AF380006

Piezoelectric Ceramics

edited by
Ernesto Suaste Gómez

Piezoelectric Ceramics
Edited by Ernesto Suaste Gómez

This is a reprint of the book published in 2010 by Sciyo

Republished by InTech
Janeza Trdine 9, 51000 Rijeka, Croatia

Copyright © 2010 InTech

All chapters are distributed under the Creative Commons Non Commercial Share Alike Attribution 3.0 license, which permits to copy, distribute, transmit, and adapt the work in any medium, so long as the original work is properly cited. After this work has been published by Sciyo, authors have the right to republish it, in whole or part, in any publication of which they are the author, and to make other personal use of the work. Any republication, referencing or personal use of the work must explicitly identify the original source.

Statements and opinions expressed in the chapters are these of the individual contributors and not necessarily those of the editors or publisher. No responsibility is accepted for the accuracy of information contained in the published articles. The publisher assumes no responsibility for any damage or injury to persons or property arising out of the use of any materials, instructions, methods or ideas contained in the book.

Publishing Process Manager Iva Lipovic

Technical Editor Goran Bajac

Cover Designer Martina Sirotic

Additional hard copies can be obtained from orders@intechopen.com

Piezoelectric Ceramics, Edited by Ernesto Suaste Gómez
p. cm.
ISBN 978-953-307-122-0

We are IntechOpen, the world's leading publisher of Open Access books

Built by scientists, for scientists

4,000+
Open access books available

116,000+
International authors and editors

120M+
Downloads

151
Countries delivered to

Our authors are among the
Top 1%
most cited scientists

12.2%
Contributors from top 500 universities

Selection of our books indexed in the Book Citation Index
in Web of Science™ Core Collection (BKCI)

Interested in publishing with us?
Contact book.department@intechopen.com

Numbers displayed above are based on latest data collected.
For more information visit www.intechopen.com

Meet the editor

Ernesto Suaste-Gomez received a doctorate degree in Biomedical Engineering in 1997 at the Center for Research and Advanced Studies, Cinvestav-IPN, Mexico. He has been a visiting scholar at several institutions including UC Berkeley, CSU Long Beach, NIH Bethesda MD, and Technical University of Ilmanau, Germany. Currently, he is a professor at the Electronic Engineering Department, Cinvestav-IPN. He teaches graduate courses in biomedical engineering and his research domain involves areas related to Human Vision, Eye Behavior and Biomedical Instrumentation in Variability Cardiac and Pupil Reflex. He is also interested in developing and applying new materials such as ferroelectric, piezoelectric, piezopolymers, pyroelectrics in the design of sensors and transducer. In addition, he also works on applications of sensors, actuators, transducers for ultrasonic imaging, positioning systems, energy harvesting and microelectronic devices associate at Biomedical Engineering.

Contents

Preface

This book reviews the importance and new opportunities for piezoelectric ceramics; to achieve it, it is necessary to investigate new materials, material combinations, structures, damages and porosity effects. In this context, it is necessary to know the trend analysis and to examine recent developments in different fields of applications. These significant aspects are discussed in the chapters of the book.

Chapter 1 reports on novel synthesis techniques for the deposition of Strontium-doped lead zirconate titanate piezoelectric thin films on silicon substrates enabling microfabrication of sensors, actuators and transducers. The thin film deposition was carried out by RF magnetron sputtering, with conditions optimized based on the outcome required – island-structures (650-700 °C) or low temperature deposition (300 °C). Furthermore, it exhibits the characterizations by spectroscopy and high resolution electron microscopy of these coatings. In chapter 2 the behaviors of lead metaniobate piezoceramic multilayer ultrasonic transducers with various adhesive bondlines under dynamic drive and self-heating conditions are studied. Inclusive, the respective effects from the piezoelectric material and adhesive bondlines on the performance of multilayer ultrasonic transducers are investigated and analyzed. Chapter 3 analyses the mechanism of force transfer through the bond layer into piezoelectrical patch transducers. It reviews the existing shear lag models, and presents a step-by-step derivation to integrate shear lag effect into electro-mechanical impedance formulations, both 1D and 2D. The two close form solutions are presented and the results are validated experimentally. Chapter 4 exhibits the characterization of mechanical and coupling properties and damage detection in piezoelectric ceramics. Numerical and experimental techniques to solve the mentioned problem are given. Besides, the relevant details of finite element formulation and simulation in piezoelectric ceramics are presented. Chapter 5 shows a damage analysis in piezoelectric ceramic structures. Starting from the background of this subject, the author explores the basis of its model proposed. The quantitative relation between the damages and parameters of microstructures is presented. Finally, these bases were used for the piezoelectric damage constitutive finite element model for PZT ceramics. Chapter 6 discusses the main processing routes necessary to obtain PZT-based materials with tailored porosity. The main topics presented are: the influence of the porosity on the electrical properties of the piezoelectric ceramics; porous piezoelectric ceramics produced by sacrificial template technique; the present status on functionally graded piezoelectric ceramics; and porous materials produced by tape casting technique for the production of multilayer ceramic packages.

Chapter 7 considers a genetic algorithm used for identification of equivalent electrical circuit RLC parameters in parallel piezoelectric ceramic (sandwich) transducers for ultrasonic systems. In chapter 8 the use of the bismuth titanate-based and potassium sodium niobate-based lead-free piezoelectric ceramics as the driving elements of ultrasonic wirebonding transducers is reported. Moreover, the fabrication and characterization of the ceramics, in the form of ring, and the wirebonding transducers are presented. Chapter 9 examines the energy harvesting problems in case of broadband vibration. Principally due to piezoelectrics, they operate in resonant mode, with the outcome that only a small part of available energy can be harvested. Solution presented here addresses this problem based on bistable electromechanical structures and compared to usual resonators, showing that energy harvesting is efficiently achievable over a large frequency bandwidth. At last, it discusses the outlines of future trends, challenges and development perspectives for piezoelectric energy harvesting. Chapter 10 describes fundamentals of electrostrictive polymer generators for energy harvesting and theoretical development. Even more practical considerations such as optimal energy harvesting circuit topologics and materials are described.

Chapter 11 gives a detailed novel device based on piezoelectric PLZT and Barium titanate ceramic with a platinum wire implanted for biomedical applications. Characterization of principal properties of this device called ceramic-controlled piezoelectric is included to show the advantages of this implant. Additionally, the potentiality and versatility for applications in different fields such as optical, acoustical, mechanical, thermal and electrical are proposed in several demonstrative experiments. Chapter 12 is focused on surface acoustic waves propagates on PZT ceramic substrate. At the same time, bio- compatibility studies between piezoceramic material and biological cell suspension are exposed. Chapter 13 presents the concept of microgrippers and concise introduction to categories of piezoelectric materials and micro-force detection technologies used in microgrippers. Also, a more detailed explanation of the basic structures is given, the corresponding mathematical equations, amplification mechanisms, among others subjects that describe in depth the behavior of this micro-structures. Thereby, a practical example, a bimorph piezoelectric ceramic microgripper integrating microforce detecting and feedback, is developed. Chapter 14 illustrates systematic studies of applications of ferroelectric materials in pulsed power technology and engineering. These devices are based on the shock depolarization of ferroelectrics to form a new class of miniature autonomous pulsed power sources. In this sense, a comparative systematic study is exposed between depolarization shock wave compression and heating method.

The editor is grateful to all the authors for their excellent contributions, as well as the members of the International Scientific Advisory Committee and other colleagues who helped to review the work published in this book.

Editor

Ernesto Suaste Gómez
Centro de Investigación y de Estudios Avanzados del Instituto Politécnico Nacional
México

Piezoelectric thin film deposition: novel self-assembled island structures and low temperature processes on silicon

Sharath Sriram, Madhu Bhaskaran and Arnan Mitchell
Microplatforms Research Group, RMIT University
Australia

1. Introduction

Lead zirconate titanate $Pb(Zr_{1-y}Ti_y)O_3$ is a material that is known for its high piezoelectric and ferroelectric properties. Thick and thin films of lead zirconate titanate (PZT) have been repeatedly used in a variety of piezoelectricity or piezoresistivity based sensors such as bio-sensing cantilevers (Park et al., 2005a) and pressure sensors (Maluf & Williams, 2004). PZT is one of many ceramic compounds with a general ABO_3 representation. Zirconium (Zr) and titanium (Ti) are B-site dopants of the compound and the ratio of Zr/Ti can alter the film properties such as the dielectric tenability (Yu et al., 2004). Certain A-site dopants (in place of a fraction of the lead atoms present in A-site locations) have been found to increase certain properties and introduce other properties. Examples of these are (i) lanthanum-doped PZT (PLZT) is known for enhanced electro-optic behaviour (Tunaboylu et al., 1998); (ii) stannum-doped PZT and niobium-doped PZT (PNZT) are known to exhibit shape memory behaviour, a property not exhibited by PZT (Yu & Singh, 2000); and (iii) the compound of interest in this chapter strontium-doped PZT (PSZT) has been reported to have improved piezoelectric, ferroelectric, dielectric tunability, and pyroelectric (infra-red radiation sensors by thermal effects) behaviour (Bedoya et al., 2000; Zheng et al., 2002; Araujo & Eiras, 2003; Yu et al., 2004).

Published literature on studies of PSZT material focuses on pellets and powders (Bedoya et al., 2000; Yu et al., 2004). However, the general consensus is that PZT and other doped-PZT compounds (e.g. PLZT and PNZT) can be deposited as thick or thin films under similar conditions, whether by sputtering or sol-gel techniques, as the percentage concentration of A-site dopants is generally very small. To exhibit piezoelectricity, ferroelectricity, and other properties the films need to be crystalline and not amorphous. The desirable crystalline phases of these films, termed perovskite, are generally rhombohedral or tetragonal (Randall et al., 1998; Yu et al., 2004).

RF magnetron sputter deposition of thin films of PZT and its doped variations (such as PLZT) have been reported in published literature (Kim et al., 1999; Wasa et al., 2004). Most published literature report deposition at temperatures of 500 °C to 700 °C (Kim et al., 1999; Wasa et al., 2004), followed in many cases by annealing at temperatures between 1000-1300 °C (Wasa et al., 2004). Such high temperature processing is undesirable when working with metal, oxide, and nitride films on silicon, which is typically the case with most microsystem devices. This increases thermal stress between layers and can deform microstructures such as cantilevers, bridges, and membranes. Reports in published literature also discuss the difficulty in controlling orientation by sputtering (Park et al., 2005b).

This chapter reports on novel synthesis techniques for the deposition of piezoelectric thin films on to silicon substrates, which allows the use of microfabrication in the realisation of sensors, actuators, and transducers. It demonstrates an intermetallic reaction driven self-assembly process for formation of islands of piezoelectric and the use of lattice guiding by a metal layer to perform low temperature deposition of piezoelectric thin films. These coatings have been characterised extensively by spectroscopy and high resolution electron microscopy. For the first case, PSZT thin films were deposited on platinum (Pt) coated silicon substrates as platinum has been the conventional choice for deposition of PZT compounds. Utilising intermetallic reactions on silicon, preferential deposition sites were defined for island-structured deposition for optical sensing. The second case focussed on the challenge of low temperature crystallization of complex oxide thin films. Deposition of such materials is often carried out at elevated temperatures in excess of 600 °C. This chapter demonstrates one of the first instances of deposition of preferentially oriented strontium-doped lead zirconate titanate thin films at a relatively low temperature of 300 °C. This was achieved by carrying out deposition on gold-coated silicon substrates which exert a guiding influence on thin film growth due to similarity in lattice parameters.

2. Experimental Details

2.1 Bottom electrode deposition

Silicon (100) substrates were dipped in hydrofluoric acid to remove the native oxide layer. The desired bottom electrode configuration was deposited by electron beam evaporation (at room temperature and under vacuum of 1×10^{-7} Torr). The four bottom electrode configurations used for this study were (i) platinum thin films with titanium adhesion layer (Pt 200 nm on Ti 20 nm on Si), (ii) platinum thin films with titanium dioxide adhesion layer (Pt 200 nm on TiO_2 20 nm on Si), (iii) gold thin films with titanium adhesion layer (Au 150 nm on Ti 15 nm on Si), and (iv) gold thin films with titanium adhesion layer and a silicon dioxide intermediate layer (Au 150 nm on Ti 15 nm on SiO_2 200 nm on Si). All bottom electrode layers for each type of sample were sequentially deposited without breaking vacuum.

2.2 Thin film deposition

Thin films of strontium-doped lead zirconate titanate (PSZT) of composition $(Pb_{0.92}Sr_{0.08})(Zr_{0.65}Ti_{0.35})O_3$ (700 nm or 1.6 µm thick) were deposited on metal coated silicon substrates by RF magnetron sputtering in a 10% oxygen balance argon atmosphere at a pressure of 10 mTorr (Sriram et al., 2009a) from a 100 mm diameter target of composition $(Pb_{0.92}Sr_{0.08})(Zr_{0.65}Ti_{0.35})O_3$. A 3-inch resistive substrate heater was used, which was compatible with deposition in an oxygen atmosphere, and samples were placed in direct contact with the heater. Very accurate control of temperature was achieved using a Model 808 temperature controller programmer (Eurotherm Controls, Inc.). The post-deposition cooling rate was found to influence the degree of perovskite orientation in the thin films (Sriram et al., 2006a) and so, a cooling rate of 5 °C/min was chosen.

2.3 Transmission electron microscopy (TEM) analysis

The transmission electron microscopy (TEM) analysis was carried out at an accelerating voltage of 200 kV on a JEOL 2010F TEM equipped with a Gatan Imaging Filter (GIF2001) and an EmiSpec E Vision energy dispersive X-ray analysis (EDX) system. Plan view specimens for *in situ* TEM analysis were prepared by mechanically grinding away the backing silicon from the film, with the aid of a calibrated disc grinder. The specimens were then ion milled to electron transparency at room temperature using 4 kV argon ions incident at 5°, with dual beam milling from the back surface of the specimen. A TEM holder with a programmable heating element, and temperature accuracy of ±0.5 °C, was used. Cross-sectional TEM specimens were prepared by gluing coated samples face-to-face with backing silicon, using high strength epoxy resin. A core (2.3 mm in diameter) was then ultrasonically machined with the interface of interest at the centre of the core. The core was then glued into a brass tube of 3 mm external diameter. This was then sawed into 500 μm thick sections, ground, and ion milled (Ar^+ at 3.5-5 keV at 4-7°, with double beam modulation) to produce electron transparent cross-sectional specimens. Knowing the silicon wafer normal and by appropriate tilting of the silicon specimen in the transmission electron microscope (TEM), it was possible to ensure that the film was viewed edge-on.

2.4 Surface enhanced Raman scattering (SERS) measurements

In order to carry out surface enhanced Raman scattering (SERS) measurements, samples with PSZT thin films were rinsed with isopropyl alcohol, air-dried, and coated with silver. Silver thin films (thickness of 60 nm) were coated at normal incidence under rotation using an Emitech K975X thermal evaporator. The samples were then immersed for 10 minutes in the analyte (10 mM thiophenol diluted in ethanol), subsequently rinsed in pure ethanol, and air-dried. SERS measurements were carried out using a Jobin Yvon Horiba TRIAX320 spectrometer with a thermoelectrically cooled CCD detector. An objective lens of 50x, wavelength of 532 nm, and power of 1.1 mW were used. Data was collected in the backscatter mode with an acquisition time of 10 seconds.

2.5 Secondary ion mass spectrometry (SIMS) analysis

The SIMS analysis was carried out using caesium ions (Cs+) with a primary accelerating voltage of 7.5 keV and a sampling voltage of 4.5 keV (Cameca 5f dynamic SIMS instrument). Considering the insulating nature of the film, the sample was biased with an offset voltage of +50 V to overcome charging effects. In view of the low concentration of strontium in the PSZT thin film samples, the two major isotopes – ^{86}Sr and ^{88}Sr – were studied to confirm the presence of strontium. The plotted data has been normalized with respect to the caesium signal measured.

2.6 X-ray diffraction (XRD) analysis

X-ray diffraction (XRD) analysis was carried out using a Scintag X-ray diffractometer operating with a cobalt X-ray source (at a wavelength of 0.179020 nm). The scans were carried out for a 2θ range of 20° to 60° with steps of 0.02°. The collected data were shifted to correspond to the copper $k\alpha$ wavelength (0.154056 nm) for comparison with the International Centre for Diffraction Data powder diffraction pattern files available.

3. Self-Assembled Island Structures

3.1 Island structured dielectric thin films

The deposition of platinum on silicon substrates typically requires the use of an intermediate adhesion layer. Inert titanium dioxide is commonly used in conjunction with pure silicon substrates (Pt/TiO$_2$/Si) (Bhaskaran et al., 2008), whereas titanium is the preferred layer for oxidized silicon surfaces (Pt/Ti/SiO$_2$/Si) (Kim et al., 1999; Million et al., 2003; Wasa et al., 2004). These two configurations prevent any reaction of the platinum with the silicon. However, in the synthesis process that we propose here, it is precisely this metal-silicon interaction which is the underlying mechanism.

Fabrication began by coating silicon substrates with a 20 nm titanium adhesion layer followed by 200 nm of platinum using electron beam evaporation. These samples were subsequently coated with dielectric thin films of strontium-doped lead zirconate titanate (PSZT) using RF magnetron sputtering.

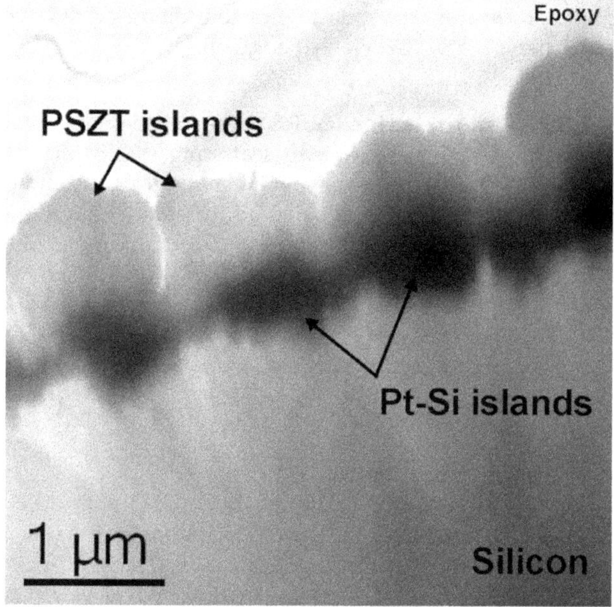

Fig. 1. Cross-sectional transmission electron micrograph analysis of PSZT thin films deposited on Pt/Ti/Si substrates. The distribution of Pt-Si islands and the preferential growth of PSZT on these islands can be observed. (Reprinted with permission from Sriram et al., 2009b)

The temperature of deposition initiates a reaction between the platinum and silicon through the thin titanium layer, resulting in the formation of islands of platinum silicide (Lee et al., 1997; Firebaugh et al., 1998). The PSZT thin film layer that is subsequently deposited prefers to grow on these platinum silicide islands. The affinity that PSZT-type films have for growing on metallic layers, and platinum in particular, has been established in the literature (Kim et al., 1999; Million et al., 2003; Wasa et al., 2004). The ability to form perovskite structured complex oxides on platinum has been used extensively for deposition processes. The platinum-silicon reaction and growth of sub-micron PSZT islands with columnar grains is shown in Fig. 1.

3.2 Mechanism of island structuring

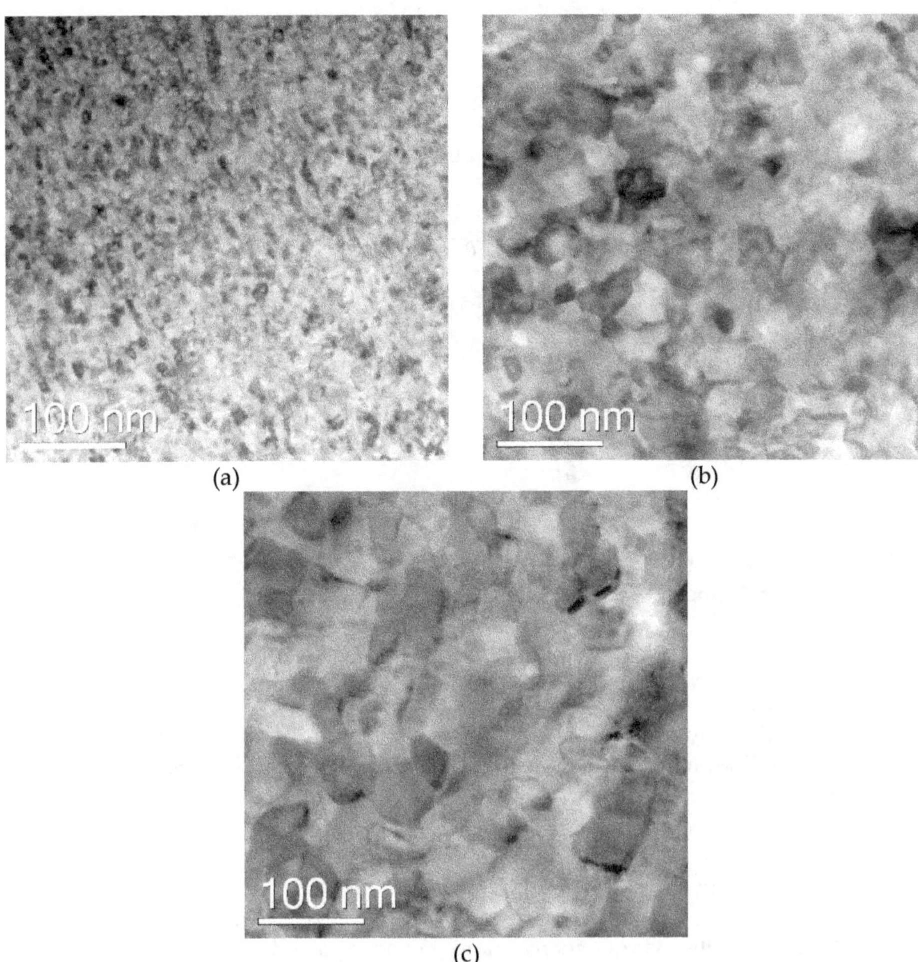

Fig. 2. TEM micrographs showing evolution of grains in the Pt/Ti/Si structure: (a) as-deposited sample, (b) at 500 °C, and (c) at 650 °C. (Reprinted with permission from Sriram et al., 2009b)

The mechanism responsible for the formation of the platinum-silicon islands was investigated using *in situ* heating during plan-view transmission electron microscopy (TEM) analysis. Silicon substrates coated with a 20 nm titanium adhesion layer and 200 nm of platinum were used. These samples were mechanically polished to create plan view TEM specimens. The samples were heated while being observed within the TEM to replicate the temperature ramp-up before the PSZT sputtering process.

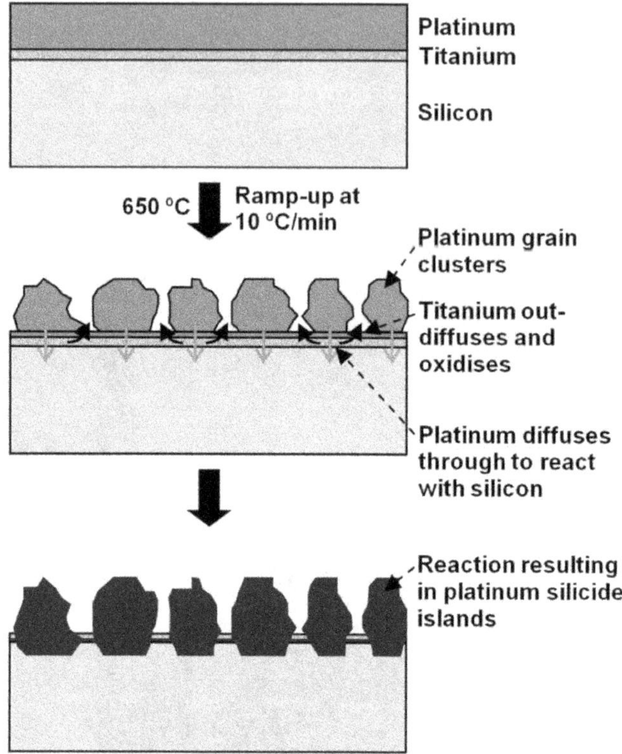

Fig. 3. Schematic depicting proposed mechanism for platinum-silicon reaction, based on *in situ* plan view TEM analysis, cross-section TEM analysis, and AES. (Reprinted with permission from Sriram et al., 2009b)

The evolution of the grain structure of platinum was studied by observing a region of the foil where the platinum layer overlapped with the silicon substrate. Grain growth with crystallization was apparent in the micrographs (Fig. 2), wherein an initial grain size of ~10 nm in the as-deposited sample increased to ~30 nm at 500 °C and ~50 nm at 650 °C. The distribution of grain sizes became more random with increasing temperature. Lattice fringes due to strong crystallization could be observed within individual grains. In the thinner regions of the specimen the formation of voids was observed, due to the formation of grain clusters. Each cluster consisted of 15-25 grains. The size and distribution of the clusters relate to the platinum-silicon islands observed in the cross-sectional micrographs (Fig. 1).

The composition of the island regions was examined by performing energy dispersive X-ray analysis (EDX), on the cross-section specimens. This analysis revealed that the island regions were composed of both platinum and silicon, while the regions between the islands had an upper surface that was rich in titanium and oxygen and a lower region composed mostly of platinum and silicon.

A possible explanation is as follows. As grain growth in platinum is initiated, outward diffusion of titanium occurs at the grain boundaries. This titanium is exposed to oxygen in the sputtering gas mixture and forms a stable oxide of titanium. As temperature increases, platinum reacts with silicon to form islands of silicide, with regions of titanium oxide present in between. This oxide presumably prevents the formation of titanium silicide; and it is notable that stable titanium silicide is formed at temperatures above 700 °C (Maex & van Rossum, 1995; Bhaskaran et al., 2007). The presence of all four elements Si, Pt, Ti, and O was confirmed by using Auger electron spectroscopy (AES), while AES depth profiling verifies the existence of Pt-Si and surface Ti-O phases (Sriram et al., 2009b). A schematic depiction of the proposed mechanism is shown in Fig. 3.

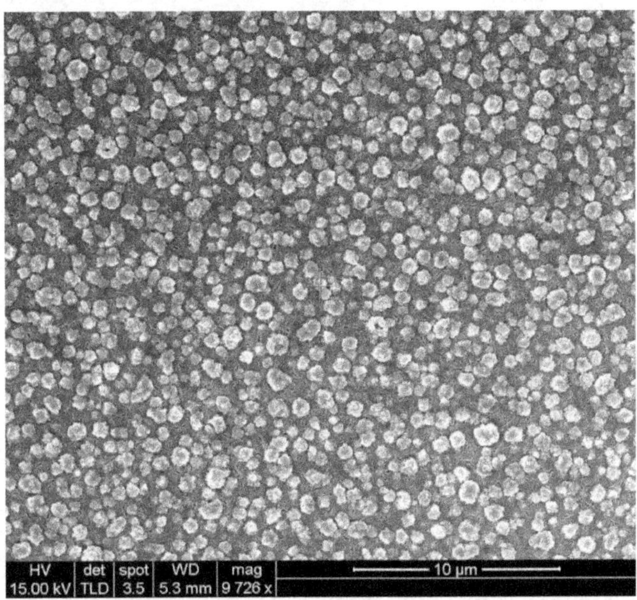

Fig. 4. Scanning electron micrograph depicting the distribution of the self-assembled Pt-Si islands with preferentially grown PSZT. (Reprinted with permission from Sriram et al., 2009b)

During the subsequent deposition of PSZT, it appears to preferentially grow on the regions with platinum, resulting in the final island-structure observed in the film, as shown in Fig. 1. The distribution and size of the islands depends on two factors that can be identified intuitively. These are the duration of thermal processing and the thickness of the platinum layer. The dependence on thermal processing duration was verified using cross-sectional TEM analysis, which showed that the average island size increased from ~500 nm for a 2 hour process to ~800 nm for a 4 hour process. The distribution of such islands on the substrate is shown in Fig. 4.

This island-structuring significantly increases the surface roughness of the samples under study. Analysis of average surface roughness (R_a) was carried out using an Ambios XP-2 surface profilometer. The Pt-Ti-Si electrode regions had a measured R_a of 24.8 nm and regions with PSZT deposited on them had R_a of 27.1 nm. This represents a significant increase over the R_a values for the continuous PSZT thin films on $Pt/TiO_2/Si$ structure of 4.9 nm, with these electrodes registering an average surface roughness lesser than 2.0 nm.

3.3 Surface-enhanced Raman scattering results

The self-assembly of island features on a substrate and their subsequent transfer to thin films can be used effectively in a wide range of sensing applications. In this section, we demonstrate that the island structures that we have synthesized in both platinum and PSZT, have potential for applications in surface-enhanced Raman scattering (SERS) sensors.

SERS is a vibrational spectroscopy technique that allows sensitive detection of chemical compounds in close proximity to a nanostructured gold, silver or copper surface (Smith, 2008). The scattering efficiency can be enhanced by a factor of 10^6 or more compared to normal Raman scattering. The effect is primarily generated by the enhanced electromagnetic field associated with the localized surface plasmon resonance that occurs around the metal nanoparticles (Schatz et al., 2006; Stewart et al., 2008). The SERS effect can most readily be generated on roughened metal surfaces, where the roughness has a characteristic scale of 10-100 nm. Examples of such substrates include electrochemically roughened electrodes (Fleischmann et al., 1974; Pettinger & Wetzel, 1982) and metal films over nanoparticles (Goudonnet et al., 1982; Kostrewa et al., 1998). As a first step towards exploiting nanostructured dielectric films for plasmonic sensors, we have used the island-structured PSZT films as a platform for fabricating silver SERS substrates.

Four substrate types were employed for these SERS experiments. These were (i) continuous platinum thin film on silicon ($Pt/TiO_2/Si$), (ii) continuous PSZT thin film deposited on continuous platinum thin films ($PSZT/Pt/TiO_2/Si$), (iii) self-assembled Pt-Si islands, starting from the Pt/Ti/Si substrates and (iv) island-structured PSZT grown preferentially on the islands of substrates of type (iii).

The thicknesses of these layers were 200nm/20nm for Pt/TiO_2 on silicon substrate, while the PSZT dielectric layers were 750-800 nm, coated under the same conditions used for the island-structuring process. All four substrates were coated with a thin layer of silver to enable surface plasmonic activity. Thiophenol was used as the reference analyte, as it is widely used for SERS validation (Sandroff & Herschbach, 1982; Viets & Hill, 1998; Haynes et al., 2005; White & Stoddart, 2005; Kostovski et al., 2009).

The SERS spectra that were collected from the continuous platinum film substrates presented very weak spectral characteristics for thiophenol [Fig. 5(a)]. The peaks expected at ~1000 cm^{-1}, ~1080 cm^{-1}, and ~1580 cm^{-1} could barely be resolved. Samples with the continuous PSZT layer [Fig. 5(b)] showed marginally better performance, albeit still unusable for SERS measurements.

SERS measurements for the island-structured substrates are shown in Figs. 5(c,d). Strong spectral signatures are present for the reference thiophenol peaks. A comparison of the intensities for the platinized substrates highlights that island-structuring [Fig. 5(c)] results in a 40x enhancement of Raman scattering over the continuous counterpart [Fig. 5(a)]. The inclusion of the PSZT layer [Fig. 5(d)] further enhances scattering counts by a factor of 2½. This represents a net 100x enhancement from the continuous metalized substrates to the island-structured PSZT substrates.

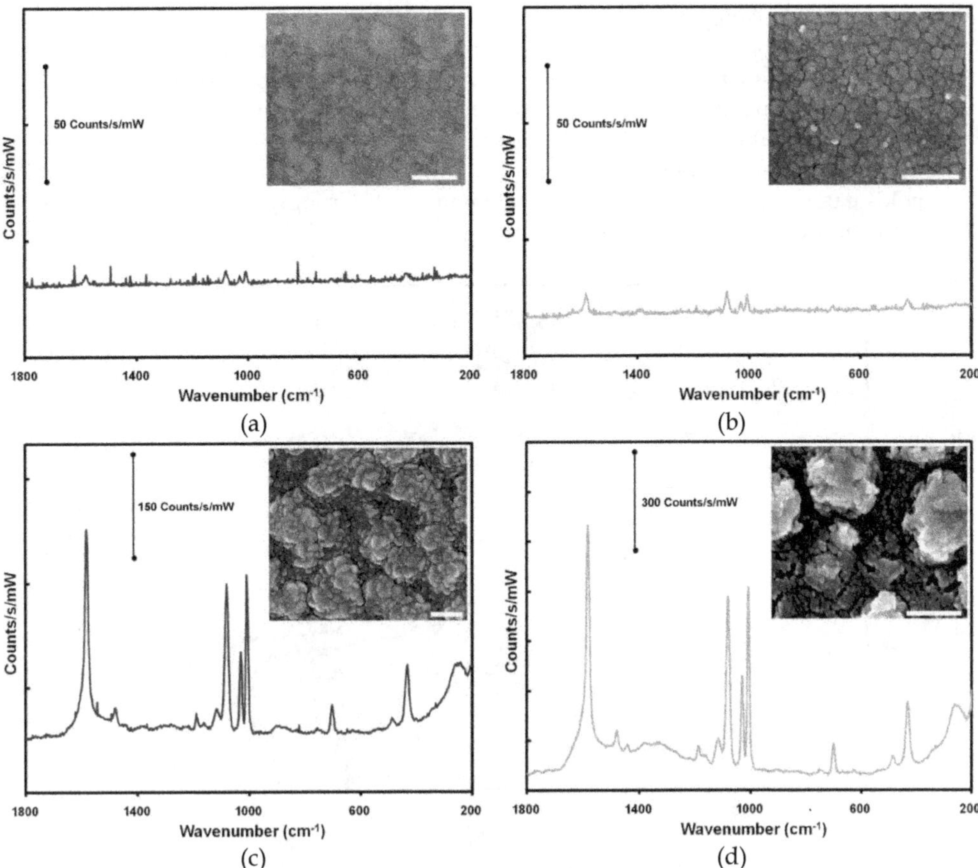

Fig. 5. SERS spectra collected for 10 mM thiophenol diluted in ethanol are presented. Spectra shown are for: (a) continuous platinum thin film on silicon; (b) continuous PSZT thin film deposited on substrate in (a); (c) self-assembled Pt-Si islands; and (d) island-structured PSZT grown preferentially on islands in (c). The interval between the y-axis markers is 50 counts/s/mW in (a) and (b) and is 150 counts/s/mW in (c) and (d). Scanning electron micrographs of the respective film surfaces coated with 60 nm of silver are shown as insets, with scale bars corresponding to 500 nm. (Reprinted with permission from Sriram et al., 2009b)

4. Low Temperature Deposition on Gold

4.1 PSZT thin films on Au/Ti/Si

Having chosen gold for having lattice spacings close to those of PSZT, thin films deposited on gold coated silicon substrates, with a titanium adhesion layer, were characterised. Secondary ion mass spectrometry (SIMS) depth profiling through the film thickness (Fig. 6) showed that it was compositionally uniform. The SIMS data also highlighted some dissolution of the titanium into the gold layer, although the peak in the titanium signal corresponded to the location of the original titanium layer (between the gold layer and the silicon substrate). This depth profile was used to explain the results of certain observations during the transmission electron microscopy analysis. Figure 7 shows a low magnification cross-sectional view of the multi-layer structure. Energy dispersive X-ray analysis (EDX) was carried out along the length and through the thickness of the film to detect gross variations in chemical composition due to second phases or other contaminants; none were found. The PSZT film was uniform in thickness (~470 nm). The dark band beneath the film is the gold-titanium-silicon layer (typically ~135 nm thick), from which prismatic ingrowths into the silicon substrate can be seen.

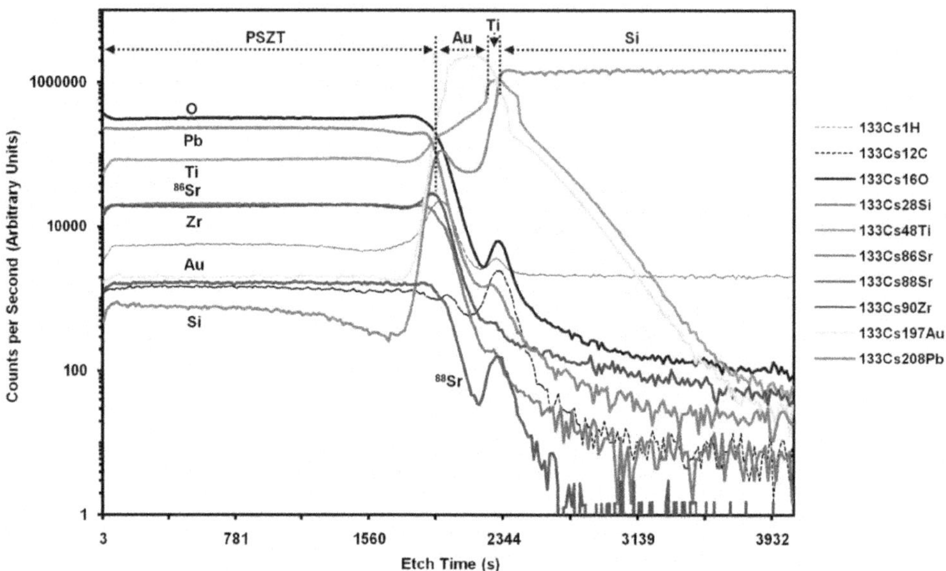

Fig. 6. Secondary ion mass spectroscopy depth profile through the thin film. (Reprinted with permission from Sriram et al., 2009c)

Detailed imaging of this interfacial region (Fig. 8) showed an amorphous layer approximately 4 nm thick. Point analysis with EDX showed that this layer was rich in silicon and oxygen, most likely SiO_2 formed by outward diffusion of silicon from the substrate through the metal layers. This is supported by previous studies on the fast outward diffusion of silicon through grain boundaries in gold (Sumida and Ikeda, 1991), with this silicon oxidising in the 10 % oxygen in argon sputtering atmosphere. This is also supported by the increase in the silicon signal at this interface in the SIMS analysis (also confirming that this is not a by-product of the specimen preparation ion milling process). The region of PSZT (~25 nm) closest to this amorphous SiO_2 layer appeared to be amorphous or at best poorly crystalline, as it exhibited strong amorphous speckle (right inset of Fig. 8). Weak lattice fringe contrast was only evident only from regions greater than 30 nm into the PSZT layer (from the SiO_2 interfacial layer, as seen in the left inset of Fig. 8).

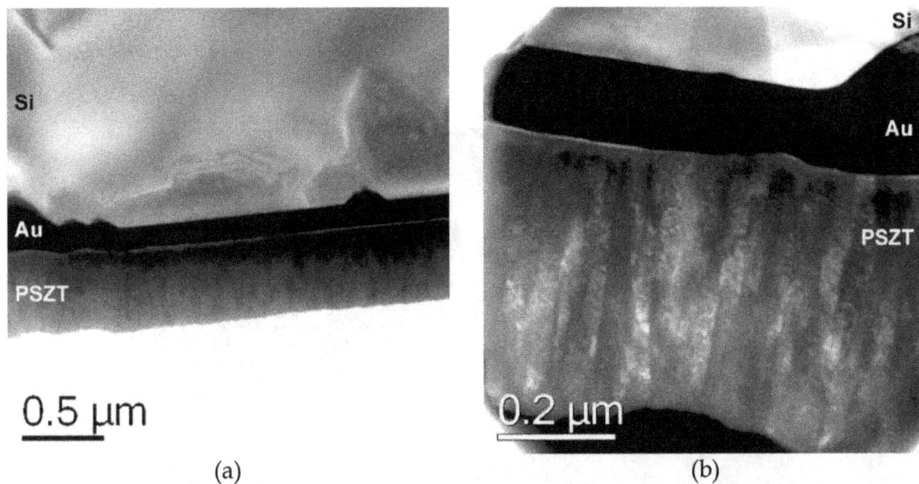

(a)   (b)

Fig. 7. XTEM analyses of a PSZT thin film deposited for a 2 hour sputtering duration: (a) Bright field diffraction contrast image of the ~470 nm thick PSZT film on a ~130 nm thick gold layer. Prismatic ingrowths of gold-silicon have formed at the interface of gold to silicon; (b) HCDF image of the PSZT film highlighting the columnar structure. *Note*: The outer region (~100 nm) of the film has been removed during ion-milling. (Reprinted with permission from Sriram et al., 2009c)

The PSZT thin film exhibited a columnar microstructure (Fig. 7). These columnar grains were perpendicular to the substrate (out of plane), and were most apparent in the energy filtered (zero loss) hollow cone dark field (HCDF) image [of a 2 hour PSZT deposition sample shown in Fig. 7(b)]. Typical columnar grain widths were 30-50 nm, and many grains appeared to span the full thickness of the film (~470 nm). The contrast due to the crystallinity was quite weak in the 25 nm thick band of PSZT closest to the gold layer, confirming the amorphous or poorly crystalline nature of this region (as observed in Fig. 8). During the initial stages of deposition, the PSZT is deposited onto a substrate covered with amorphous SiO_2, which affected crystallinity. Crystallisation of the PSZT occurs once a certain thickness of material (~25 nm) has been deposited. Growth stress may contribute to the onset of crystallisation (Mitchell et al., 2003). X-ray diffraction analysis of these films showed that the films exhibited preferential [104] orientation (discussed later).

Increasing the deposition duration from 2 to 4 hours increased the columnar grain width (measured from hollow cone dark field images) from 30-50 nm to ~90 nm (not shown). It also resulted in an increase in the average surface roughness (R_a) from 7.8 nm to 9.5 nm (Sriram et al., 2006b and Sriram et al., 2007). This was attributed to the development of marked faceting. Fig. 7(a) shows a cross-section of the relatively flat surface of the film deposited for 2 hours. The dimension of the weak faceting is similar to that of the columnar grain width (30-50 nm). For the film deposited for 4 hours, very pronounced prismatic faceting has occurred (Fig. 9). This is much coarser than seen at the shorter duration and reflects the grain coarsening taking place in the film in longer duration depositions. The increase in roughness is a consequence of grain coarsening, which can be attributed to the variations in growth rates as a function of crystal orientation as well as grain coalescence. At the growing surface, those grains in orientations promoting faster growth can dominate neighbouring grains and eventually isolate them. With continued growth, grains may then develop prismatic faceting which causes increased surface roughening as the grain width increases (Mitchell et al., 2003).

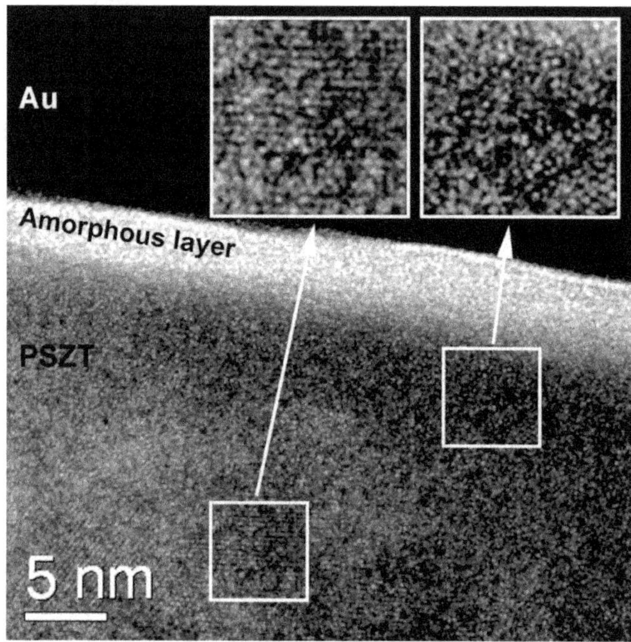

Fig. 8. High resolution energy filtered cross-sectional TEM image of the amorphous layer at the interface of PSZT and gold. (Reprinted with permission from Sriram et al., 2009c)

The eutectic point of gold and silicon is 363 °C, but at the deposition temperature of 300 °C interdiffusion of gold and silicon (enabled by outward silicon diffusion through gold grain boundaries) through the 15 nm thin titanium adhesion layer has been observed. This reaction was verified using secondary ion mass spectroscopy (SIMS) depth profiling (Fig. 6). The peak titanium concentration occurred between the gold and silicon – being the original location of the ~15 nm thick titanium layer. On moving through the gold layer towards the PSZT (a distance of 150 nm), the titanium signal showed a gradual decrease. This indicated that titanium diffused into the gold. Both titanium and gold diffusion into silicon was also apparent. The profiles for these two elements were very similar and gradually decreased over a considerable distance into the silicon. Due to the very different sputtering rates of the various phases present, it was not possible to meaningfully estimate the depth over which the gold and titanium diffused into the silicon substrate. However, the cross-sectional TEM imaging provided an understanding of the origin of the SIMS profile shapes for these elements.
The gold-(titanium)-silicon reaction can be observed in the form of prismatic crystallites growing into silicon along the metal-silicon interface. This is shown in Figs. 7 and 9 and in detail in Fig. 10. EDX analysis confirmed that these crystallites were a reaction product of gold and silicon – with titanium also present (Sriram et al., 2009c). The remnant of the

original titanium layer could be seen as a ~10 nm wide band running across the base of the crystallite in Fig. 10. The crystallites were variable in size, the largest being about 200 nm into silicon. During SIMS depth profiling (Fig. 6), the floor of the ion milling crater would break through the main (flat) part of the metal layer into the silicon substrate, and the gold and titanium signals would be expected to drop abruptly. However, the presence of prismatic crystallites rich in gold, with traces of titanium, projecting into the silicon substrate, would cause the signal for these elements to taper off gradually as the depth profiling progressed, dropping to zero when the profile passed through the apex of the largest crystallites.

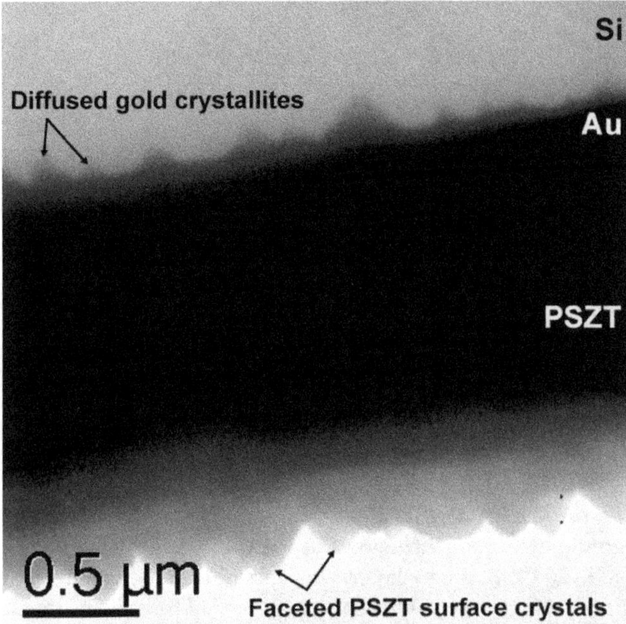

Fig. 9. Cross-section energy filtered TEM image showing the highly faceted film surface with crystallites at the gold-silicon interface. (Reprinted with permission from Sriram et al., 2009c)

The diffusion of gold into the (100) silicon results in crystallites bounded by the dense (111) silicon planes (Fig. 10). As expected the angle between these bounding planes and the silicon (100) surface was found to be 54.7° in Fig. 10, which is oriented with the electron beam parallel to silicon [110]. The resulting crystallites varied in size and the density of such crystallites was much higher in longer duration deposition (4 h) specimen. Such diffusion or dissolution of gold into silicon-based materials through titanium has been observed (Wenzel et al., 1998) and can be related to the high reactivity of titanium with silicon (Ti_xSi_y compounds are formed even at slightly elevated temperatures) and the thickness (15 nm) of the titanium layer. It has also been concluded that the titanium layer needs to have a sufficient thickness, depending on process temperature and duration, in order to serve as a barrier to prevent the gold-silicon reaction (Kanamori and Sudo, 1982), rather than serve only as an adhesion layer.

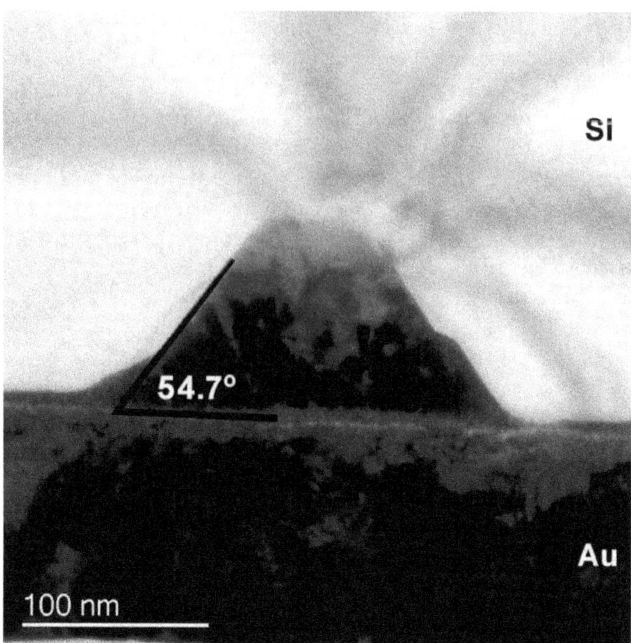

Fig. 10. Cross-sectional energy filtered TEM image of a crystallite bounded by dense (111) silicon planes. The angle between the silicon (100) surface and the crystallite edge is 54.7° being the silicon (100)^(111) interplanar angle. *Note*: The electron beam is parallel to silicon [110]. (Reprinted with permission from Sriram et al., 2009c)

4.2 PSZT thin films on Au/Ti/SiO$_2$/Si

To overcome the gold-silicon interdiffusion observed in the initial samples, an intermediate silicon dioxide (SiO$_2$, 200 nm thick) layer was introduced between the metal layers and silicon, and PSZT deposition was carried out at 300 °C. The silicon dioxide layer included between the metal layers and the silicon substrate prevents metal-silicon reactions, and uniform thin film layers can be observed in the cross-sectional transmission electron micrograph [Fig. 11(a)]. The resulting PSZT films had a nanocolumnar grain structure. The PSZT deposition temperature initiates crystal growth in the gold layer (which was nanocrystalline when deposited), resulting in significant changes to the texture of gold [Fig. 11(b)]. In the absence of gold-silicon interactions, the gold layer is able to reorient so that the most densely packed planes (111) are parallel to the surface, and thus the surface energy is minimized. The formation of the amorphous layer is also prevented. As a result, the highly textured gold layer was able exert a strong guiding effect on the subsequently deposited PSZT thin films (Fig. 12), which developed a pronounced (104) texture (discussed later). This guiding effect results in the (104) PSZT planes being aligned parallel to (111) gold planes (Fig. 13); this is a feature at different points along the interface interspersed with regions of random orientation.

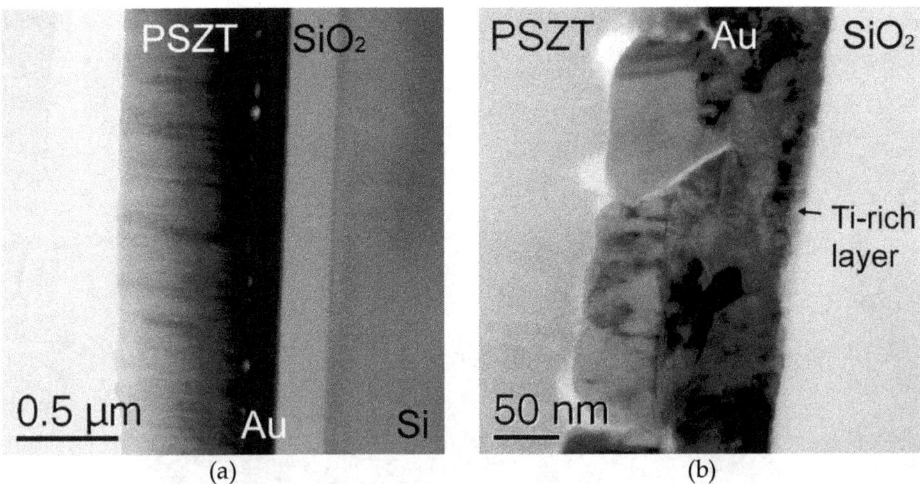

Fig. 11. Cross-sectional transmission electron microscopy results: (a) uniform thin film layers can be observed in the cross-sectional transmission electron micrograph and (b) grain growth in gold with preferential (111) texturing (lattice spacing of 0.236 nm). (Reprinted with permission from Sriram et al., 2010)

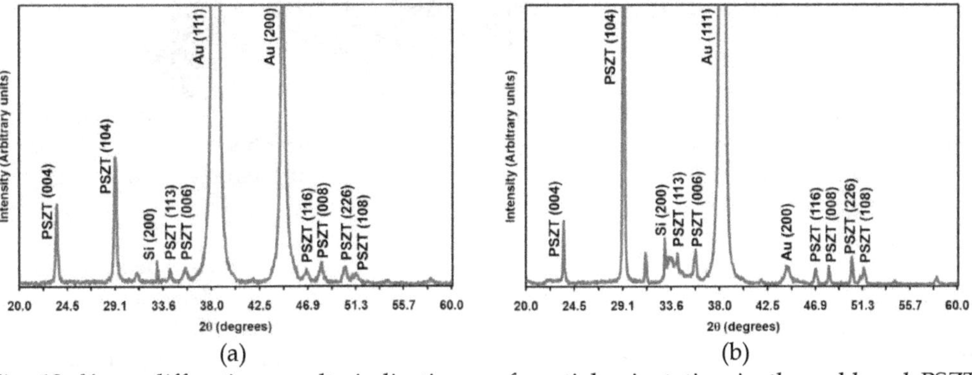

Fig. 12. X-ray diffraction results indicating preferential orientation in the gold and PSZT layers: (a) PSZT on Au/Ti/Si and (b) PSZT on Au/Ti/SiO₂/Si. (Reprinted with permission from Sriram et al., 2010)

The ratio of the relative intensities of the (111) and (200) gold X-ray diffraction peaks was ~100 after PSZT deposition at 300 °C on a sample with the buffer layer of silicon dioxide [Fig. 12(b)]; this corresponds to an increase in preferential orientation in the gold layer by a factor of 6 (compared to a sample without the silicon dioxide layer) [Fig. 12(a)].

The pronounced (111) orientation in the gold layer improves the preferential orientation of the PSZT thin film – the ratio of the X-ray diffraction relative intensities of the two major peaks corresponding to (004) and (104) PSZT orientations increased from 1.57 to 8.52 with the inclusion of the silicon dioxide layer (Fig. 12). All PSZT thin film peaks were shifted from expected peak positions, while substrate (gold and silicon) peaks were exactly at expected peak positions. The PSZT rhombohedral unit cell parameters a and c were both about 5.29%

larger than expected. This diffraction analysis also showed that the unit cells of the PSZT thin films have a 16% greater volume than theoretically expected (Bhaskaran et al., 2008).

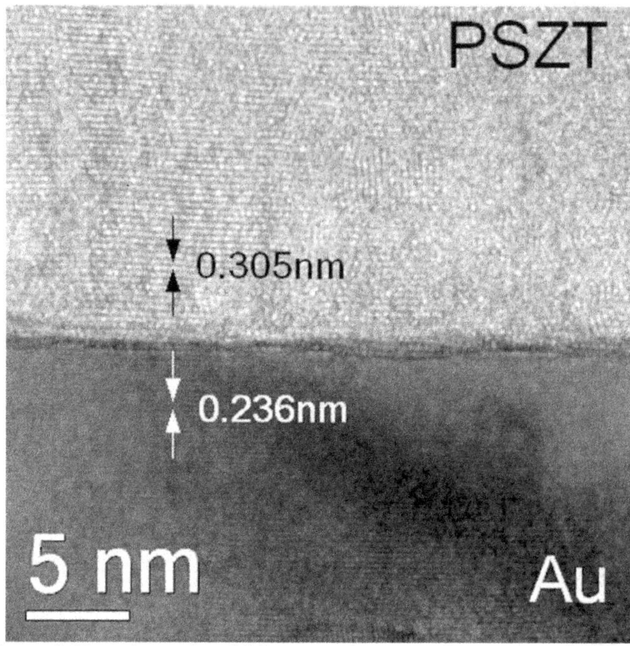

Fig. 13. The preferentially oriented gold layer appears to have a guiding effect on the PSZT thin film and this high resolution transmission electron micrograph shows the (104) planes of PSZT (with lattice spacing of 0.305 nm) preferentially aligning parallel to the (111) gold planes. (Reprinted with permission from Sriram et al., 2010)

Crystal structure simulations carried out using these modified unit cell parameters highlight the small lattice mismatches between the gold (111) and PSZT (104) planes as depicted in Fig. 14. From the overlap of atoms of these planes it can be concluded that gold can exert a guiding effect on PSZT. The simulation for gold was based on a lattice constant a of 0.236 nm and that for PSZT was based on the modified rhombohedral unit cell parameters a of 0.604 nm and c of 1.508 nm.

Based on the lattice spacing parameters (Sriram et al., 2010), a guiding effect by gold (111) on PSZT (202) might have been expected. According to the powder diffraction reference,[§] the PSZT (104) is five times more intense than the (202) orientation, indicating that the former is the preferred equilibrium state. This is the most likely reason for the observed guiding effect of gold (111) on PSZT (104). It should also be noted that the proximity of the d-spacings of the gold (111) and PSZT (202) peaks thwarts structural comparison using XRD, with no related guiding effects observed in the XTEM results. Importantly, the crystal structure simulations presented in Fig. 14 verifies the atomic overlap and the ability of gold (111) to exert a guiding effect on PSZT (104).

[§] Powder Diffraction Pattern Files, International Centre for Diffraction Data (ICDD, formerly the Joint Committee for Powder Diffraction Studies), Newtown Square, PA 19073, Card 04-002-5985.

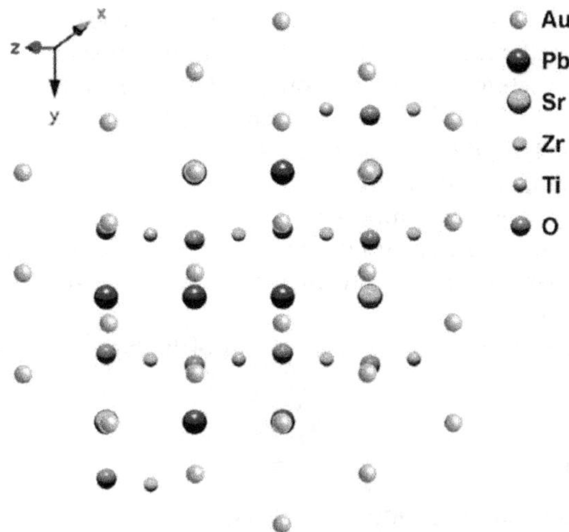

Fig. 14. Crystal structure simulation showing the superposition of the gold (111) plane on the PSZT (104) plane. From the overlap of atoms of these planes it can be concluded that gold can exert a guiding effect on PSZT. (Reprinted with permission from Sriram et al., 2010)

5. Conclusion

Extensive materials characterisation of PSZT thin films deposited on platinum and gold coated silicon substrates has been performed. The focus has been on silicon substrate based studies, in order to enable future incorporation of PSZT thin films in silicon-based devices. Spectroscopy and microscopy techniques were used, often in combination, to study the microstructure of the deposited thin films.

A synthesis process for realizing island-structured textures was proposed by inciting a reaction between the silicon substrate and the platinum thin film layer deposited on it. This process results in a distribution of sub-micron and nanoscale islands of platinum silicide, with the process scalability dependent on thermal duration and metal thickness parameters. A mechanism for this structuring process is proposed, combining results from microscopy and spectroscopy analyses. The ability of these textures to serve as SERS-active media is demonstrated, with 100x enhancement of detection sensitivity demonstrated for thiophenol as compared to the planar substrates. Further enhancement of the SERS sensitivity could be expected by optimising the silicide island morphology through control of the annealing conditions and through engineering of the nano-scale column structure of the complex oxide to best accommodate a specific analyte.

A combination of transmission electron microscopy and X-ray diffraction has been used to show that gold exerts a guiding effect on the RF magnetron sputter deposited PSZT thin films. PSZT thin films deposited on gold-coated silicon substrates using titanium as an adhesion layer showed presence of inter-diffusion. PSZT films exhibited a uniform composition through their thickness. A silicon-rich amorphous layer was formed at the interface of the PSZT thin films and gold, and was most likely to be silicon dioxide. Prismatic crystallites of a gold-silicon compound developed at the interface of the metal layers and the silicon substrate. This was caused by gold diffusing through the underlying titanium metal layer; the crystallites being bounded by the dense (111) silicon planes. On the introduction of a silicon dioxide barrier layer, the resulting films have a nanocolumnar grain structure with strong preferential orientation. This is one of the first instances of low temperature deposition of preferentially oriented piezoelectric thin films, with this deposition on silicon making the process favourable to microfabrication processes. This has the potential for enabling incorporation of complex functional oxides into devices with processing at a relatively lower temperature of 300 °C, allowing more versatile device designs and fabrication schemes. The guiding effect of (111) Au on (104) PSZT was observed by high resolution transmission electron microscopy and verified using crystal structure simulations.

These results pave the way for the next generation of high performance piezoelectric thin film devices on silicon.

Acknowledgements

The authors acknowledge funding support from the Australian Research Council (Discovery Project DP1092717) and the Australian Institute of Nuclear Science and Engineering. The authors also thank researchers who contributed to this work – A. S. Holland, G. Kostovski, D. R. G. Mitchell, K. T. Short, and P. R. Stoddart.

6. References

Araujo, E. B. & Eiras, J. A. (2003). Effects of crystallization conditions on dielectric and ferroelectric properties of PZT thin films. *Journal of Physics D: Applied Physics*, 36, 16, 2010-2013, 0022-3727

Bedoya, C.; Muller, C.; Baudour, J.-L.; Madigou, V.; Anne, M. & Roubin, M. (2000). Sr-doped $PbZr_{1-x}Ti_xO_3$ ceramic: Structural study and field-induced reorientation of ferroelectric domains. *Materials Science Engineering B*, 75, 1, 43-52, 0921-5107

Bhaskaran, M.; Sriram, S.; Short, K. T.; Mitchell, D. R. G.; Holland, A. S. & Reeves, G. K. (2007). Characterisation of C54 titanium silicide thin films by spectroscopy, microscopy and diffraction. *Journal of Physics D: Applied Physics*, 40, 17, 5213-5219, 0022-3727

Bhaskaran, M.; Sriram, S.; Mitchell, D. R. G.; Short, K. T. & Holland, A. S. (2008). Effect of multi-layered bottom electrodes on the orientation of strontium-doped lead zirconate titanate thin films. *Thin Solid Films*, 516, 22, 8101-8105, 0040-6090

Firebaugh, S. L.; Jensen, K. F. & Schmidt, M. A. (1998). Investigation of high-temperature degradation of platinum thin films with an in situ resistance measurement apparatus. *Journal of Microelectromechanical Systems*, 7, 1, 128-135, 1057-7157

Fleischmann, M.; Hendra, P. J. & McQuillan, A. J. (1974). Raman spectra of pyridine adsorbed at a silver electrode. *Chemical Physics Letters*, 26, 2, 163-166, 0009-2614

Goudonnet, J. P.; Begun, G. M. & Arakawa, E. T. (1982). Surface-enhanced raman scattering on silver-coated teflon sphere substrates. *Chemical Physics Letters*, 92, 2, 197-201, 0009-2614

Haynes, C. L.; Yonzon, C. R.; Zhang, X. & Van Duyne, R. P. (2005). Surface-enhanced Raman sensors: early history and the development of sensors for quantitative biowarfare agent and glucose detection. *Journal of Raman Spectroscopy*, 36, 6-7, 471-484, 0377-0486

Kanamori, S. & Sudo, H. (1982). Effects of titanium layer as diffusion barrier in Ti/Pt/Au beam lead metallization on polysilicon. *IEEE Transactions on Components Hybrids and Manufacturing Technology*, CHMT-5, 318–321, 0148-6411

Kim, T. S.; Kim, D. J. & Jung, H. J. (1999). N_2O reactive gas effect on rf magnetron sputtered $Pb(Zr_{0.52}Ti_{0.48})O_3$ thin films. *Journal of Applied Physics*, 86, 7024-7028, 0021-8979

Kostovski, G.; White, D. J.; Mitchell, A.; Austin, M. W. & Stoddart, P. R. (2009). Nanoimprinted optical fibres: Biotemplated nanostructures for SERS sensing. *Biosensors and Bioelectronics*, 24, 5, 1531-1535, 0956-5663

Kostrewa, S.; Hill, W. & Klockow, D. (1998). Silver films on diamond particles as substrates for surface-enhanced Raman scattering. *Sensors and Actuators B: Chemical*, 51, 1-3, 292-297, 0925-4005

Lee, C.-K.; Hsieh, C.-D. & Tseng, B.-H. (1997). Effects of a titanium interlayer on the formation of platinum silicides. *Thin Solid Films*, 303, 1-2, 232-237, 0040-6090

Maex, K. & van Rossum, M. (1995). In: *Properties of Metal Silicides*, INSPEC, 0863417760, London

Maluf, N. & Williams, K. (2004). *An Introduction to Microelectromechanical Systems Engineering*, Artech House, 9781580535915, Norwood

Millon, C.; Malhaire, C.; Dubois, C. & Barbier, D. (2002). Control of the Ti diffusion in Pt/Ti bottom electrodes for the fabrication of PZT thin film transducers. *Materials Science in Semiconductor Processing*, 5, 2-3, 243-247, 1369-8001

Mitchell, D. R. G.; Attard, D. J. & Triani, G. (2003). Transmission electron microscopy studies of atomic layer deposition TiO_2 films grown on silicon. *Thin Solid Films*, 441, 1-2, 85-95, 0040-6090

Park, C.-S.; Kim, S.-W.; Park, G.-T.; Choi, J.-J. & Kim, H.-E. (2005b). Orientation control of lead zirconate titanate film by combination of sol-gel and sputtering deposition. *Journal of Materials Research*, 20, 1, 243-246, 0884-2914

Park, J. H.; Kwon, T. Y.; Yoon, D. S.; Kim, H. & Kim, T. S. (2005a). Fabrication of microcantilever sensors actuated by piezoelectric $Pb(Zr_{0.52}Ti_{0.48})O_3$ thick films and determination of their electromechanical characteristics. *Advanced Functional Materials*, 15, 12, 2021-2028, 1616-3028

Pettinger, B. & Wetzel, H. (1982). In: *Surface-Enhanced Raman Scattering*, Chang, R. K., Furtak, T. E., Eds.; 293, Plenum Press, 9780306409073, New York

Randall, C. A.; Kim, N.; Kucera, J.-P.; Cao, W. & Shrout, T. R. (1998). Intrinsic and extrinsic size effects in fine-grained morphotropic-phase-boundary lead zirconate titanate ceramics. *Journal of the American Ceramic Society*, 81, 3, 677-688, 0002-7820

Sandroff, C. J. & Herschbach, D. R. (1982). Surface-enhanced Raman study of organic sulfides adsorbed on silver: facile cleavage of sulfur-sulfur and carbon-sulfur bonds. *Journal of Physical Chemistry*, 86, 17, 3277-3279

Schatz, G. C.; Young, M. A. & Van Duyne, R. P. (2006). In: *Surface-Enhanced Raman Scattering: Physics and Applications*, Kneipp, K., Moskovits, M., Kneipp, H., Eds.; 19-45, Springer-Verlag, 3540335668, Heidelberg

Smith, W. E. (2008). Practical understanding and use of surface enhanced Raman scattering/surface enhanced resonance Raman scattering in chemical and biological analysis. *Chemical Society Reviews*, 37, 955-964, 0306-0012

Sriram, S.; Bhaskaran, M. & Holland, A. S. (2006a). The effect of post-deposition cooling rate on the orientation of piezoelectric $(Pb_{0.92}Sr_{0.08})(Zr_{0.65}Ti_{0.35})O_3$ thin films deposited by RF magnetron sputtering. *Semiconductor Science and Technology*, 21, 9, 1236-1243, 0268-1242

Sriram, S.; Bhaskaran, M. & Holland, A. S. (2006b). Surface morphology and stress analysis of piezoelectric strontium doped lead zirconate titanate thin films, *Proceedings of the SPIE (Micro- and Nanotechnology: Materials, Processes, Packaging, and Systems III)*, p. 64150J, 9780819465238, Adelaide, December 2006, SPIE, Bellingham

Sriram, S.; Bhaskaran, M.; Holland, A. S.; Short, K. T. & Latella, B. A. (2007). Measurement of high piezoelectric response of strontium-doped lead zirconate titanate thin films using a nanoindenter. *Journal of Applied Physics*, 101, 104910, 0021-8979

Sriram, S.; Bhaskaran, M.; du Plessis, J.; Short, K. T.; Sivan, V. P. & Holland, A. S. (2009a). Influence of oxygen partial pressure on the composition and orientation of strontium-doped lead zirconate titanate thin films. *Micron*, 40, 1, 104-108, 0968-4328

Sriram, S.; Bhaskaran, M.; Kostovski, G.; Mitchell, D. R. G.; Stoddart, P. R.; Austin, M. W. & Mitchell, A. (2009b). Synthesis of Self-Assembled Island-Structured Complex Oxide Dielectric Films. *Journal of Physical Chemistry C*, 113, 38, 16610-16614, 1932-7447

Sriram, S.; Bhaskaran, M.; Mitchell, D. R. G.; Short, K. T.; Holland, A. S. & Mitchell, A. (2009c). Microstructural and Compositional Analysis of Strontium-Doped Lead Zirconate Titanate Thin Films on Gold-Coated Silicon Substrates. *Microscopy and Microanalysis*, 15, 30-35, 1431-9276

Sriram, S.; Bhaskaran, M.; Mitchell, D. R. G. & Mitchell, A. (2010). Lattice Guiding for Low Temperature Crystallization of Rhombohedral Perovskite-Structured Oxide Thin Films. *Crystal Growth & Design*, 10, 2, 761-764, 1528-7483

Stewart, M. E.; Anderton, C. R.; Thompson, L. B.; Maria, J.; Gray, S. K.; Roger, J. A. & Nuzzo, R. G. (2008). Nanostructured Plasmonic Sensors. *Chemical Reviews*, 108, 2, 494-521, 1520-6890

Sumida, N. & Ikeda, K. (1991). Cross-sectional observations of gold-silicon reaction on silicon substrate in situ in the highvoltage electron microscope. *Ultramicroscopy*, 39, 1-4, 313-320, 0304-3991

Tunaboylu, B.; Harvey, P. & Esener, S. C. (1998). Microstructure of magnetron sputtered PLZT thin films on sapphire. *Integrated Ferroelectrics*, 19, 1-4, 11-32, 1058-4587

Viets, C. & Hill, W. (1998). Comparison of fibre-optic SERS sensors with differently prepared tips. *Sensors and Actuators B: Chemical*, 51, 1, 92-99, 0925-4005

Wasa, K.; Kitabatake, M. & Adachi, H. (2004). *Thin Film Materials Technology: Sputtering of Compound Materials*, Springer-Verlag, 9783540211181, Heidelberg

Wenzel, R.; Goesmann, F. & Schmid-Fetzer, R. (1998). Diffusion barriers in gold-metallized titanium-based contact structures on SiC. *Journal of Materials Science: Materials in Electronics*, 9, 109-113, 0957-4522

White, D. J. & Stoddart, P. R. (2005). Nanostructured optical fiber with surface-enhanced Raman scattering functionality. *Optics Letters*, 30, 6, 598-600, 0146-9592

Yu, Y. & Singh, R. N. (2000). Effect of composition and temperature on field-induced properties in the lead strontium zirconate titanate system. *Journal of Applied Physics*, 88, 12, 7249-7257, 0021-8979

Yu, Y.; Tu, J. & Singh, R. N. (2001). Phase stability and ferroelectric properties of lead strontium zirconate titanate ceramics. *Journal of the American Ceramic Society*, 84, 2, 333-340, 0002-7820

Zheng, H.; Reaney, I. M.; Lee, W. E.; Jones, N. & Thomas, H. (2002). Surface decomposition of strontium-doped soft $PbZrO_3$-$PbTiO_3$. *Journal of the American Ceramic Society*, 85, 1, 207-212, 0002-7820

Investigation of elevated temperature effects on multiple layer piezoelectric ultrasonic transducers with adhesive bondlines by self-heating

Zhengbin Wu[1,4], Sandy Cochran[2] and Bo Wu[3]
[1]Shenzhen Institutes of Advanced Technology, Chinese Academy of Sciences
[3]Huazhong University of Science and Technology
[4]The Chinese University of Hong Kong
China
[2]University of Dundee
UK

1. Introduction

Ultrasound is referred to as sound signal with frequencies above the upper limit of human audible range, i.e. 20 kHz. The modern study and usage of ultrasound is believed to originate from the underwater detection of Paul Langevin in 1917 (Mason, 1976). Since then, ultrasound has taken an important role in a broad variety of applications including underwater sonar (Foote, 2008), non-destructive testing (NDT) and evaluation (Thompson, 1985), biomedical diagnosis and therapy (Duck, 1998), and process industry (Hoyle, 1994), and so on.

The generation of ultrasound waves for the first time by electromechanical transduction was based on the piezoelectric effect, which was discovered by Pierre and Jacques Curie in 1880 (Ballato, 1995). With the piezoelectric effect, mechanical energy and electrical energy can be converted into each other. As one of the most commonly used mechanisms, the piezoelectric effect has been intensively put into practical applications to produce ultrasound with piezoelectric transducers. Such transducers have a great flexibility of either piezoelectric materials or structural modes. In the 20th century, metal oxide-based piezoelectric ceramics (Jaffe et al., 1971) and other man-made materials including single crystals (Ritter et al., 2000), polymers (Fukada, 2000), composites (Akdogan et al., 2005) and thin films (Lakin, 2005) have enabled designers to employ the piezoelectric effect in increasingly wide applications based on their specific properties.

As one of the important structural types, multiple layer or stacked transducers have been widely used for their specific benefits (Cochran & Hayward, 1999). In these multiple layer piezoelectric ultrasonic transducers, layers are connected mechanically in series and electrically in parallel, which can generate high strain through the establishment of high electric fields at relatively low operating voltages with their lower electrical impedance compared to single layer structure. Fig. 1 is the schematic diagram of an adhesively-bonded three-layer piezoelectric ultrasonic transducer.

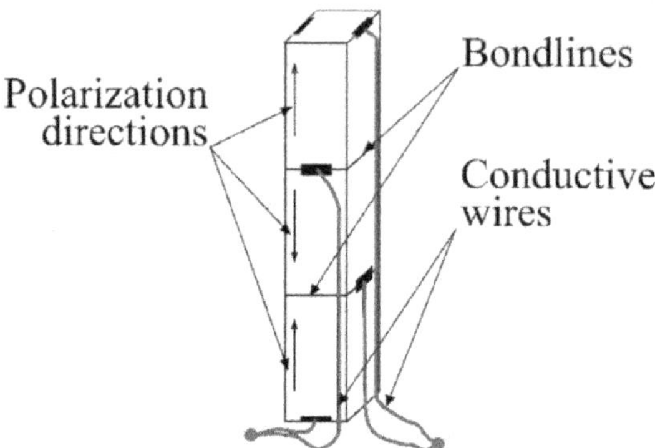

Fig. 1. Schematic diagram of a three-layer piezoelectric ultrasonic transducer

Fundamental work on multilayer structures has been carried out by several researchers (Sittig, 1967; Hossack & Auld, 1993). The applications of multilayer structures have also been investigated in the area of medical ultrasound imaging. Multilayer piezoelectric ceramic structures for 2D array transducers were reported in 1994 (Goldberg & Smith, 1994). It was found that the stacked structure has a better electrical impedance match and signal noise ratio (SNR) than the single layer. A multilayer PZT/polymer composite hybrid array was investigated in 1999 and it was concluded that the multilayer structure improved SNR and bandwidth (Mills & Smith, 1999). Cofired resonators with the layer number, N, up to 20 was reported in 1996 (Greenstein & Kumar, 1996). It was found the magnitude of the electrical impedance followed the relation $1/N$ of single layer, and the coupling coefficient k_t remained relatively constant below 15 layers. Systematic studies of multilayer ultrasonic transducers with piezocomposites for low frequency applications have been carried out with investigating various prototypes and simulations (Cochran & Hayward, 1999; Cochran et al., 1997; Cochran et al., 1998). In these work, the authors highlighted the advantages and the fabrication difficulties of the multilayer composite structures in specific low frequency applications, one of the difficulties being the bondlines. Multilayer ultrasonic transducers with non-uniform layer thicknesses have been studied with theoretical optimisation to improve their bandwidth (Abrar & Cochran, 2007).

Excited by high electric field signals at their resonant frequencies, various losses in piezoelectric materials resulting in heat and temperature increase normally limit the practical applications of multilayer devices (Yao et al., 2000). In practical high power operation, the dominant type of transducers are mass-spring-mass models, i.e. Langevin/Tonpilz transducers. These structures are free of adhesive bondlines subjected to tensile stress because adhesive bonds have low thermal conductivity, low Q_m value, and low maximum operating temperature. Different aspects such as loss effect, temperature rise, and self-heating mechanisms of piezoelectric devices have been studied on this topic, both on practical experiments (Robertson & Cochran, 2002; Dubus et al., 2002; Wehrsdorfer et al., 1995; Zheng et al., 1996) and theoretical analyses (Wang et al., 1996; Flint et al., 1995; Ye & Sun, 1997). Heat generation in various types of multilayer PZT-based actuators was reported in 1996 and a simplified analytic method to evaluate the temperature rise was also developed (Zheng et al., 1996). The phenomenology of mechanical, piezoelectric and dielectric losses was analyzed in 2001 (Uchino & Hirose, 2001). It was also reported that how to measure these losses separately. A nonequilibrium thermodynamic model was developed in 2004, which couples irreversible mechanical, electric, and thermal processes, all contributing to dissipation (Xia & Hanagud, 2004). It was verified that the dissipation nonlinearly depends on operating frequency, related to characteristic relaxation of irreversible processes. Generally, however, it is still difficult to accurately simulate the thermal property of multilayer piezoelectric devices for their complicated structural and material behaviors under dynamic excitation conditions. Experimental investigations into the dynamic and thermal properties of several Langevin/Tonpilz structures have been conducted from different aspects (Ochi et al., 1985; Yao et al., 2000; Zheng et al., 1996; Dubus et al., 2002; Pritchard et al., 2004). For such typical multilayer piezoelectric devices, the frequency bandwidth under high level dynamic conditions is seriously limited. This disadvantage of traditional structures may be overcome to use a multilayer piezoelectric composite structure. But it has been found that this is infeasible for its practical manufacture procedure issues (Robertson & Cochran, 2002).

To solve this problem, therefore, adhesively-bonded multilayer ultrasonic transducers have been suggested to be used for high power applications at relatively low frequencies. It is desired to investigate the thermal and mechanical behaviors of bondlines with newly developed adhesives in multilayer transducers under dynamic drive and self-heating conditions.

In this chapter, the behaviour of multilayer piezoelectric ultrasonic transducers with two typical adhesive bondlines under self-heating conditions is to be studied. The material characteristics of high Curie temperature piezoelectric cermaic, which was used to manufacture the multilayer structures, under different temperatures will be investigated. The effects of self-heating on temperature rise and mechanical output characteristics of adhesively-bonded multilayer ultrasonic transducers under various dynamic drive conditions are to be reported. Further investigation is also to be presented into the loss characteristics derived from measured time-temperature behaviour of multilayer piezoelectric ultrasonic transducers with adhesive bondlines, corresponding to stable temperature rises. In particular, the effect of the resultant higher temperature on the complete structures including the piezoelectric material and the adhesive bondlines will be considered. Various drive electric fields and the corresponding temperatures at resonant frequency will be reported too.

2. Material and Experimental Characterizations

Ultrasonic transducers and other smart devices made from piezoelectric materials operated under self-heating or elevated temperature desire high Curie temperature piezoelectric materials. Lead metaniobate piezoelectric ceramic (PZ35, Ferroperm Piezoceramics A/S, Hejreskovvej 18A, DK-3490 Kvistgård, Denmark) has a higher Curie temperature around 500 °C and therefore is approriate to elevated temperature applications. It also has very low width vibration mode. Its high loss properties compared to lead zirconate titanate (PZT) piezoceramic allow it to be manufactured into broadband ultrasonic transducers.

Two two-part epoxy adhesives were selected to bond multilayer structures here, RX771/HY1300 and silver-loaded epoxy. RX771/HY1300 (Robnor Resins Ltd, Swindon, UK) has commonly been used as the polymer in piezocomposite materials. It has low viscosity of 0.6-1.2 Pa s at 25°C prior to curing and is able to produce bondlines with thickness in the range of several microns (Wu el al., 2005). Silver-loaded epoxy (Part No. 186-3616, RS Components Ltd, Corby, UK) is a highly viscous gray paste before curing, and often used to establish electrical contacts in electronic devices. The glass transition temperature T_g of RX771/HY1300 is 60 °C after cured while silver-loaded epoxy has a maximum useful temperature of 150 °C after set, which offers a potential to overcome the disbond issue of multilayer structures resulted from the weak thermal characteristics of epoxy adhesive bondlines under elevated temperature.

The electrical impedance characteristics of the piezoelectric ultrasonic transducers were measured by Impedance Analyzer HP4192A (Agilent, Winnersh Triangle, Wokingham, Berkshire, UK). The surface displacement of the transducers was measured with a laser vibrometer system (Laser Vibrometer Controller OFV-2700-2 and Differential Fiber Optic Unit OFV-512, Polytec GmbH, Waldbronn, Germany).

3. Effects of Elevated Temperature on Piezoelectric Material Parameters

In oder to predict the effects of self-heating or elevated temperature on the multilayer piezoelectric transducers made from it, it is fundamentally crucial to consider its material constant variations with increased temperature beforehand.

For the characterization of piezoelectric materials, the standard method is normally used to analyze the electrical impedance resonant characteristics (ANSI/IEEE, 1987). The Piezoelectric Resonance Analysis Program (PRAP, TASI Technical Software Inc., Kingston, Ontario, Canada) offers a quick convergent fitting and analysis process and a more accurate characterization of lossy piezoelectric materials such as lead metaniobate or piezoelectric polymers (Kwok et al., 1997). Elevated temperature may enhance the loss mechanisms of piezoelectric materials (Sherrit et al., 1999). Here the behaviour of lead metaniobate piezoelectric ceramic from ambient temperature to higher temperatures was investigated. Material parameters were precisely calculated by theoretical fitting and analysis with PRAP.

A lead metaniobate disc under thickness extensional vibration mode was studied here. The diameter and thickness are 12 mm and 0.7 mm respectively. When the measurements going on, the lead metaniobate sample was put inside a temperature-controllable oven and connected outside with the HP4192A Impedance Analyzer through a short braided coax cable, which has 50 Ohm impedance and low attenuation over the studied frequency domain. A digital thermocouple meter was used to record the inside temperature of the oven. The electrical impedance resonant characteristics of the lead metaniobate sample were measured at 25 °C, 40 °C, 60 °C, 80 °C, and 100 °C, respectively.

The basic theoretical model of the PRAP analysis package was developed (Smits, 1976; Sherrit et al., 1992). The initial value of the elastic stiffness constant under constant displacement $C_{33}{}^{D*}$ is calculated from Equations (1) and (2).

$$f_p^* = f_p\left(1 - iw\right)^{-1/2} \tag{1}$$

$$C_{33}^{D*} = 4\rho t^2 f_p^{*2} \tag{2}$$

Here f_p is the frequency where the real part of $fZ(f)$ has a maximum, ρ and t are the density and thickness of the sample respectively, w is the ratio of the frequency difference between where the imaginary part of $fZ(f)$ has the minimum and the maximum to f_p, and * represents complex variants.

The imaginary part of all the material parameters represents the respective loss. Two arbitrary points at frequencies either side of f_p along with the initial value for the elastic constant $C_{33}{}^{D*}$ are used to calculate the electromechanical coupling constant K_t^* and the permittivity under constant strain $\varepsilon_{33}{}^{S*}$ of the sample following Equation (3)

$$Z(f)^* = \frac{t}{i2\pi f \varepsilon_{33}^{S*} A}\left[1 - \frac{K_t^{*2}\tan\left(\pi f t\sqrt{\rho/C_{33}^{D*}}\right)}{\pi f t\sqrt{\rho/C_{33}^{D*}}}\right] \tag{3}$$

where, f and A are frequency and the sample electrode area respectively. The obtained values K_t^*, $\varepsilon_{33}{}^{S*}$ and f_p^* can be used to calculated a new $C_{33}{}^{D*}$ following the relationship of Equation (3). The process continues iteratively until the calculated results becoming convergent. The complex piezoelectric constant e_{33}^* can be calculated from Equation (4).

$$e_{33}^* = \sqrt{K_t^{*2} C_{33}^{D*} \varepsilon_{33}^{S*}} \tag{4}$$

PRAP was used to fit and analyze the measured electrical impedance resonant characteristics. The magnitude and phase of the measured data and fitted results at 25 °C, 60 °C, and 100 °C are shown in Fig. 2. From Fig. 2, it is found that there is a good agreement between the fitted results and measured data.

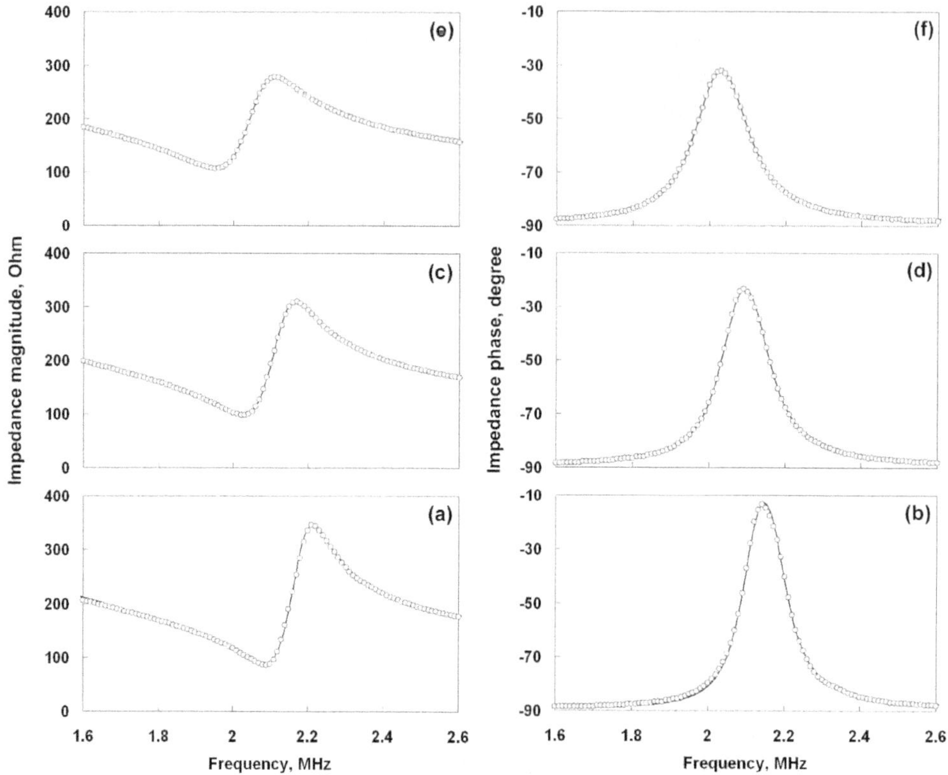

Fig. 2. Magnitude (a, c, and e) and phase (b, d, and f) impedance resonant characteristics of lead metaniobate at 25 °C (a and b), 60 °C (c and d), and 100 °C (e and f) (circle represents measured value, and solid line calculated data.)

Temperature	25 °C	40 °C	60 °C	80 °C	100 °C
f_p, MHz	2.19	2.16	2.13	2.10	2.07
Imag(f)	0.0543	0.0585	0.0643	0.0708	0.0766
C_{33}^D, 10^{10} N/m²	5.00	4.89	4.77	4.61	4.48
Imag(C)	0.248	0.266	0.288	0.311	0.332
K_t	0.296	0.297	0.300	0.302	0.304
Imag(K)	0.00678	0.00416	0.00216	0.00160	0.00188
ε^S_{33}, 10^{-9} F/m	2.40	2.44	2.50	2.56	2.62
Imag(ε)	-0.0326	-0.0505	-0.0400	-0.0504	-0.0548
e_{33}, C/m²	3.248	3.247	3.278	3.283	3.298
Imag(e)	0.133	0.100	0.096	0.095	0.108

Table 1. Some material parameters of lead metaniobate at various temperatures

Table 1 shows the analyzed material property data at 25 °C, 40 °C, 60 °C, 80 °C, and 100 °C, respectively. The loss component of each parameter is represented by the imaginary part value. Increasing the temperature from 25 °C to 100 °C alters f_p by more than 5% from 2.19

MHz to 2.07 MHz. Furthermore, Imag(f), which reflects the elastic damping characteristic, becomes greater with the higher temperatures. These are also reflected by the data in Fig. 2. Because the thermal expansion effect of piezoelectric ceramics in the temperature range studied here is negligible (Takahashi et al., 1974), the shift of f_p is attributed to the thermal variation of sound velocity of lead metaniobate. This hypothesis is supported by the 10% reduction of $C_{33}{}^D$ over the same temperature range. The sound velocity of lead metaniobate is a function of the elastic stiffness coefficient and density. Whilst the density was found to be temperature invariant, it can be seen from Table 1 that $C_{33}{}^D$ has a value of 5.00×10^{10} N/m^2 at 25 °C but only 4.48×10^{10} N/m^2 at 100 °C. Increased temperature also caused increasing Imag(C). The sound velocity and the elastic damping changes are thought to result from the characteristic variation of the sound wave propagation in the material at various temperatures. The analyzed elastic constants of lead metaniobate presented here have a different trend from those reported data of PZT ceramics (Sherrit et al., 1999), wherein the elastic compliance coefficients shifted downwards under higher temperatures. However the dielectric behaviour follows similar trends to that reported for PZT (Sherrit et al., 1999) and the authors observed that $\varepsilon_{33}{}^S$ increases by 7% from 2.40×10^{-9} F/m to 2.62×10^{-9} F/m as temperature increases from 25 °C to 100 °C. K_t and e_{33} have changes less than 3% with the elevating temperature.

4. Self-heating Effects

Now that the stable elevated temperature effects on the piezoelectric material properties have been studied, in this section, the effects of self-heating and resulted stable temperature rise on multilayer piezoelectric transducers with adhesive bondlines are to be investigated. Two two-layer lead metaniobate transducers bonded with RX771/HY1300 and silver-loaded epoxy respectively were prepared. Each active layer of two-layer thickness extensional mode transducers has a lateral dimension of 20 mm square and thickness of 1.55 mm. After fabrication, the bondlines of RX771/HY1300 and silver-loaded epoxy have mean thicknesses of 10.4 μm and 104 μm respectively, which were determined by an acoustical fitting process (Wu et al., 2003). A single-layer lead metaniobate transducer with the same lateral dimensions as the two layer transducers and the thickness of 1.49 mm was also prepared and tested. The thickness extensional vibration mode single layer transducer was used to compare to the multilayer structures, which is able to be used to justify various loss effects from the piezoelectric material and adhesive bondlines on the multilayer transducers.

A thermocouple was mounted near the centre of one major working surface of those samples, as shown in Fig. 3, to measure its surface temperature, which connected to a digital meter (0.1 °C resolution). The effect of this thermocouple on the transducer is considered to have been very minor and is neglected here. In each experiment, a stable temperature value was recorded two minutes after the electrical signal was applied.

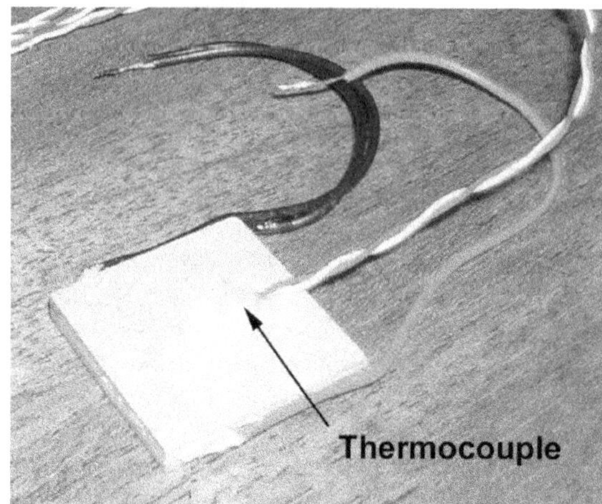

Fig. 3. Prototype of a two-layer lead metaniobate transducer

The surface displacement frequency response of the two multilayer transducers at ambient temperature was measured. The normalized surface displacement responses in frequency domain were obtained by the absolute displacement values dividing the maximum, 233 nm, of the two transducers under 50-cycle tone burst sinusoidal driving signal at fixed 200 V peak-peak level without self-heating, which are shown in Fig. 4. The thickness extension within the frequency domain between 400 and 500 kHz is the dominant vibration mode for the both transducers, while the much lower peaks around 330 kHz represent the width-vibrational modes. Those around 1.45 MHz are the third harmonies of the thickness extension vibration mode of the multilayer transducers. Therefore the latter two vibration modes in the frequency range up to 1.8 MHz can simply be ignored. The multilayer transducer with silver-loaded epoxy bondline has the lower thickness extensional vibration mode resonant frequency, 419 kHz, than the one with RX771/HY1300, 496 kHz for their different bondlines and longitudinal ultrasound signal propagation time along the thickness direction.

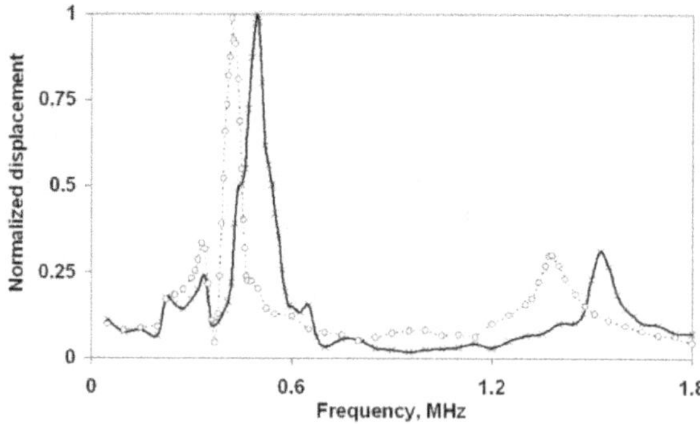

Fig. 4. Surface displacement frequency spectrum of the two-layer lead metaniobate transducers bonded with RX771/HY1300 (solid line and cross) and silver-loaded epoxy (dashed line and circle), respectively

For a working ultrasonic transducer, the total electrical power supplied by the driver P_T can be calculated as equation (5).

$$P_T = \frac{(E \cdot t_E)^2}{R_E} \cdot \eta \tag{5}$$

Here E is root of mean square (RMS) value of applied electric field, t_E is the transducer thickness under applied electric field, η is the driving duty cycle, and R_E the transducer's electrical resistance.

For the transducers studied here, operating in air, the radiated acoustic power is approximately zero because of the mismatch of acoustic impedance. Therefore P_T is equal to the sum of the power losses, which comprise mechanical power loss, P_M, and dielectric power loss, P_D, resulting in heating and temperature rises. The power losses are a function of the stable temperature rise in Kelvin of the piezoelectric transducers under constant (zero) stress condition, δT, and the effective heat transfer constant, k_H. Their relationships are shown as equation (6).

$$P_T = P_M + P_D = k_H \cdot \delta T \tag{6}$$

Here, $\delta T = T - T_A$. T is the equilibrium temperature at the surface of the transducer, T_A the ambient temperature. At resonance the power dissipation of a transducer is thought to be primarily due to the mechanical loss (Sherrit, 2001). Under free load conditions, the relationships between P_M, the mechanical loss resistance R_M, the RMS displacement of the working surface, d, the driving frequency, f, and η is described as equation (7).

$$P_M = R_M \cdot (2\pi f \cdot d)^2 \cdot \eta \tag{7}$$

All the transducers reported in this chapter were tested in air by tone burst sinusoidal signals with different duty cycles. The drive frequencies of all the transducers were their measured resonant frequencies under longitudinal thickness extensional vibration mode without self-heating. The detailed excitation conditions are listed in Table 2.

	Frequency (kHz)	Lower duty cycle		Higher duty cycle	
		Cycle number	Duty cycle	Cycle number	Duty cycle
Two-layer transducer with RX771/HY1300	496	50	1.0%	1000	20%
Two-layer transducer with silver-loaded epoxy	419	50	1.2%	1000	24%
Single-layer transducer	1021	100	1.0%	2000	20%

Table 2. Excitation conditions for the lead metaniobate transducers

Excited by low duty cycles at resonance, the measured temperature and resonant frequency of all the transducers remain almost unchanged due to the low power and limited amplitude electric field. As shown in Fig. 5, under no more than 1.2% duty cycle excitation, the resonant displacement of all the transducers has a nearly linear relationship with the driving electric field within the range up to around 160 V mm^{-1}. According to the relationships shown in equations (5) and (7), this indicates that the mechanical power loss is approximately proportional to the low amplitude driving power. It is also found that the displacement responses of the two layer transducers are approximately twice that of the single layer at a given electric field level indicating that under low power and ambient temperature conditions, the adhesive bondlines support the resonant vibration of multilayer structures very well.

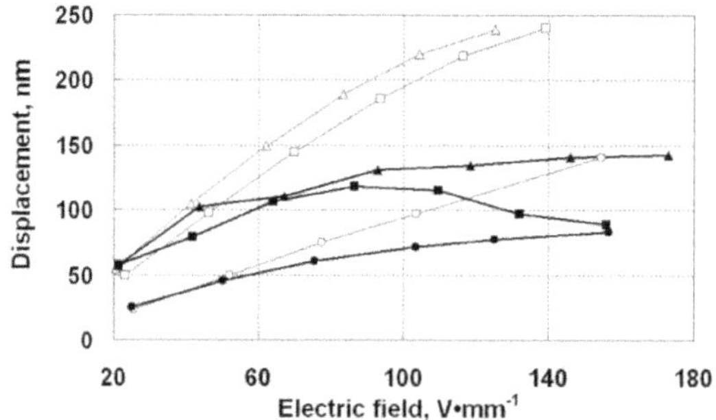

Fig. 5. Measured displacement amplitude under various driving electric fields of the transducers (solid marks and thick lines, high duty cycle; hollow marks and thin lines, low duty cycle; circle, single layer transducer; triangle, two-layer transducer bonded with RX771/HY1300; square, two-layer transducer bonded with silver-loaded epoxy)

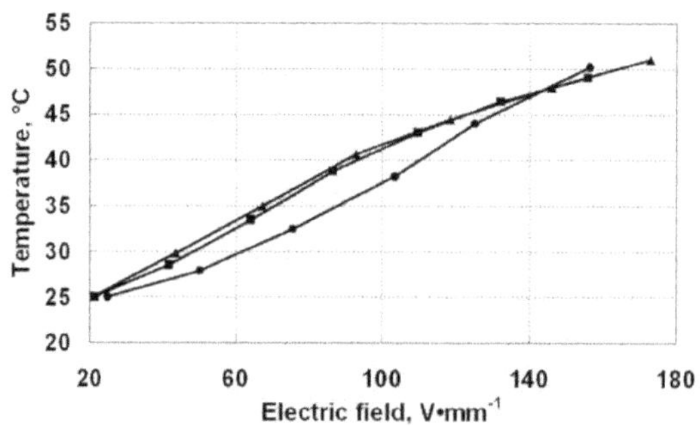

Fig. 6. Measured temperature characteristics under various driving electric fields of the transducers (solid marks and thick lines, high duty cycle; hollow marks and thin lines, low duty cycle; circle, single layer transducer; triangle, two-layer transducer bonded with RX771/HY1300; square, two-layer transducer bonded with silver-loaded epoxy)

With the duty cycle increased to above 20%, as shown in Fig. 6, the temperature of the transducers rises from ambient to around 50 °C with electric field increasing from below 30

V mm^{-1} to above 150 V mm^{-1}. According to equation (6), the elevated temperature is resulted from the higher excitation and power losses. For the single layer ceramic sample, as shown in Fig. 5, the displacement response presents a decreased gradient with the high duty cycle excitation electric field increasing from approximately 50 V mm^{-1} to 155 V mm^{-1}, indicating that, at a given electric field, the displacement at high duty cycle is lower than at low duty cycle excitation. This is attributed to the variation of electrical impedance frequency characteristics of this ceramic with the elevated temperature resulting from high duty cycle driving as presented in the previous section. The peak displacement of the two-layer transducer with silver-loaded epoxy under high duty cycle excitation corresponds to an electric field of 86.5 V mm^{-1}. Above this value, the displacement decreases with the increased electric field and resulting in higher temperature above 38.8 °C, which is due to the reduced elastic coefficient of the bondline. For the two layer transducer with RX771/HY1300, although its surface displacement increases with increasing electric field in the whole range covered in the present investigation, the value is reduced and is less than twice that of the single layer transducer under high duty cycle drive conditions. This is attributed to the lower elastic coefficient of RX771/HY1300 with increasing temperature, below its T_g, and higher excitation amplitudes and power losses (Wu & Cochran, 2008).

5. Loss Characteristics of Self-heating

In the previous section, the self-heating effects on the performance of multilayer piezoelectric ultrasonic transducers with adhesive bondlines have been investigated. In this section, further study is to be carried out to analyze the loss characteristics of self-heating.

The time-temperature characteristics of the single and two layer transducers, which have been introduced in the previous section, at resonances under different drive electric fields were measured and shown in Fig. 7. As shown in Fig. 7, the temperature rise of all transducers became stable less than five minutes after excitations applied. The stable temperature rise increases with the enhanced electric fields for all the piezoelectric ultrasonic transducers. The time-temperature behaviour of a piezoelectric device can be described by Equation (8) (Zheng et al., 1996)

$$\Delta T(t_s) = \delta T (1 - e^{-t_s/\alpha})\qquad(8)$$

where $\Delta T(t_s)$ is the temperature rise of the piezoelectric device at the instantaneous time t_s, and α the time constant. Using Equation (8), the theoretical exponential time-temperature characteristics of all the transducers under different electric fields were calculated with measured δT and fitted α and shown in Fig. 7. Generally, the experimental results agree the theoretical data well but it is also found that the measured results rise steeper than theory at the beginning of the experiments. This indicates that more than one time constants are involved in the self-heating processes. Further accurate characterization of the whole self-heating process is proposed to be studied in the future work.

The stable temperature rise δT, which is the function of loss, can be approximately expressed as equation (9) (Zheng et al., 1996).

$$\delta T = \frac{u f \eta v_e}{k(T) A_t} \tag{9}$$

Here u the total loss of the actuator per unit volume and driving cycle, and v_e the active volume of the actuator. A_t and $k(T)$ are the total surface area of the device and the overall heat transfer coefficient respectively.

The time constant α in Equation (8) is defined as

$$\alpha = \frac{v \rho c}{k(T) A_t} \tag{10}$$

where v is the total volume of the device. c is the specific heat capacity of the actuator. In the present work, because the volume of the relatively thin bondlines was negligible compared to the piezoelectric material and no other material was present, $v_e = v$ then the function below is built up by dividing Equation (9) to Equation (10)

$$\frac{\delta T}{\alpha} = \frac{f \eta}{\rho} \cdot \beta \tag{11}$$

where the loss constant $\beta = u/c$. Since c is a constant for each specific material, β is proportional to u and is used to characterize the loss of drive signal in the piezoelectric devices, here relating only to the relevant piezoceramic used in this work. Using Equation (11), with measured δT, fitted α, density of lead metaniobate ρ, and driving signal frequency f, and duty cycle η, the loss constant β of the piezoelectric ultrasonic transducers is calculated to characterize their loss. The temperature and the normalized loss constant for each transducer versus surface displacement are shown in Fig 8. The normalized loss constant is calculated by loss constant β over the maximum of all the transducers under different drives. It is found that the normalized loss constant corresponds well with the temperature for all transducers as shown in Fig.8.

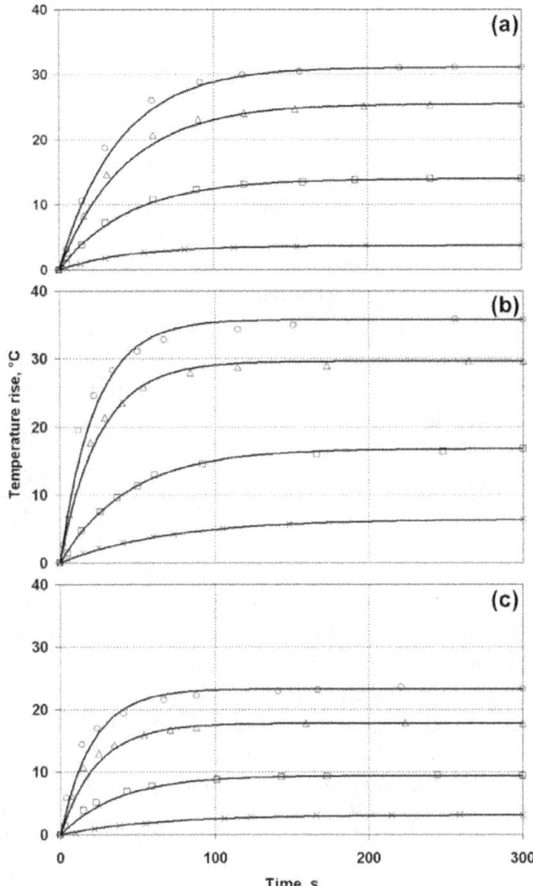

Fig. 7. Comparison of time–temperature characteristics of the single-layer (a) and two-layer transducers bonded with RX771/HY1300 (b) and with silver-loaded epoxy (c). Experimental results under various electric fields are shown (markers) along with theoretically fitted data (solid lines). [(a) cross, 50.1 V·mm^{-1}; rectangle, 103.4 V·mm^{-1}; triangle, 156.4 V·mm^{-1}; and circle, 183.2 V·mm^{-1}; (b) cross, 43.2 V·mm^{-1}; rectangle, 92.4 V·mm^{-1}; triangle, 191.8 V·mm^{-1}; and circle, 233.9 V·mm^{-1}; (c) cross, 45.7 V·mm^{-1}; rectangle, 97.0 V·mm^{-1}; triangle, 198.0 V·mm^{-1}; and circle, 249.2 V·mm^{-1}]

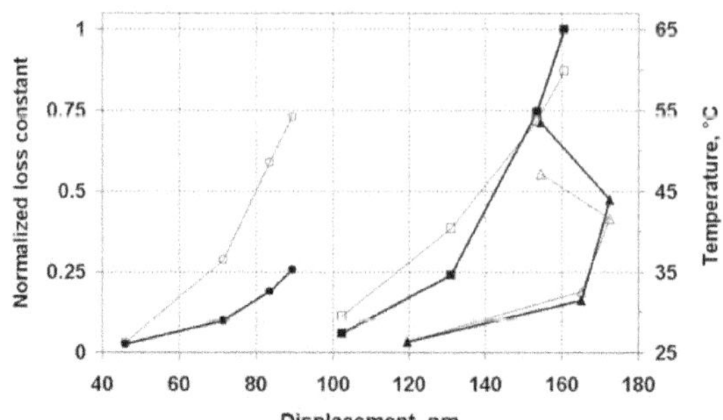

Fig. 8. The normalized loss constant (solid markers and solid lines) and the temperature (hollow markers and thin lines) of the sing-layer and two-layer transducers versus measured displacement. (circle, single-layer transducer; square, two-layer transducer with RX771/HY1300; triangle, two-layer transducer with a silver-loaded epoxy)

Under resonant drive conditions, the total loss is primarily due to the mechanical losses, although the ferroelectric hysteresis loss may become dominant under off-resonance drive conditions (Uchino & Hirose, 2001). For the single-layer transducer and the transducer bonded with RX771/HY1300, the normalized loss constant and the temperature are increased with the enhanced surface displacement in the temperature range below 60 °C, i.e. T_g of RX771/HY1300. The total loss and the temperature variation of these transducers are positive functions of the mechanical loss, related directly with the surface displacement. However, for the two-layer transducer with silver-loaded epoxy, the curves have turning points where the normalized loss constant and temperature are 0.47 and 41.6 °C, respectively. Above these values, the surface displacement (mechanical loss) drops down although the total loss and temperature are still increasing. The temperature has significantly affected the mechanical properties and the resonant frequency of this transducer. When the temperature above 41.6 °C, the transducer started to shift to off-resonance condition and the total loss characteristics started to vary off from resonance condition, which the authors believe this is because the silver-loaded bondline starts to soften though it can still bond the piezoceramics at even higher temperature (Wu & Cochran, 2010).

6. Conclusion

In this chapter, the behaviour of adhesively-bonded multilayer piezoelectric ultrasonic transducers with two typical adhesives has been investigated under self-heating conditions. The material characteristic variation of high Curie temperature lead metaniobate piezoelectric cermaic, used to manufacture the multilayer structures, with different temperature were studied. The temperature rise and mechanical output characteristics of adhesively-bonded multilayer ultrasonic transducers resulted from self-heating under dynamic drive conditions were presented. Investigation has also been carried out into the loss characteristics derived from measured time-temperature behaviour of multilayer piezoelectric ultrasonic transducers with adhesive bondlines. It is concluded that elevated temperature changes the characteristics of the piezoelectric material in the same way, the dramatically different effects of self-heating and the elevated temperature on the performances of the multilayer transducers are still presented with the respective effects on the elastic characteristics of each adhesive bondline, and in turn the resonant characteristics of the whole structure to different degrees.

7. Acknowledgement

Zhengbin Wu was sponsored by a University of Paisley (University of the West of Scotland) Studentship and Sandy Cochran was supported by a UK EPSRC Advanced Fellowship during the period of the work described in this chapter. Zhengbin Wu is also supported by the National Natural Science Foundation of China (NSFC, No. 10804075) and Shenzhen Science and Technology Research and Development Foundation 2008 (Approval no. 343).

8. References

Abrar, A. & Cochran, S. (2007). Mathematical optimization of multilayer piezoelectric devices with nonuniform layers by simulated annealing. *IEEE Transactions on Ultrasonics, Ferroelectrics, and Frequency Control*, Vol. 54, No. 10, October 2007, pp. 1920-1929, ISSN 0885-3010

Akdogan, E. K.; Allahverdi, M. & Safari, A. (2005). Piezoelectric composites for sensor and actuator applications. *IEEE Transactions on Ultrasonics, Ferroelectrics, and Frequency Control*, Vol. 52, No. 5, May 2005, pp. 746-775, ISSN 0885-3010

ANSI/IEEE Std 176-1987 (1987). *IEEE Standard on Piezoelectricity*, IEEE, New York

Ballato, A. (1995). Piezoelectricity: old effect, new thrusts. *IEEE Transactions on Ultrasonics, Ferroelectrics, and Frequency Control*, Vol. 42, No. 5, September 1995, pp. 916-926, ISSN 0885-3010

Cochran, A.; Reynolds, P. & Hayward, G. (1997). Multilayer piezocomposite transducers for applications of low frequency ultrasound, *Proceedings of IEEE International Ultrasonics Symposium*, pp. 1013-1016, ISBN 0-7803-4153-8, Toronto, Canada, October 1997, IEEE, New York

Cochran, A.; Reynolds, P. & Hayward, G. (1998). Progress in stacked piezocomposite ultrasonic transducers for low frequency applications. *Ultrasonics*, Vol. 36, No. 10, October 1998, pp. 969-977, ISSN 0041-624X

Cochran, A. and Hayward, G. (1999). Multilayer piezocomposite ultrasonic transducers operating below 50 kHz, *Proceedings of IEEE International Ultrasonics Symposium*, pp. 953-956, ISSN 1051-0117, Nevada, USA, October 1999, IEEE, New York

Dubus B.; Haw, G.; Granger, C. & Ledez, O. (2002). Characterization of multilayered piezoelectric ceramics for high power transducers. *Ultrasonics*, Vol. 40, No. 1-8, May 2002, pp. 903–906, ISSN 0041-624X

Duck, F. A.; Baker, A. C. & Starritt, H. C. (1998). *Ultrasound in Medicine*, Institute of Physics Publishing, ISBN 0-75-030593-8, London

Flint, E.; Liang, C. & Rogers, C. A. (1995). Electromechanical analysis of piezoelectric stack active member power consumption. *Journal of Intelligent Material Systems and Structures*, Vol. 6, No. 1, January 1995, pp. 117–124, ISSN 1045-389X

Foote, K. G. (2008). Underwater acoustic technology: review of some recent developments, *Proceedings of Oceans 2008 Conference*, pp. 1327-1332, ISSN 0197-7385, Quebec, Canada, September 2008, IEEE, New York

Fukada, E. (2000). History and recent progress in piezoelectric polymers. *IEEE Transactions on Ultrasonics, Ferroelectrics, and Frequency Control*, Vol. 47, No. 6, November 2000, pp. 1277-1290, ISSN 0885-3010

Goldberg, R. L. & Smith, S. W. (1994). Multilayer piezoelectric ceramics for two-dimensional array transducers. *IEEE Transactions on Ultrasonics, Ferroelectrics, and Frequency Control*, Vol. 41, No. 5, September 1994, pp. 761-771, ISSN 0885-3010

Greenstein, M. & Kumar, U. (1996). Multilayer piezoelectric resonators for medical ultrasound transducers. *IEEE Transactions on Ultrasonics, Ferroelectrics, and Frequency Control*, Vol. 43, No. 4, July 1996, pp. 620-622, ISSN 0885-3010

Hossack, J. A. & Auld, B. A. (1993). Improving the characteristics of a transducer using multiple piezoelectric layers. *IEEE Transactions on Ultrasonics, Ferroelectrics, and Frequency Control*, Vol. 40, No. 2, March 1993, pp. 131-139, ISSN 0885-3010

Hoyle, B. S. & Luke, S. P. (1994). Ultrasound in the process industries. *Engineering Science and Education Journal*, Vol. 3, No. 3, June 1994, pp. 119-122, ISSN 0963-7346

Jaffe, B.; Cook, W. R. & Jaffe, H. (1971). *Piezoelectric Ceramics*, Academic Press Inc., ISBN 0-12-379550-8, London

Kwok, K. W.; Chan, H. L. W. & Choy, C. L. (1997). Evaluation of the material parameters of piezoelectric materials by various methods. *IEEE Transactions on Ultrasonics, Ferroelectrics, and Frequency Control*, Vol. 44, No. 4, July 1997, pp. 733–742, ISSN 0885-3010

Lakin, K. M. (2005). Thin film resonator technology. *IEEE Transactions on Ultrasonics, Ferroelectrics, and Frequency Control*, Vol. 52, No. 5, May 2005, pp. 707-716, ISSN 0885-3010

Mason, W. P. (1976). Sonics and ultrasonics: early history and applications. *IEEE Transactions on Sonics and Ultrasonics*, Vol. SU-23, No. 4, July 1976, pp. 224-232, ISSN 0018-9537

Mills, D. M. & Smith, S. W. (1999). Multi-layered PZT/polymer composites to increase signal-to-noise ratio and resolution for medical ultrasound transducers. *IEEE Transactions on Ultrasonics, Ferroelectrics, and Frequency Control*, Vol. 46, No. 4, July 1999, pp. 961-971, ISSN 0885-3010

Ochi, A.; Takahashi, S. & Tagami, S. (1985). Temperature characteristics for multilayer piezoelectric ceramic actuator. *Japanese Journal of Applied Physics*, Vol. 24, Supplement 24-3, 1985, pp. 209-212, ISSN 0021-4922

Pritchard, J.; Ramesh, R. & Bowen, C.R. (2004). Time-temperature profiles of multi-layer actuators, *Sensors and Actuators A: Physical*, Vol. 115, No. 1, September 2004, pp. 140-145, ISSN 0924-4247

Ritter, T.; Geng, X.; Shung, K. K.; Lopath, P. D.; Park, S-E. & Shrout, T. R. (2000). Single crystal PZN/PT-polymer composites for ultrasound transducer applications. *IEEE Transactions on Ultrasonics, Ferroelectrics, and Frequency Control*, Vol. 47, No. 4, July 2000, pp. 792-800, ISSN 0885-3010

Robertson, D. & Cochran, S. (2002). Experimental investigation of alternative pre-stress components for a 3-1 connectivity multilayer piezoelectric-polymer composite ultrasonic transducer. *Ultrasonics*, Vol. 40, No. 1-8, May 2002, pp. 913-919, ISSN 0041-624X

Sherrit, S.; Wiederick, H. D. & Mukherjee, B. K. (1992). Non-iterative evaluation of the real and imaginary material constants of piezoelectric resonators. *Ferroelectrics*, Vol. 134, No. 1-4, September 1992, 134, pp. 111-119, ISSN 0015-0193

Sherrit, S.; Yang, G.; Wiederick, H. D. & Mukherjee, B. K. (1999). Temperature dependence of the dielectric, elastic and piezoelectric material constants of lead zirconate titanate ceramics, *Proceedings of International Conference on Smart Material, Structures and Systems*, pp. 121-126, Bangalore, India, July 1999

Sherrit, S.; Bao, X.; Sigel, D. A.; Gradziel, M. J.; Askins, S. A.; Dolgin, B. P. & Bar-cohen, Y. (2001). Characterization of transducers and resonators under high drive levels, *Proceedings of IEEE International Ultrasonics Symposium*, pp. 1097-1100, ISBN 0-7803-7177-1, Atlanta, USA, October 2001, IEEE, New York

Sittig, E. K. (1967). Transmission parameters of thickness-driven piezoelectric transducers arranged in multilayer configurations, *IEEE Transactions on Sonics and Ultrasonics*, Vol. SU14, No. 4, October 1967, pp. 167-174, ISSN 0018-9537

Smits, J.G. (1976). Iterative method for accurate determination of the real and imaginary parts of the materials coefficients of piezoelectric ceramics. *IEEE Transactions on Sonics and Ultrasonics*, Vol. SU-23, No. 6, November 1976, pp. 393–402, ISSN 0018-9537

Takahashi, M.; Yamauchi, F. & Takahashi, S. (1974). Stabilization of resonance frequencies in piezoelectric ceramic resonators against sudden temperature change, *Proceedings of IEEE Frequency Control Symposium*, pp. 109–116, New Jersey, USA, May 1974, IEEE, New York

Thompson, R. B. & Thompson, D. O. (1985). Ultrasonics in nondestructive evaluation. *Proceedings of the IEEE*, Vol. 73, No. 12, December 1985, pp. 1716-1755, ISSN 0018-9219

Uchino, K. & Hirose, S. (2001). Loss mechanisms in piezoelectrics: how to measure different losses separately. *IEEE Transactions on Ultrasonics, Ferroelectrics, and Frequency Control*, Vol. 48, No. 1, January 2001, pp. 307-321, ISSN 0885-3010

Wang, X.; Ehlers, C. & Neitzel, M. (1996). Electro-mechanical dynamic analysis of the piezoelectric stack. *Smart Materials and Structures*, Vol. 5, No. 4, August 1996, pp. 492–500, ISSN 0964-1726

Wehrsdorfer, E.; Borchhardt, G. & Karthe, W. (1995). Large signal measurements on piezoelectric stacks. *Ferroelectrics*, Vol. 174, No. 1, December 1995, pp. 259–275, **ISSN** 1563-5112

Wu, Z.; Abrar, A.; McRobbie, G.; Gallagher, S. & Cochran, S. (2003). Implementation of multilayer ultrasonic transducer structures with optimised non-uniform layer thicknesses, *Proceedings of IEEE International Ultrasonics Symposium*, pp. 1175-1178, ISBN 0-7803-7922-5, Hawaii, USA, October 2003, IEEE, New York

Wu, Z.; Cochran, S.; McRobbie, G. & Kirk, K. J. (2005). Nondestructive and destructive investigation of bondlines for high-power multilayer ultrasonic transducers for underwater sonar, *Proceedings of IEEE International Ultrasonics Symposium*, pp. 2219-2222, ISBN 0-7803-9382-1, Rotterdam, Netherlands, September 2005, IEEE, New York

Wu, Z. & Cochran, S. (2008). Characterisation of self-heating effects on multilayer ultrasonic transducers with adhesive bondlines. *Electronics Letters*, Vol. 44, No. 22, October 2008, pp. 1333-1334, ISSN 0013-5194

Wu, Z. & Cochran, S. (2010). Loss effects on adhesively-bonded multilayer ultrasonic transducers by self-heating. *Ultrasonics*, Vol. 50, No. 4-5, April 2010, pp. 508-511, ISSN 1874-9968"

Xia, L. & Hanagud, S.V. (2004). Extended irreversible thermodynamics modeling for selfheating and dissipation in piezoelectric ceramics. *IEEE Transactions on Ultrasonics, Ferroelectrics, and Frequency Control*, Vol. 5, No. 12, December 2004, pp. 1582-1592, ISSN 0885-3010

Yao, K.; Uchino, K.; Xu, Y.; Dong, S. & Lim, L. (2000). Compact piezoelectric stacked actuators for high power applications. *IEEE Transactions on Ultrasonics, Ferroelectrics, and Frequency Control*, Vol. 47, No. 4, July 2000, pp. 819-825, ISSN 0885-3010

Ye, D. & Sun, J. Q. (1997). Dynamic analysis of piezoelectric stack actuators including thermal and pyroelectric effects, *Proceedings of SPIE International Symposium on Smart Structures and Materials: Mathematics and Control in Smart Structures*, pp. 619-629, ISBN 0-8194-2452-8, San Diego, USA, March 1997, SPIE, Bellingham

Zheng, J.; Takahashi, S.; Yoshikawa, S.; Uchino, K. & deVries, J. W. C. (1996). Heat generation in multilayer piezoelectric actuators. *Journal of American Ceramic Society*, Vol. 79, No. 12, December 1996, pp. 3193–3198, ISSN 0002-7820

3

Modelling shear lag phenomenon for adhesively bonded piezo - transducers

Suresh Bhalla and Ashok Gupta

*Department of Civil Engineering, Indian Institute of Technology Delhi, Hauz Khas,
New Delhi 110016, India*

1. Introduction

During the last one and a half decades, the electro-mechanical impedance (EMI) technique has emerged as a universal cost-effective technique for structural health monitoring (SHM) and non destructive evaluation (NDE) of all types of engineering structures and systems (Sun et al., 1995; Ayres et al., 1998; Soh et al., 2000; Park et al., 2000, 2001; Giurgiutiu and Zagrai, 2000, 2002; Bhalla & Soh, 2003, 2004a, 2004b, 2004c). In this technique, a lead zirconate titanate piezo-electric ceramic (PZT) patch, surface bonded to the monitored structure, employs ultrasonic vibrations (typically in 30-400 kHz range), to derive a characteristic electrical 'signature' of the structure (in frequency domain), containing vital information concerning the phenomenological nature of the structure. Electro-mechanical admittance, which is the measured electrical parameter, can be decomposed and analyzed to extract the mechanical impedance parameters of the host structure (Bhalla & Soh, 2004b, 2004c). In this manner, the PZT patch, acting as 'piezo-impedance transducer', enables structural identification, health monitoring and NDE (Bhalla, 2004).

The PZT patches are made up of 'piezoelectric' materials, which generate surface charges in response to mechanical stresses and conversely undergo mechanical deformations in response to electric fields. In the EMI technique, the bonded PZT patch is electrically excited by applying an alternating voltage using an impedance analyzer. This produces deformations in the patch as well as in the local area of the host structure surrounding it. The response of this area is transferred back to the PZT wafer in the form of admittance (the electrical response), comprising of the conductance (the real part) and the susceptance (the imaginary part). Hence, the same PZT patch acts as an actuator as well as a sensor concurrently. Any damage to the structure manifests itself as a deviation in the admittance signature, which serves as an indication of the damage (assuming that the integrity of the PZT patch is granted).

The EMI technique has been shown to possess far greater sensitivity to structural damages than the conventional global vibration techniques. It is typically of the order of the local ultrasonic techniques. The EMI technique employs low-cost PZT patches, which can be permanently bonded to the structures and unlike the ultrasonic techniques, can be interrogated without removal of the finishes or rendering the monitored structure out of service. In addition, no complex data processing or expensive hardware is warranted since the data is directly acquired in frequency domain. The limited sensing area of the piezo-impedance transducers helps in isolating the effects of far field changes, such as mass loading and normal operational vibrations, thereby enabling damage localization (Park et al., 2000).

The PZT patches are normally bonded to the surface of the monitored component using adhesives, which introduce the so-called 'shear lag effect'. This chapter is primarily focused on development of analytical models for considering the shear lag effect inherent in the adhesively bonded PZT patches for direct use in SHM/ NDE via the EMI technique. The chapter covers a review of the modelling strategies since the 1980s and presents a detailed description of two shear lag models specifically developed for the EMI technique, the first one by Bhalla & Soh in 2004 and the second one, a simplified version, by Bhalla, Kumar, Gupta and Datta in 2009.

2. Impedance Modelling of PZT-structure Interaction

The PZT patches, which play the key role in the EMI technique, typically develop surface charges under mechanical stresses; and conversely undergo mechanical deformations when subjected to electric fields, expressed mathematically by (IEEE standard, 1987)

$$D_i = \overline{\varepsilon_{ij}^T} E_j + d_{im} T_m \tag{1}$$

$$S_k = d_{jk} E_j + \overline{s_{km}^E} T_m \tag{2}$$

where D_i is the electric displacement, S_k the mechanical strain, E_j the electric field and T_m the mechanical stress. $\overline{\varepsilon_{ij}^T}$ denotes the complex electric permittivity of the PZT material at constant stress, d_{im} and d_{jk} the piezoelectric strain coefficients (or constants) and $\overline{s_{km}^E}$ the complex elastic compliance at constant electric field. The superscripts 'T' and 'E' indicate that the quantity has been measured at constant stress and constant electric field respectively.

During the last one and half decades, several attempts have been made to model the PZT-structure electromechanical interaction. The beginning was made by Crawley and de Luis (1987) in the form of 'static approach', later substituted by the 'impedance approach' of Liang, et al. (1994). Liang and coworkers modelled the host structure as mechanical impedance Z_s connected to the PZT patch at the end, as shown in Fig. 1(a), with the patch undergoing axial vibrations under an alternating electric field E_3. Mathematically, Z_s is related to the force F and the velocity $\dot{u}$ by

$$F_{(x=l)} = -Z_S \dot{u}_{(x=l)} \tag{3}$$

Solution of the governing 1D wave Eq. resulted in following expression for the complex electromechanical admittance for the system of Fig. 1(a)

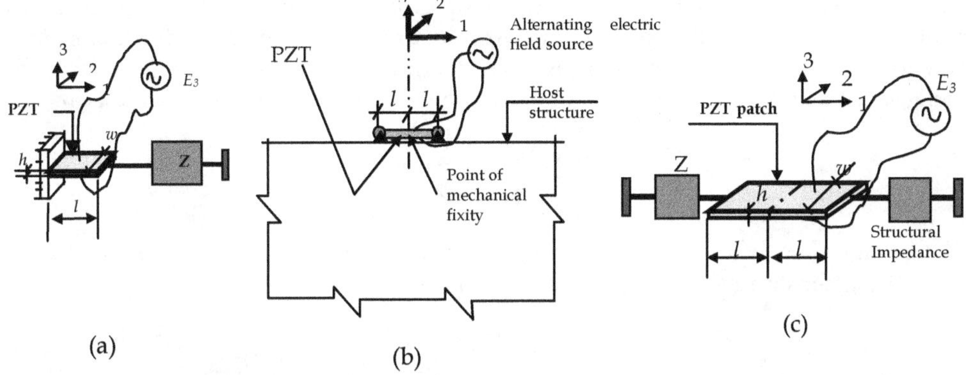

Fig. 1. (a) Liang's 1D impedance model (Liang et al. 1994).
(b) A PZT patch surface-bonded to a structure.
(c) Impedance model for the system shown in (b).

$$\overline{Y} = G + Bj = \omega j \frac{wl}{h}\left[(\overline{\varepsilon_{33}^T} - d_{31}^2 \overline{Y^E}) + \left(\frac{Z_a}{Z_S + Z_a} \right) d_{31}^2 \overline{Y^E} \left(\frac{\tan \kappa l}{\kappa l} \right) \right] \tag{4}$$

where w, l and h represent the PZT patch's dimensions (see Fig. 1a), d_{31} the piezoelectric strain coefficient for the 1-3 axes and ω the angular frequency. $\overline{Y^E} = Y^E(1 + \eta j)$ is the complex Young's modulus of the PZT patch (at constant electric field) and $\overline{\varepsilon_{33}^E} = \varepsilon_{33}^E(1 - \delta j)$ the complex electric permittivity (at constant stress), with the symbols η and δ denoting the mechanical loss factor and the dielectric loss factor respectively. Z_a represents the mechanical impedance of the PZT patch (in short circuited condition), given by

$$Z_a = \frac{\kappa wh}{\tan \kappa l} \frac{\overline{Y^E}}{(j\omega)} \tag{5}$$

where κ, the wave number, is related to the density ρ and the Young's modulus $\overline{Y^E}$ of the patch by

$$\kappa = \omega \sqrt{\frac{\rho}{\overline{Y^E}}} \tag{6}$$

In real-life applications, where the PZT patch is surface-bonded on a structure, as shown in Fig. 1(b), the nodal plane passes through the centre line of the patch. The structure can be represented as a set of two impedances Z_s connected on the either side of the patch, as illustrated in Fig. 1(c). For this scenario, l would be the half-length of the patch and Eq. (4) needs to be modified as

$$\overline{Y} = G + Bj = 2\omega j \frac{wl}{h}\left[(\overline{\varepsilon_{33}^T} - d_{31}^2 \overline{Y^E}) + \left(\frac{Z_a}{Z_S + Z_a} \right) d_{31}^2 \overline{Y^E} \left(\frac{\tan \kappa l}{\kappa l} \right) \right] \tag{7}$$

Zhou et al. (1996) extended the formulations of Liang to model the a PZT element coupled to a 2D host structure. The related physical model is schematically illustrated in Fig. 2. Zhou and coworkers replaced the single term Z_s by a matrix consisting of the direct impedances Z_{xx} and Z_{yy}, and the cross impedances Z_{xy} and Z_{yx}, related to the planar forces F_1 and F_2 (along axes 1 and 2 respectively) and the corresponding planar velocities $\dot{u}_1$ and $\dot{u}_2$ by

$$\begin{bmatrix} F_1 \\ F_2 \end{bmatrix} = -\begin{bmatrix} Z_{xx} & Z_{xy} \\ Z_{yx} & Z_{yy} \end{bmatrix}\begin{bmatrix} \dot{u}_1 \\ \dot{u}_2 \end{bmatrix} \tag{8}$$

Considering dynamic equilibrium along the two principal axes in conjunction with piezoelectric constitutive relations (Eqs. 1 and 2), they derived

$$\bar{Y} = j\omega\frac{wl}{h}\left[\overline{\varepsilon_{33}^T} - \frac{2d_{31}^2\overline{Y^E}}{(1-v)} + \frac{d_{31}^2\overline{Y^E}}{(1-v)}\left\{\frac{\sin\kappa l}{l}\quad\frac{\sin\kappa w}{w}\right\}\begin{bmatrix} \kappa\cos(\kappa l)\left\{1-v\frac{w}{l}\frac{Z_{xy}}{Z_{axx}}+\frac{Z_{xx}}{Z_{axx}}\right\} & \kappa\cos(\kappa w)\left\{\frac{l}{w}\frac{Z_{yx}}{Z_{ayy}}-v\frac{Z_{yy}}{Z_{ayy}}\right\} \\ \kappa\cos(\kappa l)\left\{\frac{w}{l}\frac{Z_{xy}}{Z_{axx}}-v\frac{Z_{xx}}{Z_{axx}}\right\} & \kappa\cos(\kappa w)\left\{1-v\frac{l}{w}\frac{Z_{yx}}{Z_{ayy}}+\frac{Z_{yy}}{Z_{ayy}}\right\} \end{bmatrix}^{-1}\begin{Bmatrix} 1 \\ 1 \end{Bmatrix}\right] \tag{9}$$

where κ, the 2D wave number, is given by

$$\kappa = \omega\sqrt{\frac{\rho(1-v^2)}{Y^E}} \tag{10}$$

Z_{axx} and Z_{ayy} are the two components of the mechanical impedance of the PZT patch along the two principal directions, given by Eq. 5. Although the analytical derivations of Zhou and co-workers are accurate in themselves, the experimental difficulties prohibit their direct application for the inverse problem, i.e. the extraction of host structure's mechanical impedance. Using the EMI technique, one can experimentally obtain two parameters- G and B for a surface-bonded PZT patch. If complete information about the structure is desired, Eq. (9) needs to be solved for 4 complex unknowns- Z_{xx}, Z_{yy}, Z_{xy}, Z_{yx} (or 8 real unknowns). Hence, the model could not be employed for the experimental determination of the drive point mechanical impedance from measurements alone.

To alleviate these shortcomings, the concept of 'effective impedance' was introduced by Bhalla & Soh (2004b). The related physical model is shown in Fig. 3 for a square-shaped PZT patch of half-length l. Bhalla & Soh (2004b) represented the PZT-structure interaction in the form of boundary traction f per unit length, varying harmonically with time. The 'effecive mechanical impedance', $Z_{a,eff}$, of the patch was defined as

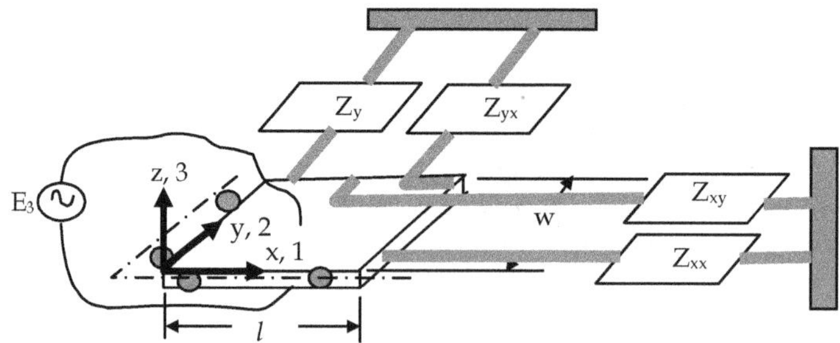

Fig. 2. 2D impedance model of Zhou et al. (1996).

$$Z_{a,eff} = \frac{F_{eff}}{\dot{u}_{eff}} = \frac{\oint_S \vec{f}.\hat{n}\,ds}{\dot{u}_{eff}} = \frac{2h\overline{Y^E}}{j\omega(1-v)\overline{\overline{T}}} \tag{11}$$

where F_{eff} is the overall planar force (or the effective force) causing area deformation of the PZT patch and $\hat{n}$ the unit vector normal to the boundary. $u_{eff} = \delta A/p_o$ is the 'effective displacement', with δA denoting the change in the patch's area and p_o its original undeformed perimeter. Differentiation of the effective displacement with respect to time yields the effective velocity, $\dot{u}_{eff}$. The effective drive point impedance of the host structure can be similarly defined, by applying a force on the surface of the host structure, along the boundary of the proposed location of the PZT patch. The term $\overline{T}$ is the complex tangent ratio, theoretically equal to $[tan(\kappa l)\,/\kappa l]$. However, in actual situations, it needs correction to realistically consider the deviation of the PZT patch from the ideal behavior, to accommodate which Bhalla and Soh (2004b) introduced correction factors as

$$\overline{T} = \frac{1}{2}\left(\frac{\tan C_1\kappa l}{C_1\kappa l} + \frac{\tan C_2\kappa l}{C_2\kappa l}\right) \tag{12}$$

The correction factors C_1 and C_2 can be determined from the experimentally obtained conductance and susceptance signatures of the PZT patch in 'free-free' conditions before bonding it on the host structure. It has been demonstrated that this 'updating' enables much more accurate reults. Solution of the governing 2D wave Eq. for this system yielded following expression for the complex electro-mechanical admittance $\overline{Y}$

$$\overline{Y} = G + Bj = 4\omega\ j\frac{l^2}{h}\left[\overline{\varepsilon_{33}^T} - \frac{2d_{31}^2\overline{Y^E}}{(1-v)} + \frac{2d_{31}^2\overline{Y^E}}{(1-v)}\left(\frac{Z_{a,eff}}{Z_{s,eff}+Z_{a,eff}}\right)\overline{\overline{T}}\right] \tag{13}$$

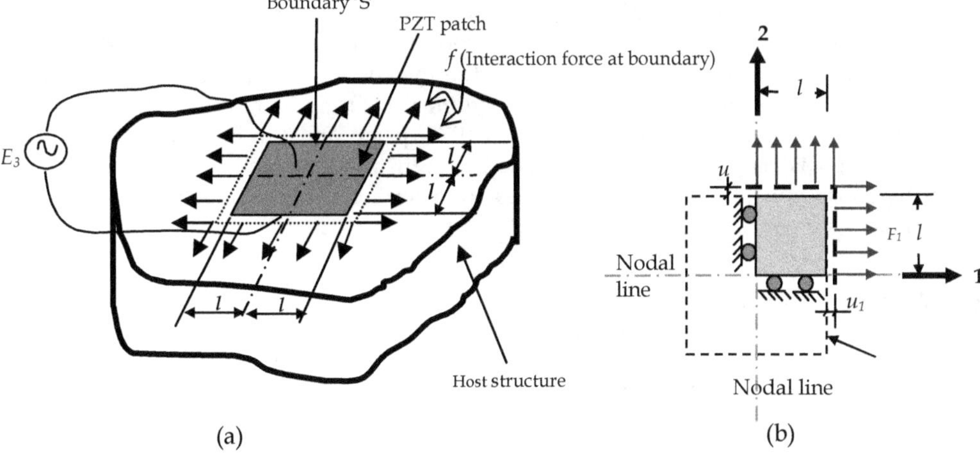

Fig. 3. Effective impedance model of Bhalla & Soh (2004a).
(a) PZT bonded to host structure. (b) Interaction forces at boundary.

Here, a single complex term for $Z_{s,eff}$ (rather than four terms as in Zhou's model) accounts for the 2D mechanical interaction of the patch with the host structure. This makes the resulting equation. simple enough to solve the inverse problem, i.e. to extract $Z_{s,eff}$ (Bhalla & Soh, 2004c), to be directly utilized for SHM/ NDE. No modelling is required for the host structure and the necessary data is directly obtainable from experimental measurements. Further, the corrected actuator effective impedance, $Z_{a,eff}$, can be expressed as

$$Z_{a,eff} = \frac{2h\overline{Y^E}}{j\omega(1-v)\overline{\overline{T}}} \tag{14}$$

All the above models ignore the fact that the mechanical interaction between the PZT patch and the host structure occurs through a finitely thick layer of adhesive sandwiched between the PZT patch and the host structure, which introduces the so-called 'shear lag effect' through its elastic deformation. Presented in the following sections is a detailed review of the shear lag mechanism inherent in adhesively bonded PZT patches and its rigorous integration in 1D and 2D impedance models, as proposed by Bhalla & Soh (2004d). Further, a new simplified model model proposed by Bhalla et al. (2009), which is especially suitable for solving the inverse problem (of extracting Z_s), considering the presence of bond layer, is also described.

3. Shear Lag Effect

Crawley and de Luis (1987) and Sirohi and Chopra (2000) respectively modelled the actuation and sensing of a generic beam element by an adhesively bonded PZT patch. The typical configuration of the system is shown in Fig. 4. The patch has a half-length l, width w_p and thickness h_p, while the bonding layer has a thickness h_s. The beam has depth h_b and width w_b. Let T_p be the axial stress in the PZT patch and τ the interfacial shear stress. The system is under quasi-static equilibrium and the beam is actuated in pure bending mode, with the bending strain linearly distributed across the cross section. Further, the PZT patch is in a state of pure 1D axial strain and the bond layer in pure shear, with the shear stress independent of 'y'. The ends of the segmented PZT actuator/ sensor are stress free, implying a uniform strain distribution across the thickness of the patch. A more detailed deformation profile is shown in Fig. 5 for the symmetrical right half of the system. Let u_p be the displacement at the interface between the PZT patch and the bond layer, and u the corresponding displacement at the interface between the bond layer and the beam. The following subsections briefly review the shear transfer mechanism for sensor and actuator respectively.

3.1 PZT patch as sensor
Let the PZT patch be instrumented only to sense the strain on the beam surface, and hence, no external electric field be applied across it. Considering static equilibrium of the differential element of the PZT patch in the x-direction, as shown in Fig. 4, we can derive

$$\tau = \frac{\partial T_p}{\partial x} h_p \tag{15}$$

At any cross section of the beam, within the portion containing the PZT patch, the bending moment is given by

$$M = T_p w_p h_p (0.5h_b + h_S + 0.5h_p) \tag{16}$$

Using Euler-Bernoulli's beam theory and assuming $(h_p+2h_s)<<h_b$, we can reduce Eq. (16) to

$$T_b + \left(\frac{3T_p w_p h_p}{w_b h_b} \right) = 0 \tag{17}$$

where T_b denotes the bending stress on the beam surface. Differentiating with respect to x, and substituting Eq. (15), we get

$$\frac{\partial T_b}{\partial x} + \left(\frac{3w_p}{w_b h_b} \right) \tau = 0 \tag{18}$$

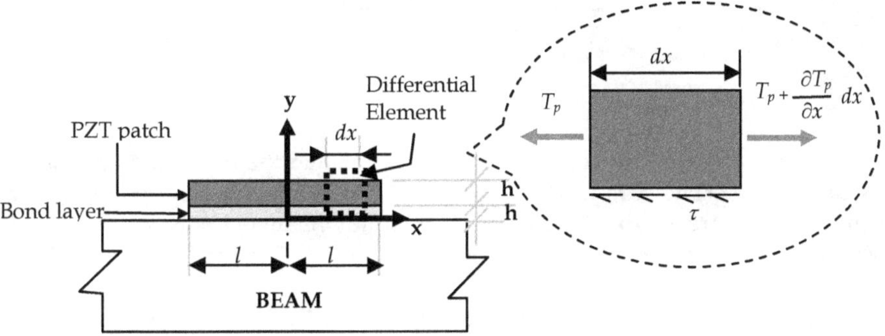

Fig. 4. A PZT patch bonded to a beam using adhesive bond layer.

Further, substituting $T_b = Y_b S_b$, $T_p = Y^E S_p$ and $\tau = G_s \gamma = G_s (u_p - u)/u$ into Eq.s (15) and (18), and differentiating with respect to x, we respectively obtain

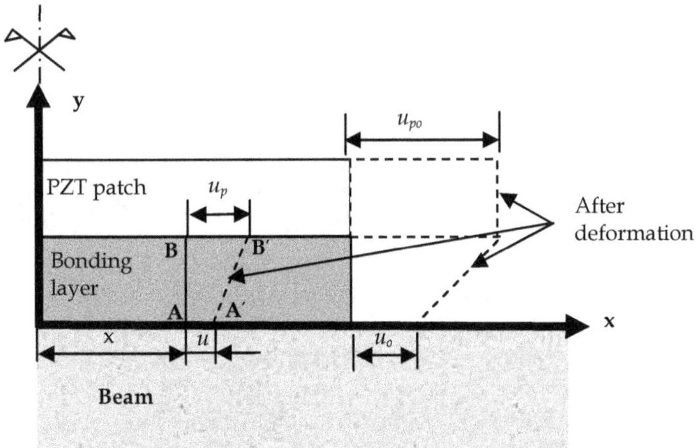

Fig. 5. Deformation in bonding layer and PZT patch.

$$\frac{\partial^2 S_P}{\partial x^2} = \left(\frac{G_s S_b}{Y_P h_S h_P}\right)\xi \tag{19}$$

$$\frac{\partial^2 S_b}{\partial x^2} = -\left(\frac{3w_P G_s S_b}{Y_b w_b h_b h_P}\right)\xi \tag{20}$$

where Y_b is the Young's modulus of elasticity of the beam, S_b and S_P the beam and PZT strains, G_s the shear modulus of the bond layer, γ the shear strain in the bond layer and $\xi = \left(S_p / S_b - 1\right)$. Subtracting Eq. (20) from Eq. (19), we get

$$\frac{\partial^2 \xi}{\partial x^2} - \Gamma^2 \xi = 0 \tag{21}$$

Where
$$\Gamma^2 = \left(\frac{G_s}{Y_P h_s h_P} + \frac{3G_s w_P}{Y_b w_b h_b h_P}\right) \tag{22}$$

This phenomenon of the difference in the PZT strain and the host structure's strain is called the *shear lag effect*. The parameter Γ (unit m^{-1}) is called the *shear lag parameter*. The ratio ξ, which is a measure of the differential PZT strain relative to surface strain of the host substrate, is called the *strain lag ratio*. The general solution for Eq. (21) is

$$\xi = A\cosh\Gamma x + B\sinh\Gamma x \tag{23}$$

Since no external electric field is applied across the PZT patch, the free PZT strain, $d_{31}E_3 = 0$. Thus, at $x = -l$, $S_P = 0 \Rightarrow \xi = -1$. Similarly, at $x = +l$, $\xi = -1$. Applying these boundary conditions, we can obtain

$$\xi = -\frac{\cosh(\Gamma x)}{\cosh(\Gamma l)} \tag{24}$$

Since
$$\xi = \left(S_p / S_b - 1\right), $$

$$\frac{S_P}{S_b} = \left[1 - \frac{\cosh(\Gamma x)}{\cosh(\Gamma l)}\right] \tag{25}$$

Fig. 6 shows a plot of the strain ratio (S_P/S_b) across the length of a PZT patch ($l = 5$mm) for typical values of $\Gamma = 10, 20, 30, 40, 50$ and 60 (cm^{-1}). It is observed that the strain ratio (S_P/S_b) is less than unity near the ends of the PZT patch. The length of this zone depends on Γ, which in turn depends on the stiffness and thickness of the bond layer (Eq. 22). As G_s increases and h_s reduces, Γ increases, the shear lag phenomenon diminishes and the shear is effectively transferred over very small zones near the ends of the PZT patch.

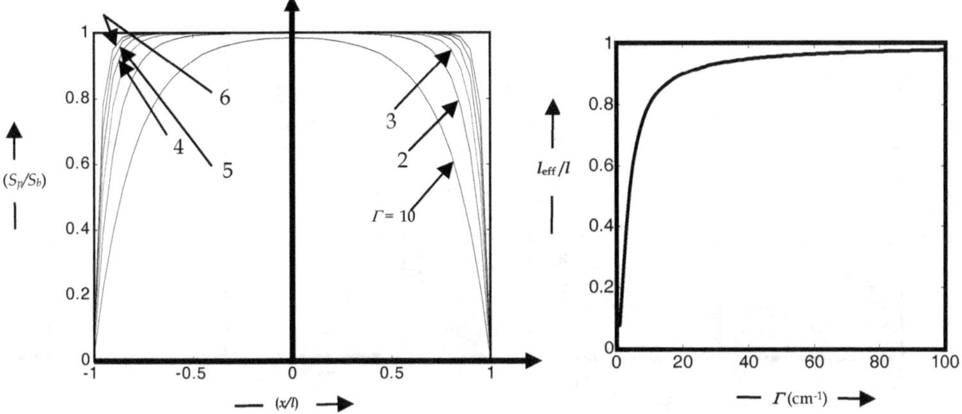

Fig. 6. Strain distribution across the length of PZT patch for various values of Γ.

Fig. 7. Variation of effective length with shear lag factor.

This analysis shows that if the PZT patch is employed as a sensor, it would develop less voltage across its terminals (than for perfectly bonded conditions) and hence underestimate the strain in the substructure. In order to quantify the effect of shear lag, we can compute the effective length, l_{eff}, of the sensor, as defined by (Sirohi and Chopra, 2000)

$$\frac{l_{eff}}{l} = \frac{1}{l}\int_{x=0}^{x=l}(S_p/S_b)dx = 1 - \frac{\tanh(\Gamma l)}{\Gamma l} \tag{26}$$

which is nothing but the area under the curve (Fig. 6) between $x/l = 0$ and $x/l = 1$. Fig. 7 shows a plot of the effective length (Eq. 26) for various values of the shear lag parameter Γ. From this figure, it can be observed that typically, for $\Gamma > 30 \text{cm}^{-1}$, ($l_{eff}/l$) is very large, typically greater than 93%, suggesting that shear lag effect can be ignored for relatively high ($> 30 \text{ cm}^{-1}$) values of Γ.

3.2 PZT patch as actuator

If the same PZT patch is employed as an actuator for a beam structure, it can be shown (Crawley and de Luis, 1987) that the strains S_p and S_b are given by

$$S_p = \frac{3\Lambda}{(3+\psi)} + \frac{\Lambda\psi\cosh\Gamma x}{(3+\psi)\cosh\Gamma l} \quad \text{and} \quad S_b = \frac{3\Lambda}{(3+\psi)} - \frac{3\Lambda\cosh\Gamma x}{(3+\psi)\cosh\Gamma l} \tag{27}$$

where $\Lambda = d_{31}E_3$ is the free piezoelectric strain, and $\psi = (Y_b h_b/Y^E h_p)$ the product of modulus and thickness ratios of the beam and the PZT patch. Fig. 8 shows the plots of (S_p/Λ) and (S_b/Λ) along the length of the PZT patch ($l = 5$mm) for $\psi = 15$. It is observed that as in the case of sensor, as Γ increases, the shear is effectively transferred over small zone near the two ends of the patch. Typically, for $\Gamma > 30\text{cm}^{-1}$, the strain energy induced in the substructure by PZT actuator is within 5% of the perfectly bonded case.

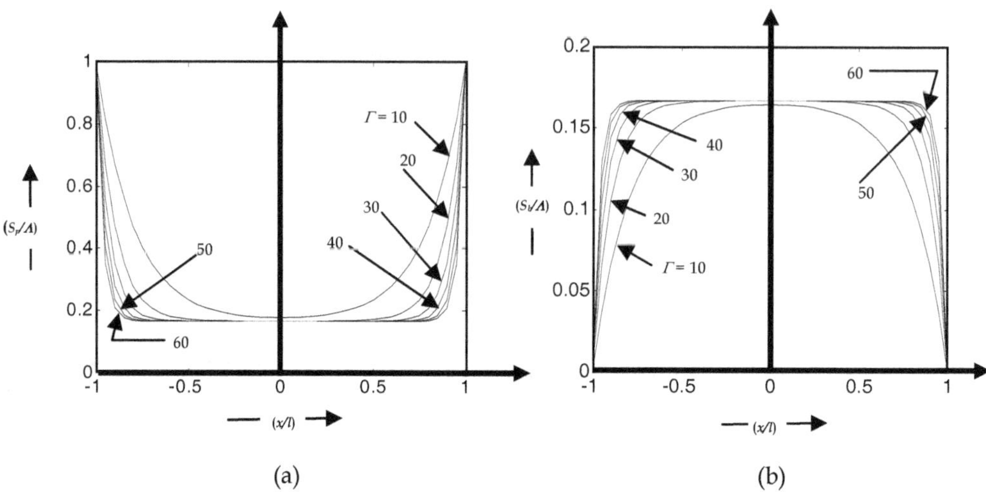

(a)   (b)

Fig. 8. Distribution of piezoelectric and beam strains for various values of Γ.
(a) Strain in PZT patch. (b) Beam surface strain.

4. Shear Lag Effect in Electro-Mechanical Impedance Formulations

It is observed in the preceding section that when acting as an actuator and/ or a sensor, there is shear lag phenomenon associated with force transmission between the PZT patch and the host structure through the adhesive bond layer. However, this aspect has not been thoroughly investigated for the EMI technique, in which the same patch concurrently serves both as a sensor as well as an actuator. Abe et al. (2002) encountered large errors in their stress prediction methodology using EMI technique, which were attributed to imprecise modeling of the interfacial bonding layer. This highlights the importance of modelling the shear lag mechanism accurately.

Xu and Liu (2002) proposed a modified 1D impedance model in which the bonding layer was modelled as a single degree of freedom (SDOF) system connected in between the PZT patch and the host structure, as shown in Fig. 9. The bonding layer was assumed to possess a dynamic stiffness $\overline{K_b}$ (or mechanical impedance, $Z_b = \overline{K_b}/j\omega$) and the structure a dynamic stiffness $\overline{K_S}$ (or mechanical impedance, $Z_s = \overline{K_S}/j\omega$). Hence, the resultant mechanical impedance for this series system can be determined as (Hixon, 1988)

$$Z_{res} = \frac{Z_b Z_s}{Z_b + Z_s} = \left(\frac{\overline{K_b}}{\overline{K_b} + \overline{K_S}} \right) Z_s = \zeta Z_S \tag{28}$$

where
$$\zeta = \frac{1}{1 + \left(\overline{K_S} / \overline{K_b} \right)} \tag{29}$$

The coupled electromechanical admittance, as measured across the terminals of the PZT patch and expressed earlier by Eq. (4), can therefore be modified as

$$\overline{Y} = 2\omega j \frac{wl}{h}\left[(\overline{\varepsilon_{33}^{T}} - d_{31}^{2}\overline{Y^{E}}) + \left(\frac{Z_a}{Z_a + \zeta Z_s}\right)d_{31}^{2}\overline{Y^{E}}\left(\frac{\tan \kappa l}{\kappa l}\right)\right] \tag{30}$$

$\zeta = 1$ implies infinitely stiff bond layer where as $\zeta = 0$ implies a free PZT patch. Xu and Liu (2002) demonstrated numerically that for a SDOF system, as ζ decreases (i.e. as the bond quality degrades), the PZT system shows an increase in the associated structural resonant frequencies. It was stated that $\overline{K_b}$ depends on the bonding process and the thickness of the bond layer. However, no closed form solution was presented to quantitatively determine $\overline{K_b}$ and hence ζ (From Eq. 29). Also, no experimental verification was attempted.

Ong et al. (2002) integrated the shear lag effect into impedance modelling using the analysis presented by Sirohi and Chopra (2000). The PZT patch was assumed to possess an effective length l_{eff} (Eq. 26) instead of the actual length. However, since the effective length was determined by considering sensor effect only, the method considered the associated shear lag only partially. Also, the resulting formulations are valid for beam type structures only and are not generic in nature. In addition, since frequencies of the order of 30-400 kHz are involved in the EMI technique, quasi-static approximation (for calculating l_{eff}) is strictly not valid. The next section presents the model of Bhalla & Soh (2004d) to alleviate all the above shortcomings.

5. Rigorous Shear Lag Model (Bhalla & Soh, 2004d)

5.1 1D shear lag model
This section presents a detailed step-by-step analysis for including the shear lag effect, first into 1D model (Liang et al., 1994) and then its extension into 2D effective impedance based model (Bhalla & Soh, 2004b).

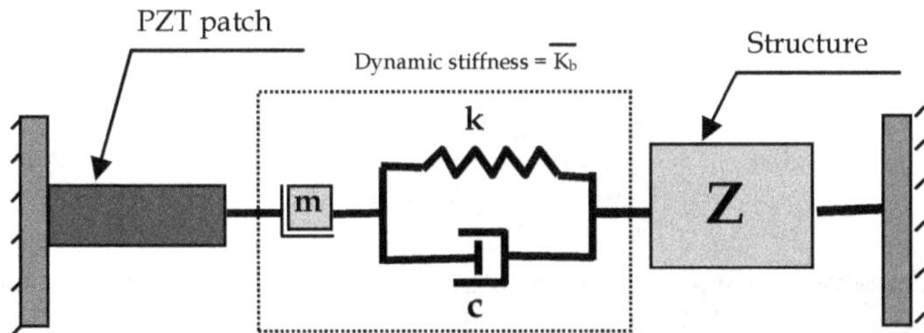

Fig. 9. Modified impedance model of Xu and Liu (2002) including bond layer.

Consider the PZT patch, shown in Figs. 4 and 5, to be driven by an alternating voltage source and let it be attached to any host structure (not necessarily beam). Also, we assume the PZT patch to be infinitesimally small as compared to the host structure. This means that the host structure has constant mechanical impedance all along the points of attachment of the patch. By D'Alembert's principle, the dynamic equilibrium of an infinitesimal element of the patch leads to

$$\tau w_p dx + (dm)\frac{\partial^2 u_p}{\partial t^2} = \frac{\partial T_p}{\partial x} h_p w_p dx \tag{31}$$

where dm is the infinitesimal mass of the element considered. Due to the dominance of the shear stress term, the inertial term can be neglected, which reduces Eq. (31) to Eq. (15). It should be noted that the inertial force term has been separately considered in impedance formulations (Liang et al., 1994), where, as a matter of fact, the shear lag effect has been ignored. Hence, the two effects are independently considered and will be finally combined. Assuming pure shear in the bond layer,

$$\tau = \frac{\overline{G_s}(u_p - u)}{h_s} \tag{32}$$

where $\overline{G_s} = G_s(1+\eta' j)$ is the complex shear modulus of the bond layer and η' is the associated mechanical loss factor. From PZT constitutive relation, Eq. (2), the axial stress in the PZT patch is given by

$$T_p = \overline{Y^E}(S_p - \wedge) = \overline{Y^E}(u'_p - \wedge) \tag{33}$$

where $S_p = u'_p$ is the PZT strain and $\wedge = E_3 d_{31}$ is the free piezoelectric strain. Substituting Eq.s (32) and (33) into Eq. (15) and simplifying, we get

$$u_p - u = \left(\frac{\overline{Y^E} h_p h_s}{\overline{G_s}}\right) u''_p \tag{34}$$

At any vertical section through the host structure (which includes the PZT patch), the force transmitted to the host structure is related to the drive point impedance Z_s of the host structure by

$$F = T_p w_p h_p = -Z_s u j \omega \tag{35}$$

where u is the drive point displacement at the point in question on the surface of the host structure. Since the PZT patch is infinitesimally small, Z_s is practically the same along the entire length of the PZT patch. Substituting Eq. (33), differentiating with respect to x (noting that Z_s is constant), and rearranging, we get

$$u''_p = -\left(\frac{Z_s j \omega}{w_p h_p \overline{Y^E}}\right) u' \tag{36}$$

From Eq.s (34) and (36), we derive

$$u_p - u = -\left(\frac{Z_s h_s j \omega}{\overline{G_s} w_p}\right) u' \tag{37}$$

Eq.s (34) and (37) are the fundamental Eq.s governing the shear transfer mechanism via the adhesive bonding layer. Eliminating u_p from these equations and differentiating with respect to x, we can derive

$$u''' + \bar{p}u''' - qu'' = 0 \tag{38}$$

where

$$\bar{p} = -\frac{w_p \bar{G_s}}{Z_s h_s j\omega} \tag{39}$$

and

$$q = \frac{\bar{G_s}}{Y^E h_s h_p} = \frac{G_s(1+\eta'j)}{Y^E h_s h_p(1+\eta j)} \approx \frac{G_s}{Y^E h_s h_p} \tag{40}$$

$\bar{p}$ and q are shear lag parameters, similar to the factor Γ in Eq. (22). The parameter q is equivalent to the first component of Γ and $\bar{p}$ to the second component. As seen from Eq. (40), q is directly proportional to the bond layer's shear modulus and inversely proportional to the PZT patch's Young's modulus, the PZT patch's thickness and the bond layer thickness. Examination of Eq. (39) similarly shows that $\bar{p}$ is directly proportional to the bond layer's shear modulus and the PZT patch's width. It is inversely proportional to structural mechanical impedance and the bond layer thickness. Being dynamic parameter, the frequency ω also comes into picture, influencing $\bar{p}$ inversely. Further, it should be noted that $\bar{p}$ is a complex term whereas the term q is approximated as a pure real term assuming η and η' to be very small in magnitude. Solving Eq. (41), we get the roots of the characteristic Eq. as

$$\lambda_1 = 0, \quad \lambda_2 = 0, \quad \lambda_3 = \frac{-\bar{p} + \sqrt{\bar{p}^2 + 4q}}{2}, \quad \lambda_4 = \frac{-\bar{p} - \sqrt{\bar{p}^2 + 4q}}{2} \tag{41}$$

Hence, the solution of the governing differential equation (Eq. 38) can be written as

$$u = A_1 + A_2 x + Be^{\lambda_3 x} + Ce^{\lambda_4 x} \tag{42}$$

Differentiating with respect to x, we get

$$u' = A_2 + B\lambda_3 e^{\lambda_3 x} + C\lambda_4 e^{\lambda_4 x} \tag{43}$$

From Eqs (37), (42) and (43), we obtain

$$u_p = (A_1 + \bar{n}A_2) + A_2 x + B(1 + \bar{n}\lambda_3)e^{\lambda_3 x} + C(1 + \bar{n}\lambda_4)e^{\lambda_4 x} \tag{44}$$

where
$$n = \left(-\frac{\overline{Z} h_s j\omega}{w_p \overline{G_s}} \right) \tag{45}$$

Differentiating Eq. (44) with respect to x, we obtain the strain in the PZT patch as

$$S_p = A_2 + B\lambda_3(1+\overline{n}\lambda_3)e^{\lambda_3 x} + C\lambda_4(1+\overline{n}\lambda_4)e^{\lambda_4 x} \tag{46}$$

At $x = 0$ (see Figs 4 and 5), $u = 0$, hence Eq. (42) leads to

$$A_1 = -(B + C) \tag{47}$$

Similarly, applying the boundary condition that at $x = 0$ and $u_p = 0$ to Eq. (44) leads to

$$A_2 = -(B\lambda_3 + C\lambda_4) \tag{48}$$

Hence, Eq. (46) can be modified as

$$S_p = B[\lambda_3(1+\overline{n}\lambda_3)e^{\lambda_3 x} - \lambda_3] + C[\lambda_4(1+\overline{n}\lambda_4)e^{\lambda_4 x} - \lambda_4] \tag{49}$$

The third and the fourth boundary conditions are imposed by the stress free ends of the PZT patch. That is, at $x = -l$ and at $x = +l$, the axial strain in the PZT patch is equal to the free piezoelectric strain or $\Lambda = E_3 d_{31}$ (Crawley and de Luis, 1987). This leads to constants B and C as

$$\begin{bmatrix} B \\ C \end{bmatrix} = \frac{\Lambda}{(k_1 k_4 - k_2 k_3)} \begin{bmatrix} k_4 - k_2 \\ k_1 - k_3 \end{bmatrix} \tag{50}$$

where $k_1 = \lambda_3(1+\overline{n}\lambda_3)e^{-\lambda_3 l} - \lambda_3$, $k_2 = \lambda_4(1+\overline{n}\lambda_4)e^{-\lambda_4 l} - \lambda_4$, $k_3 = \lambda_3(1+\overline{n}\lambda_3)e^{\lambda_3 l} - \lambda_3$ and
$k_4 = \lambda_4(1+\overline{n}\lambda_4)e^{\lambda_4 l} - \lambda_4$.

In general, the force transmitted to the host structure can be expressed as

$$F = -Z_s j\omega u_{(x=l)} \tag{51}$$

where $u_{(x=l)}$ is the displacement at the surface of the host structure at the end point of the PZT patch. Conventional impedance models (for example, Liang and coworkers) assume perfect bonding between the PZT patch and the host structure, i.e. the displacement compatibility $u_{(x=l)} = u_{p(x=l)}$, thereby approximating Eq. (51) as $F = -Z_s j\omega u_{p(x=l)}$. However, due to the shear lag phenomenon associated with finitely thick bond layer, $u_{(x=l)} \neq u_{p(x=l)}$. Based on the analysis presented in this section, we can obtain following relationship between $u_{(x=l)}$ and $u_{p(x=l}$ from Eq. (37)

$$\frac{u_{(x=l)}}{u_{p(x=l)}} = \frac{1}{1-\left(\dfrac{Z_s h_s j\omega}{w_p \overline{G_s}}\right)\dfrac{u'_{(x=l)}}{u_{(x=l)}}} = \frac{1}{\left(1+\dfrac{1}{\overline{p}}\dfrac{u'_o}{u_o}\right)} \tag{52}$$

The term u'_o / u_o can be determined by using Eq.s (42) and (43). Making use of this relationship, Eq. (51) can be rewritten as

$$F_S = \frac{-Z_s}{\left(1+\dfrac{1}{\overline{p}}\dfrac{u'_o}{u_o}\right)} j\omega u_{p(x=l)} = Z_{s,eq} j\omega u_{p(x=l)} \tag{53}$$

where $Z_{s,eq} = Z_s / \left(1+\dfrac{1}{\overline{p}}\dfrac{u'_o}{u_o}\right)$ is the 'equivalent impedance' apparent at the ends of the PZT

patch, taking into consideration the shear lag phenomenon associated with the bond layer. In the absence of shear lag effect (i.e. perfect bonding), $Z_{s,eq} = Z_s$. On comparing with the result of Xu and Liu (2002) i.e. Eq. (29), we find that

$$\zeta = \frac{1}{1+\dfrac{1}{\overline{p}}\dfrac{u'_o}{u_o}} \tag{54}$$

Hence, the derivation presented in this section enables quantitative prediction of the modifying term of the structural impedance, which was left undone by Xu and Liu (2002).

5.2 Extension to 2D shear lag model

The formulations derived above can be easily extended to the 2D effective impedance based electro-mechanical model introduced by Bhalla & Soh (2004b). For this derivation, it is assumed that the PZT is square in shape with a half-length equal to l. The strain distribution and the associated shear lag are determined along each principal direction and the two effects are assumed to be independent, which means that the effects at the corners are neglected.

The patch is assumed to be mechanically isotropic and piezoelectrically orthotropic in the x-y plane. The constitutive relations (Equations 1 and 2) can be thus reduced to (see Figure 3)

$$D_3 = \overline{\varepsilon_{33}^T} E_3 + d_{31}(T_1 + T_2) \tag{55}$$

$$S_1 = \frac{T_1 - \nu T_2}{\overline{Y^E}} + d_{31} E_3 \tag{56}$$

$$S_2 = \frac{T_2 - \nu T_1}{\overline{Y^E}} + d_{31} E_3 \tag{57}$$

Consider an infinitesimal element of the PZT patch, in dynamic equilibrium with the host structure, as shown in Fig. 10. Since this figure shows a planar view, the shear stresses τ_{xz} and τ_{yz} are not visible. Considering force equilibrium along x-direction, we can write (De Faria, 2003)

$$\frac{\partial T_1}{\partial x} + \frac{\partial \tau_{xy}}{\partial y} - \frac{\tau_{xz}}{h_p} = 0 \tag{58}$$

Ignoring the terms involving rate of change of shear strains (consistent with the observation by Zhou et al., 1996), we get

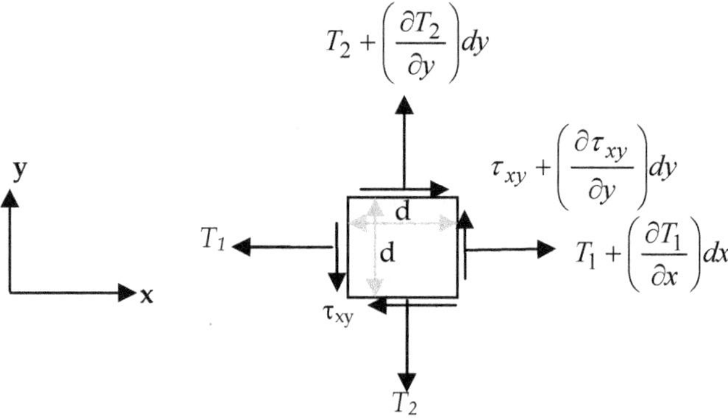

Fig. 10. Stresses acting on an infinitesimal PZT element.

$$\frac{\partial T_1}{\partial x} = \frac{\tau_{xz}}{h_p} \tag{59}$$

Further, using Eqs (56) and (57), we can derive (noting $\Lambda = E_3 d_{31}$)

$$T_1 = \frac{\overline{Y^E}}{(1-\upsilon^2)}\left[(S_1 + \upsilon S_2) - \Lambda(1+\nu)\right] \tag{60}$$

or
$$T_1 = \frac{\overline{Y^E}}{(1-\upsilon^2)}\left[(u'_{px} + \upsilon u'_{py}) - \Lambda(1+\nu)\right] \tag{61}$$

Differentiating with respect to x and ignoring the second order terms involving both x and y (Zhou et al., 1996), we get

$$\frac{\partial T_1}{\partial x} = \frac{\overline{Y^E}}{(1-\upsilon^2)} u''_{px} \tag{62}$$

Substituting Eq. (62) into Eq. (59), expanding the term τ_{xz} and rearranging, we get

$$u_{px} - u_x = \frac{\overline{Y^E} h_s h_p}{\overline{G_s}(1-\upsilon^2)} u''_{px} \tag{63}$$

Similarly, for the other direction, we get

$$u_{py} - u_y = \frac{\overline{Y^E} h_s h_p}{\overline{G_s}(1-\upsilon^2)} u''_{py} \tag{64}$$

Adding Eqs (63) and (64) and dividing by 2, we obtain, based on the definition of 'effective displacement' that is $u_{eff} = \dfrac{\delta A}{P_o} = \dfrac{u_1 l + u_2 l + u_1 u_2}{2l} \approx \dfrac{u_1 + u_2}{2}$ (see Fig. 3), we obtain

$$u_{p,eff} - u_{eff} = \left(\frac{\overline{Y^E} h_p h_s}{\overline{G_s}(1-\upsilon^2)} \right) u''_{p,eff} \approx \frac{1}{q_{eff}} u''_{p,eff} \tag{65}$$

where q_{eff} has been approximated as pure real number, as in the 1D case. Here, $u_{p,eff}$, by definition, is the effective displacement at the interface between the PZT patch and the bond layer, u_{eff} is the corresponding effective displacement at the interface between the structure and the bond layer. Further, from the definition of effective impedance, we can write, for the host structure

$$F = T_1 l h_p + T_2 l h_p = -Z_{s,eff} u_{eff} j\omega \tag{66}$$

Making use of Eqs. (56) and (57), and noting $\Lambda = E_3 d_{31}$, we get

$$\frac{\overline{Y^E} l h_p (S_{p1} + S_{p2} - 2\Lambda)}{(1-\upsilon)} = -Z_{s,eff} u_{eff} j\omega \tag{67}$$

Substituting for $S_{p1} = u'_{px}$, $S_{p2} = u'_{py}$, making use of the definition of effective displacement, and differentiating, we can derive

$$u''_{p,eff} = -\left[\frac{Z_{s,eff}(1-\upsilon) j\omega}{2\overline{Y^E} l h_p} \right] u'_{p,eff} \tag{68}$$

Eliminating $u''_{p,eff}$ from E.s (65) and (68), we get

$$u_{p,eff} - u_{eff} = -\left(\frac{Z_{s,eff} h_s j\omega}{2\overline{G_s}(1+\upsilon)l} \right) u'_{eff} = \left(\frac{1}{\overline{P_{eff}}} \right) u'_{eff} \tag{69}$$

Eqs (65) and (69) are the governing equations for 2D case. The parameters $\overline{p}_{eff}$ and q_{eff}, which are the equivalent of the 1D shear lag parameter $\overline{p}$ and q respectively, and can be expressed as

$$\overline{p}_{eff} = -\left(\frac{2\overline{G}_s(1+\upsilon)l}{Z_{s,eff}h_s j\omega}\right) \quad \text{and} \quad q_{eff} \approx \frac{G_s(1-\upsilon^2)}{Y^E h_p h_s} \tag{70}$$

The rest of the procedure is identical to the one outlined in the previous section for 1D case. Hence, the equivalent effective impedance can be expressed as

$$Z_{s,eff,eq} = \frac{Z_{s,eff}}{1+\left(\dfrac{1}{p_{eff}}\dfrac{u'_{eff,(x=l)}}{u_{eff,(x=l)}}\right)} \tag{71}$$

5.3 Experimental verification

In order to verify the derivations outlined above, two PZT patches, 10x10x0.3mm and 10x10x0.15mm, conforming to grade PIC 151 (PI Ceramic, 2003), were bonded to two aluminium blocks, each 48x48x10mm in size, conforming to grade Al 6061-T6. The experimental set-up shown in Fig. 11 was employed. The PZT patches were bonded to the blocks using RS 850-940 two-part epoxy adhesive (RS Components, 2003). Before applying the epoxy, two optical fiber pieces, 0.125mm in diameter, were laid down on the surface of the specimens parallel to each other. The layer of epoxy was then applied on the surface and the PZT patch was placed on it. Light pressure was maintained over the assembly using a small weight. The setup was left undisturbed in this condition at room temperature for 24 hours to enable full curing of adhesive. The optical fiber pieces were left permanently in the adhesive layer. This procedure ensured a uniform thickness of 0.125mm of the bond layer in both the specimens tested. The two specimens have (h_s/h_p) ratio equal to 0.417 and 0.833 respectively. During the test, the voltage level of the impedance analyzer was maintained equal to 1 volt root mean square. Each admittance reading was worked out as the average of three experimental recordings. In this way, the experimental signatures, consisting of the real part- the conductance (G) and the imaginary part- the susceptance (B) were obtained. Signatures of the five representative PZT patches were acquired in 'free-free' conditions to determine the key parameters of the patches. Table 1 lists the key averaged PZT parameters for the batch.

The numerical approach based on FEM, as outlined by Bhalla & Soh (2004b), was employed to determine the effective mechanical impedance of the host structure. The physical properties of Al 6061-T6 were considered as: Young's modulus = 68.95GPa, density = 2715 kg/m³ and Poisson's ratio = 0.33. Rayleigh damping was considered with $a = 0$ and $\beta = 3 \times 10^{-9}$. Wavelength analysis and convergence test on this model has already been reported by Bhalla & Soh (2004b). Fig. 12 shos the finite element model of a quarter of the structure considered for this purpose. The PZT patch or the bond layer need not be meshed since their stiffness, mass and damping are separately considered in the formulations. A uniformly distributed planar harmonic force was applied along the boundary of the PZT patch and the displacement response was obtained by full dynamic harmonic analysis to determine the effective drive point impedance of the structure as

$$Z_{s,eff} = \frac{F_{eff}}{j\omega u_{eff}} \tag{72}$$

The shear modulus of elasticity of the epoxy adhesive was assumed as 1.0 GPa, in accordance with Adams and Wake (1984). The mechanical loss factor of commercial adhesives shows a wide variation and is strongly dependent on temperature. It might vary from 5% to 30% at room temperature, depending upon the type of adhesive (Adams and Wake, 1984). For this study, a value of 10% was considered.

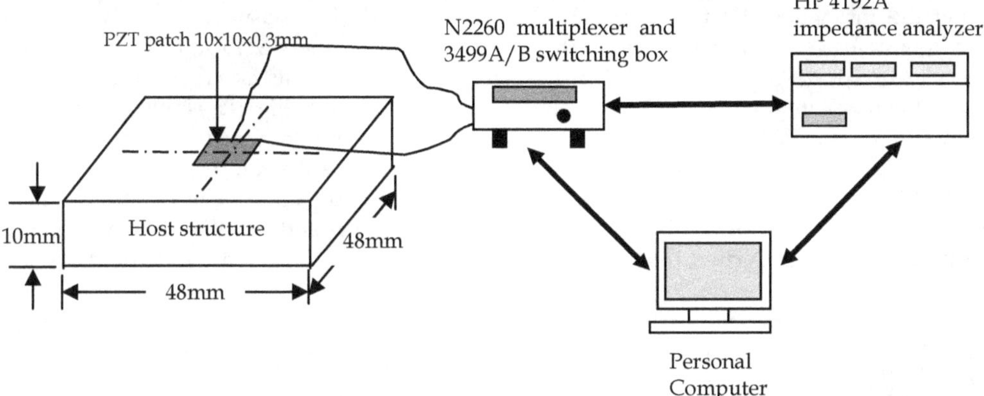

Fig. 11. Experimental set-up to verify new electro-mechanical formulations including bond layer.

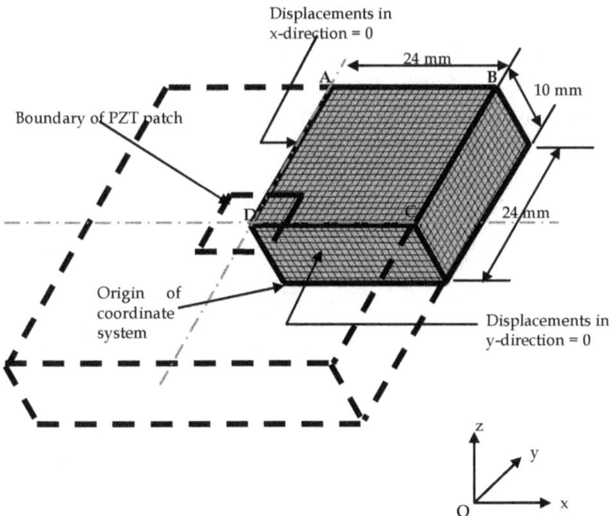

Fig. 12. 3D finite element model of one-quarter of structure.

Physical Parameter	Value
Electric Permittivity, ε_{33}^{T} (farad/m)	1.7785 x 10⁻⁸
Peak correction factor, C_f	0.898
$K = \dfrac{2d_{31}^2 Y^E}{(1-\upsilon)}$ (N/V²)	5.35x10⁻⁹
Mechanical loss factor, η	0.0325
Dielectric loss factor, δ	0.0224

Table 1. Averaged parameters of test sample of PZT patches.

Fig. 13 shows the plot of normalized conductance (Gh/L^2) worked out for the two specimens using the integrated 2D model presented in this paper. The plot for perfectly bonded condition is also shown. It is observed that on increasing thickness of the adhesive layer, the sharpness of peaks in the conductance plot tends to diminish. This fact is confirmed by the experimental plots shown in Fig. 14 for the two specimens. Fig. 15 shows the plot of normalized susceptance (Bh/L^2), worked out using the new model for three cases- no bond layer, (h_s/h_p) = 0.417 and (h_s/h_p) = 0.833. Again, it is observed that an increase in thickness tends to flatten the peaks. In addition, the average slope of the curve also reduces marginally. This is confirmed by Fig. 16, which shows the curves determined experimentally for the two specimens. Thus, the shear lag model has made reasonably accurate predictions.

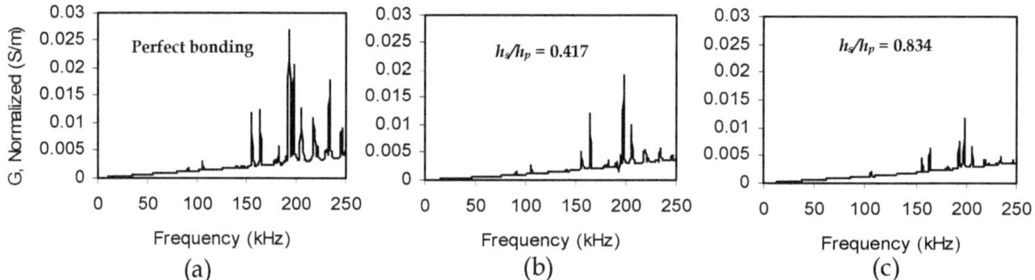

Fig. 13. Theoretical normalized conductance.
(a) Perfect bonding. (b) h_s/h_p = 0.417. (c) h_s/h_p = 0.834.

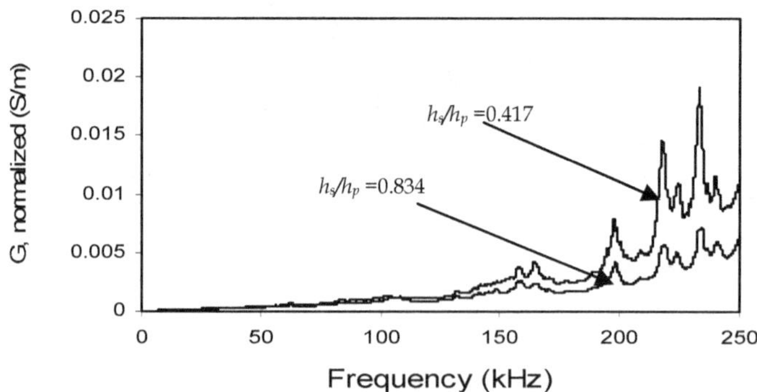

Fig. 14. Experimental normalized conductance for h_s/h_p = 0.417 and h_s/h_p = 0.834.

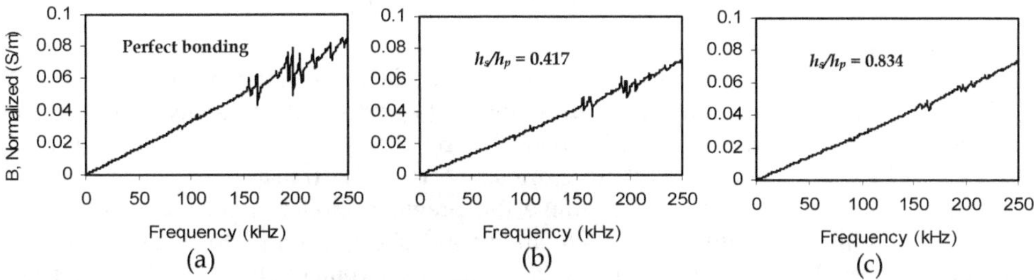

Fig. 15. Theoretical normalized susceptance.
(a) Perfect bonding. (b) $h_s/h_p = 0.417$. (c) $h_s/h_p = 0.834$.

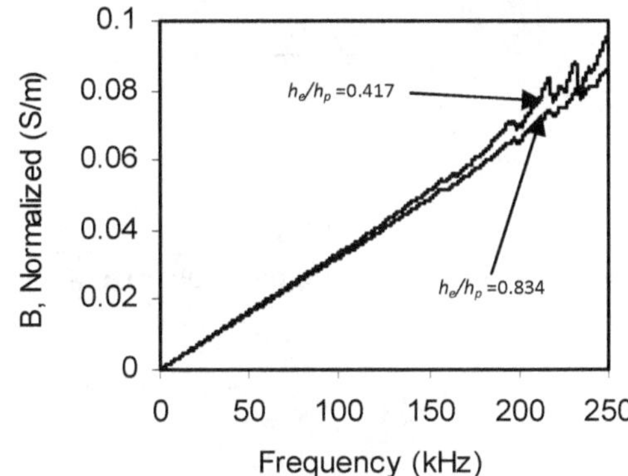

Fig. 16. Experimental normalized susceptance for $h_s/h_p = 0.417$ and $h_s/h_p = 0.834$.

This model is not only more rigorous but at the same time generic in nature. A parametric study revealed that to achieve best results, the adhesive layer should possess high shear modulus and minimum practicable thickness. A related experimental study has been reported by Qing et al. (2006). However, shortcoming of this model is visible for solving the inverse problem for NDE. In the damage quantification approach postulated by Bhalla & Soh (2004c), one needs to extract the mechanical impedance of the host structure $(Z_s = x + yj)$ from the measured admittance signature. In the presence of the adhesive layer, this would be $Z_{s,eq}$, from which it is computationally very difficult to obtain the true structural impedance Z_s, as clearly evident from Eqs. (69) to (71).

This difficulty of solving the inverse problem taking due consideration of the adhesive bond layer is very well alleviated by the simplified impedance model of Bhalla et al. (2009). In the next sections, the model is first derived first derived for 1D case and then extended to 2D situations.

6. Simplified Shear Lag Model (Bhalla et al., 2009)

6.1 1D shear lag model

Fig. 17 shows the physical aspects of the proposed simplified 1D impedance model. The PZT patch has length $2l$ with zero displacement at the mid point, which is the nodal point. Hence, only right half of the system is modelled here. The bond layer is assumed to be connected in between the PZT patch and the host structure such that it can transfer the force between the two through pure shear mechanism. Unlike the previous model of Bhalla & Soh (2004d), where shear strain varied along the patch, an average shear strain uniform along the length has been considered as a simplification. Let u_p be the displacement at the tip of the PZT patch at any point of time. Due to the shearing of the bond layer, same displacement would not be transferred to the host structure. Let u be the displacement of the host structure at a point just underneath the tip of the PZT patch. Let h_p and h_S respectively denote the thickness of the patch and the bond layer. Shear strain in the bond layer is given by

$$\gamma = \frac{u_p - u}{h_s} \tag{73}$$

which can be rearranged as,

$$u = u_p - \left(\frac{\tau}{G_s}\right) h_s \tag{74}$$

If F be the force transmitted to the host structure over the area A of one-half of the patch, Eq. (74) can be rewritten as

$$u = u_p - \left(\frac{F}{AG_S}\right) h_s \tag{75}$$

Further, in terms of the structural impedance Z_s, the force transmitted to the host structure can be expressed as

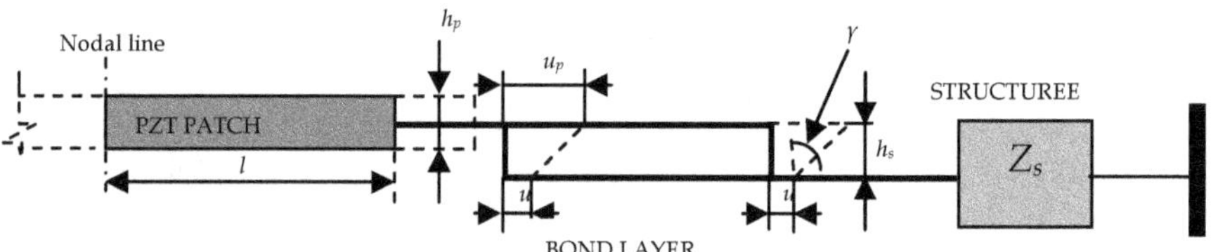

Fig. 17. Simplified 1D impedance model proposed by Bhalla et al., 2009 (showing right symmetrical half of PZT patch- structure system).

$$F = -Z_S \dot{u} = -Z_S u j \omega \tag{76}$$

Substituting u from Eq. (75) and simplifying, we get

$$F = -Z_S j \omega \left[u_p - \frac{F h_S}{AG_S} \right] \tag{77}$$

By rearranging the terms and with $A = wl$, Eq. (77) can be simplified as

$$F = -\frac{Z_S}{\left(1 - \dfrac{Z_S \omega h_s j}{wl\overline{G_s}}\right)} j\omega u_p \tag{78}$$

This can be expressed in a format similar to Eq. (76) as

$$F = -Z_{S,eq} j\omega u_p \tag{79}$$

where

$$Z_{S,eq} = \frac{Z_S}{\left(1 - \dfrac{Z_S \omega h_s j}{wl\overline{G_s}}\right)} \tag{80}$$

is the 'equivalent impedance', apparent at the ends of the PZT patch, taking into consideration the shear lag phenomenon associated with the bond layer. Replacing Z_s, by $Z_{s,eq}$ in Eq. (7), the modified expression for admittance across the PZT patch can be obtained as in the case of the model of Bhalla & Soh (2004d).

6.2 Extension to 2D

This section extends the 1D shear lag based impedance formulations derived above to 2D effective impedance-based electromechanical model. The PZT patch is assumed to be square in shape with a half-length equal to l. Again, the strain distribution and the associated shear lag are determined along each principal direction independently, invariably introducing discontinuity at the corners, which is ignored. By applying Eq. (75) along each principal direction for the configuration of Fig. 3b (for a quarter of the patch),

$$u_1 = u_{p1} - \left(\frac{F_1}{l^2 \overline{G_s}}\right) h_s \tag{81}$$

and

$$u_2 = u_{p2} - \left(\frac{F_2}{l^2 \overline{G_s}}\right) h_s \tag{82}$$

where F_1 and F_2 are the forces along each direction as shown in Fig. 3(b). Adding Eqs. (81) and (82) and dividing by 2, we get

$$\frac{u_1 + u_2}{2} = \frac{u_{p1} + u_{p2}}{2} - \left(\frac{F_1 + F_2}{2l^2 \overline{G_s}}\right) h_s \tag{83}$$

From the definition of effective displacement (Bhalla & Soh, 2004b)

$$u_{eff} = \frac{\delta A}{p_o} = \frac{u_1 l + u_2 l + u_1 u_2}{2l} \approx \frac{u_1 + u_2}{2} \tag{84}$$

Further, from Eq. (11),

$$F_{eff} = F_1 + F_2 \tag{85}$$

Thus, using Eqs. (84) and (85), Eq. (83) can be reduced to

$$u_{eff} = u_{p,eff} - \left(\frac{F_{eff}}{2l^2 \overline{G_s}} \right) h_s \tag{86}$$

From the definition of effective impedance,

$$F_{eff} = -Z_{eff} u_{eff} \, j\omega \tag{87}$$

Substituting Eq. (86) into (87) and solving, as for the 1D case, an expression for the equivalent effective impedance can be derived as

$$Z_{s,eff,eq} = \frac{Z_{s,eff}}{\left(1 - \dfrac{Z_{s,eff}\,\omega h_s j}{2l^2 \overline{G_s}} \right)} \tag{88}$$

With the above result, Eq. (13) can be modified, by replacing Z_{eff} by $Z_{s,eff,eq}$ as

$$\overline{Y} = G + Bj = 4\omega \ j\frac{l^2}{h}\left[\overline{\varepsilon_{33}^T} - \frac{2d_{31}^2 \overline{Y^E}}{(1-v)} + \frac{2d_{31}^2 \overline{Y^E}}{(1-v)}\left(\frac{Z_{a,eff}}{Z_{s,eff,eq} + Z_{a,eff}} \right)\overline{T} \right] \tag{89}$$

In order to verify the proposed new model, the same aluminium block, 48x48x10mm in size, was considered as the host structure. A PZT patch, 10x10x0.3mm in size, was assumed to be surface-bonded on this structure. The effective drive point impedance of the host structure was computed by carrying out 3D dynamic harmonic analysis, as described earlier. Final values for G and B were determined in the frequency range 0-250kHz using Eq. (89). A 0.150mm thick epoxy layer was considered with shear modulus of G_s = 1 GPa and a mechanical loss factor of η = 10%. The parameters of the PZT patch considered are listed in Table 1.

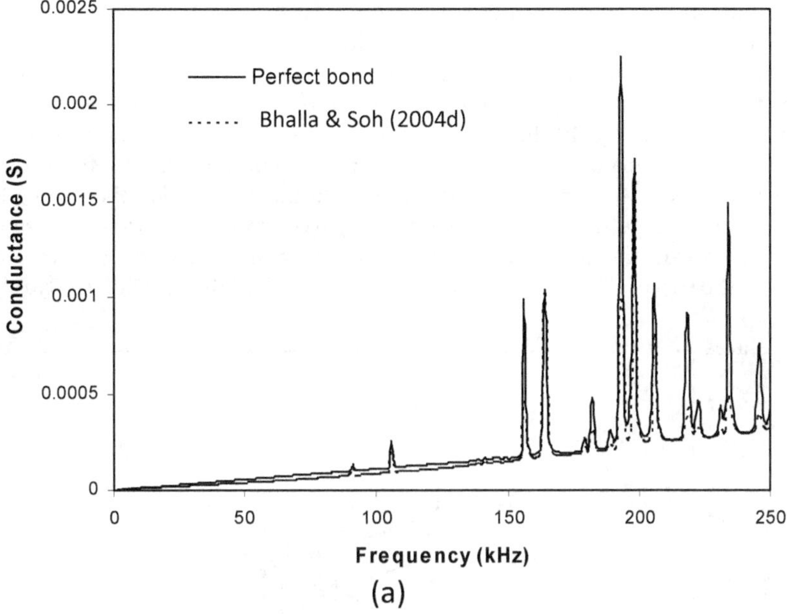

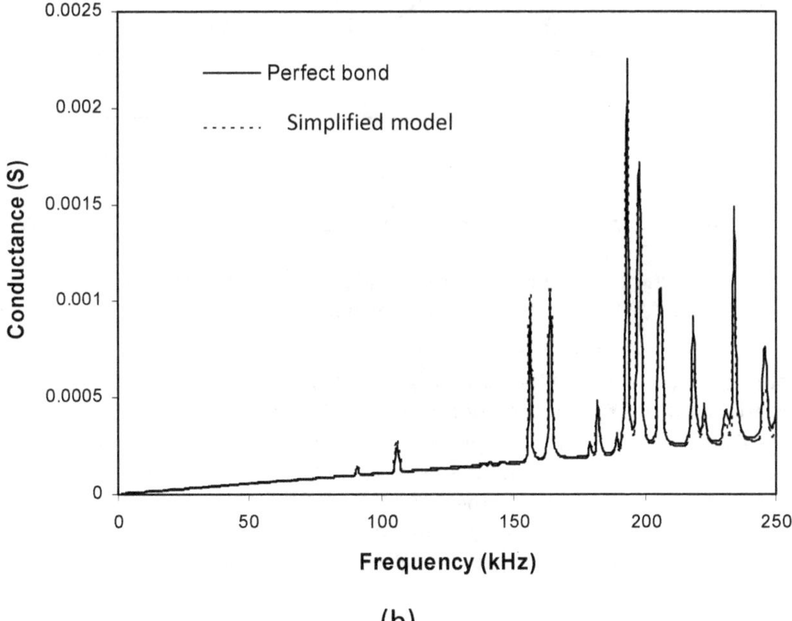

Fig. 18. Comparison of conductance signature.
(a) Bhalla & Soh's model. (b) Simplified model.

Fig. 18 shows a comparison of the variation of conductance with frequency obtained using the simplified model and also the rigorous model of Bhalla & Soh (2004d). Curves obtained by both models are plotted alongside the curves for the case of perfect bonding for comparison. The proposed simplified model predicts the conductance in consistency with the rigorous model. Both the models predict that the peaks tend to diminish down due to the presence of the bond layer. Similarly, Fig. 19 shows a comparison of the variation of susceptance for three cases- no bond layer, Bhalla & Soh (2004d) model and the simplified model. It is observed that like the previous model, the simplified model leads to the observation that peaks lose their sharpness and the average slope of the susceptance curve tends to reduce owing to the shear lag effect. However, the susceptance curve resulting from the simplified model lies intermediate of the two cases i.e perfect bond and the model of Bhalla & Soh (2004d).

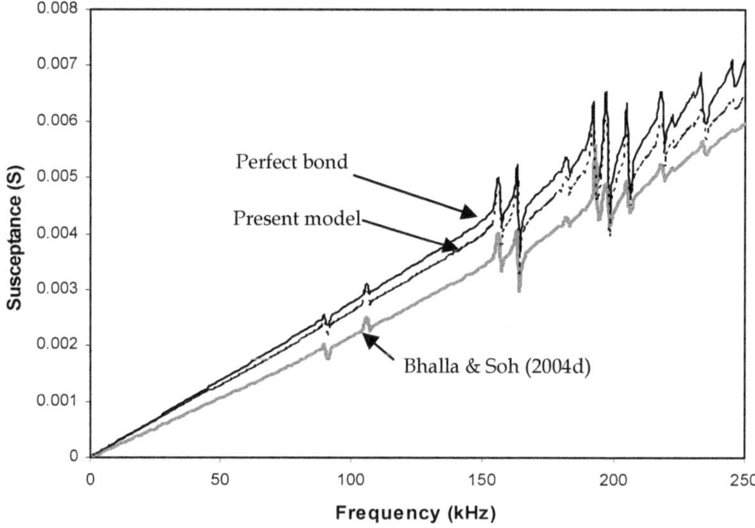

Fig. 19. Comparison of susceptance signature from new model with Bhalla & Soh (2004d) model.

The influence of important parameters on the conductance and susceptance signatures was studied by Bhalla et al. (2009) and the observations matched with the previous model.

6.3 Practical relevance of simplified shear lag model
The results presented in the previous section show that in spite of its simplicity, the new model produces results, comparable to the previous model of Bhalla & Soh (2004d) that was analytically far more complicated . The main strength of the simplified shear lag model is the simplicity of application for solving the inverse problem. As pointed out above, it is computationally very difficult to obtain the true structural impedance $Z_{s,eff}$ from $Z_{s,eff,eq}$, using the previous model for adhesively bonded PZT patches. On the other hand, using the new simplified model, the true structural impedance can be directly determined, from Eq. (48) as,

$$Z_{s,eff} = \frac{2l^2 \overline{G_s} Z_{s,eff,eq}}{2l^2 \overline{G_s} + Z_{s,eff,eq}\, \omega h_s j} \tag{90}$$

$Z_{s,eff,eq}$ can be obtained from the measured G and B directly using the equations derived by Bhalla & Soh (2004c) for use in Eq. (90) above. No modelling is required for the host structure or the bond layer The finite element modelling done in the previous section was solely for model verification purpose only and not required in the actual applications where G and B will be available through measurement. The true structural mechanical impedance can be conveniently used for SHM of structural and aerospace components using the method proposed by Bhalla & Soh (2004 c).

Fig. 20 compares the extracted structural impedance for an aluminium block 48x48x10mm, with and without considering the bond layer. $Z_{s,eff,eq}$ is derived from the experimentally obtained admittance signatures (Bhalla & Soh, 2004b, c) followed by $Z_{s,eff}$ using Eq. (90). It is observed that the ignoring the bond layer tends to overestimate the structural true impedance. This is because the bond layer offers additional impedance on account of its own stiffness, damping and inertia. Solving the inverse problem assuming perfect bond results into impedance "apparent " at the patch ends, i.e with bond layer included . On the other hand, using the proposed formulations eliminates the effect of the bond layer and hence the impedance gets reduced. To determine the true impedance using the previous model would have demanded solving 4th order differential Eq., which is circumvented by the new simplified model.

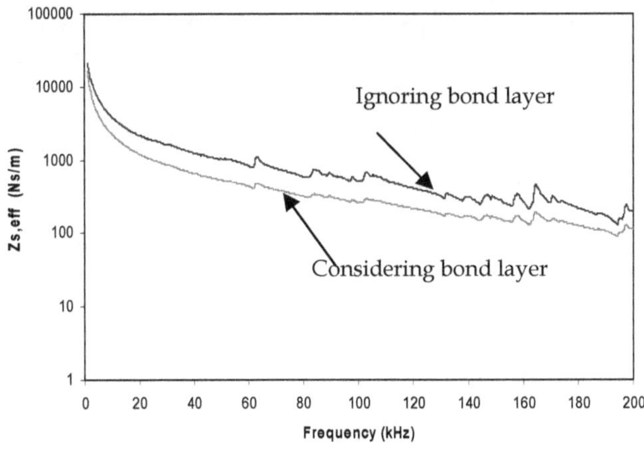

Fig. 20. Considering influence of bond layer to extract structural mechanical impedance.

7. Conclusions

This chapter has rigorously addressed the problem of incorporating the presence of adhesive layer in the electro-mechanical impedance modeling. The treatment presented is generic in nature and not restricted to beam structures alone as in the case of Crawley and de Luis (1987) and Sirohi and Chopra (2000). Besides, dynamic equilibrium of the system has been considered rather than relying on equivalent length static coefficients. The formulations have been extended to 2D effective impedance based model and have been experimentally verified. Hence, the treatment is more general, rigorous and accurate. The bond layer can significantly influence structural identification if not carefully accounted for. This chapter has also presented the development of a simplified impedance model incorporating the shear lag effect into electro-mechanical admittance formulations. The results of the model have been compared with those of the Bhalla & Soh's (2004d) shear lag impedance model. Although far simplified, the model is found to predict the conductance and the susceptance signatures in close proximity with those predicted by the model of Bhalla & Soh (2004d). The advantages of the new model are quite apparent. This model simplifies the complex shear lag phenomenon associated with the force transmission between the PZT patch and the host structure bonded to each other by the adhesive bond layer. It enables computing the true mechanical impedance of the structure from the measured experimental data alone, thus circumventing the necessity of preparing a model of the host structure or the bond layer.

8. References

Abe, M.; Park, G. & Inman, D. J. 2002. Impedance-based monitoring of stress in thin structural members. *Proceeding of 11th International Conference on Adaptive Structures and Technologies*, October 23-26, Nagoya, Japan, pp. 285-292.

Adams, R. D & Wake, W. C. 1984. *Structural adhesive joints in engineering*. Elsevier Applied Science Publishers, London.

Ayres, J. W.; Lalande, F.; Chaudhry, Z. & Rogers, C. A. 1998. Qualitative Impedance-Based Health Monitoring of Civil Infrastructures," *Smart Materials and Structures*, 7, 5, 599-605.

Bhalla, S. & Soh, C. K. 2003. Structural impedance based damage diagnosis by piezo-transducers. *Earthquake Engineering and Structural Dynamics*, 32, 12, 1897-1916.

Bhalla, S. & Soh C. K. 2004a. High frequency piezoelectric signatures for diagnosis of seismic/ blast Induced structural damages. *NDT &E International*, 37, 1, 23-33.

Bhalla, S. & Soh, C. K. 2004b. Structural health monitoring by piezo-impedance transducers I: modeling. *Journal of Aerospace Engineering*, ASCE, 17, 4, 154-165.

Bhalla, S. & Soh, C. K. 2004c. Structural health monitoring by piezo-impedance transducers II: Applications. *Journal of Aerospace Engineering*, ASCE, 17, 4, 166-175.

Bhalla, S. & Soh, C. K. 2004d. Impedance based modeling for adhesively bonded piezo-transducers. *Journal of Intelligent Material Systems and Structures*, 15, 12, 955-972.

Bhalla, S., Kumar, P., Gupta, A. & Datta, T. K. 2009. A simplified impedance model for adhesively-bonded piezo-impedance transducers. *Journal of Aerospace Engineering*, ASCE, 22, 4, 373-382.

Crawley, E. F. & de Luis, J. 1987. Use of piezoelectric actuators as elements of intelligent structures. *AIAA Journal*, 25,10, 1373-1385.

De Faria, A. R. 2003. The effect of finite stiffness bonding on the sensing effectiveness of piezoelectric patches.*Smart Materials and Structures*, 12: N5-N8.

Giurgiutiu, V. & Zagrai, A. N. 2000. Characterization of piezoelectric wafer active sensors. *Journal of Intelligent Material Systems and Structures*, 11, 12, 959-976.

Giurgiutiu, V. & Zagrai, A. N. 2002. Embedded self-sensing piezoelectric active sensors for on-line structural identification. *Journal of Vibration and Acoustics*, ASME, 124, 1, 116-125.

Hixon, E.L. 1988. Mechanical Impedance. *Shock and Vibration Handbook*, edited by C. M. Harris, 3rd ed., Mc Graw Hill Book Co., New York, 10.1-10.46.

IEEE 1987. *IEEE Standard on Piezolectricity*. Std. 176, IEEE/ ANSI.

Liang, C., Sun, F. P. & Rogers, C. A. 1994. Coupled electro-mechanical analysis of adaptive material systems- determination of the actuator power consumption and system energy transfer. *Journal of Intelligent Material Systems and Structures*, 5, 12-20.

Ong, C.W., Yang, Y., Wong, Y.T., Bhalla, S., Lu, Y. & Soh, C.K. 2002. The effects of adhesive on the electro-mechanical response of a piezoceramic transducer coupled smart system. *Proc. ISSS-SPIE International Conference on Smart Materials, Structures and Systems*, 12-14 December, Bangalore, 191-197.

Park, G., Cudney, H. H. and Inman, D. J. 2000. Impedance-based health monitoring of civil structural components. *Journal of Infrastructure Systems*, ASCE, 6(4), 153-160.

Park, G., Cudney, H. H. & Inman, D. J. 2001. Feasibility of using impedance-based damage assessment for pipeline structures. *Earthquake Engineering and Structural Dynamics*, 30, 10, 1463-1474.

PI Ceramic. 2003. *Product Information Catalogue*, Lindenstrabe, Germany, http://www.piceramic.de.

Qing, X. P., Chan, H.- L., Beard, S. J., Ooi, T. K. & Marotta, S. A. 2006. Effect of adhesive on performance of piezoelectric elements used to monitor structural health. *International Journal of Adhesion and Adhesives*, 26, 8, 622-628.

RS Components. 2003. Northants, UK, http://www.rs-components.com.

Sirohi, J. and Chopra, I. 2000. Fundamental understanding of piezoelectric strain sensors," *Journal of Intelligent Material Systems and Structures*, 11, 4, 246-257.

Soh, C. K., Tseng, K. K. H., Bhalla, S. and Gupta, A. 2000. Performance of smart piezoceramic patches in health monitoring of a RC Bridge. *Smart Materials and Structures*, 9, 4, 533-542.

Sun, F. P., Chaudhry, Z., Rogers, C. A., Majmundar, M. & Liang, C. 1995. Automated real-time structure health monitoring via signature pattern recognition. *Proc. SPIE Conference on Smart Structures and Materials*, San Diego, California, Feb.27-Mar1, 2443, 236-247.

Xu, Y. G. & Liu, G. R. 2002. A modified electro-mechanical impedance model of piezoelectric actuator-sensors for debonding detection of composite patches, *Journal of Intelligent Material Systems and Structures*, 13, 6, 389-396.

Zhou, S. W., Liang, C. & Rogers, C. A. 1996. An impedance-based system modelling approach for induced actuator-driven structures, *Journal of Vibration and Acoustics*, ASME, 118, 323-331.

4

Characterization of properties and damage in piezoelectrics

Guillermo Rus, Roberto Palma and Javier Suárez
Dept. Structural Mechanics and Hydraulic Engineering, University of Granada
Spain

1. Introduction

The aim of this chapter is to review the state of the art of the possibilities and existing knowledge on characterization of mechanical and coupling properties and damage detection in piezoelectric ceramics. To address the theoretical problem with a proper rational basis, a review of numerical and experimental techniques to solve the so-called inverse problem of characterization of properties and damage is given. Finally, the relevant details of finite element and boundary element formulations and implementation when addressing correct damage simulation in piezoelectric ceramics (Pérez-Aparicio et al. (2007)) is presented.

In recent years, the theoretical framework of inverse problems has been developed to understand and formalize the property and damage characterization problems. This theory has been applied in a variety of continuum mechanics applications, by the groups of Liu & Chen (1996), Oh et al. (2005), Pagano (1970), Rus et al. (2005) and Tarantola & Valette (1982). There is a growing body of knowledge in identification inverse problems for elasticity, like the groups of Tardieu & Constantinescu (2000) or Bonnet & Constantinescu (2005), but due to the intrinsic coupling of magnitudes, the formulation of the piezoelectric problems is more complex than in elastic problems. Despite this difficulty, the electric field and its coupling with the elastic field has the potential to be exploited in monitoring piezoelectrics and in the solution of the inverse problem. Inverse problem techniques have been applied specifically for the piezoelectric ceramics for:

Characterization of properties: The basic formulation has been established by Kaltenbacher et al. (2006) defining a cost functional as the difference between electric impedances observed in laboratory and those obtained after solving the direct problem by the finite element method. A similar cost functional was used by Ruíz et al. (2004a;b), which was minimized using genetic algorithms. On the other hand, Araújo et al. (2006); Araújo et al. (2002) proposed an inverse problem to obtain the constitutive properties of composite plate specimens with surface bonded piezoelectric patches or layers, where the cost functional was the difference between the experimental and FEM–predicted eigen–frequencies and its minimization was carried out using two strategies: a gradient–based method, and neural networks. A genetic algorithm was applied by Chou & Ghaboussi (2001) and Mares & Surace (1996) to solve the IP in elastic structures. Based on crystallographic criteria by Russell & Ghomshei (1997) a cost functional was formulated as the difference in the orientation distribution function, which provides a statistical description of the orientation.

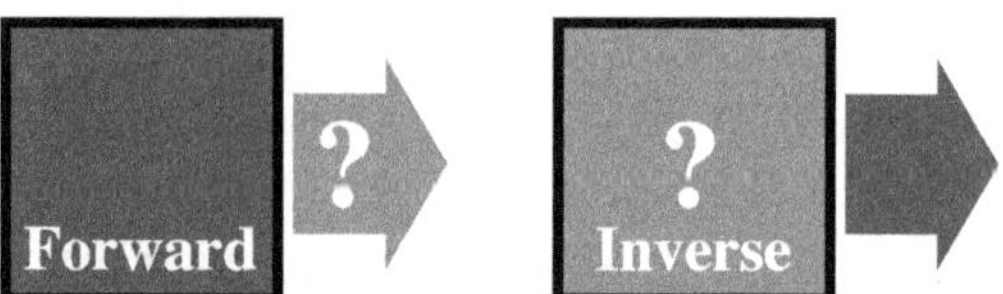

Fig. 1. Inverse problem concept

Damage detection: The group of Palma et al. (2009); Rus et al. (2009) developed the identification inverse problems to design self-diagnosing piezoceramics, by formulating and solving the defects identification problem in piezoelectric plates. They obtained the optimal experimental configuration based on the concept of the probability of detection. Their findings show which excitation-measurement combination provides the most robust self-diagnosis design, and also give criteria to determine which degree of uncertainty needs to be assured in each constitutive property to yield a robust and sensitive damage characterization. A small body of research groups is currently emerging both in the experimental and theoretical fields of monitoring damage in piezoelectrics.

2. Inverse Problems

An Inverse Problem (IP) can be defined in opposition to the forward problem. If a forward problem aims at finding the response (output, in red colour, Fig. 1) of a system given a known model (input, in blue colour), an inverse problem consists in retrieving unknown information of the model given the response as known input data. IPs have been recently applied to study and characterize not only damage or mechanical properties of piezoelectric and classical materials, but provide the general framework for reconstructing an unknown part of a system model.

The model-based IP is the most advanced approach for IP solution. Following Fig. 2, it consists in (1) obtaining a set of experimental measurements given a specific experimental design, which interrogate the system by propagating some physical magnitude that interacts with the unknown part of the system and manifests on an accessible part of it. Some physical assumptions need to be made to generate a mathematical model (2) that can be solved computationally, in which the unknown part of the model to be reconstructed is dependent on some defined parameters. This model simulates the measurements given a set of parameter values. A discrepancy (3) between experimenental and simulated measurements is defined using some metrics to define a Cost Function (CF). The IP is finally solved by finding the values of the parameters that minimize the CF (4), and thus the problem is mathematically fully stated.

Inverse problems are ill-conditioned problems, in the sense that the solution may not be unique, may not exist and may be unstable and divergent. This ill-conditioning is rooted in the physical meaning of the problem, and cannot be completely avoided by purely mathematical manipulations. Instead, some physical pieces of a priori information have to be incorporated into the formulation. This is the basis of a set of techniques called regularization that were formally introduced by Tikhonov & Arsenin (1979); Tikhonov & Arsénine (1974) and extended by Menke (1984) and Aster et al. (2005). These strategies were successfully incorporated to the model-based inverse problem by Rus, establishing a corps of knowledge to be used in this project, while dealing with a variety of mechanical parameter extraction ranging from continuum damage Lee et al. (2007; 2008); Rus & Gallego (2002); Rus & García-Martínez

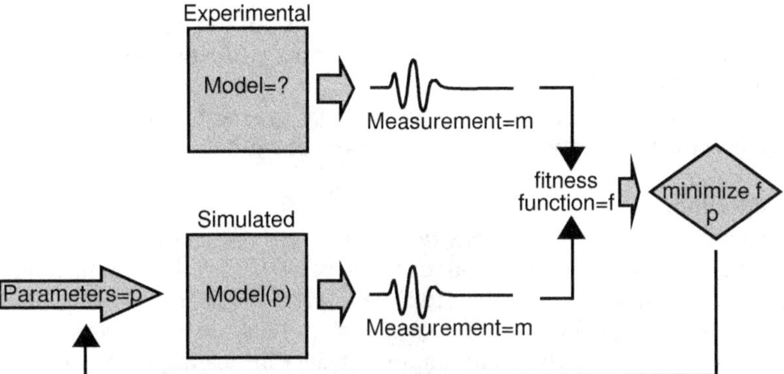

Fig. 2. Model-based inverse problem flowchart

(2007); Rus et al. (2005; 2004) to fractures Gallego & Rus (2004); Rus & Gallego (2005; 2006); Rus et al. (2007a;b) in advanced materials, and using either advanced finite element (FEM) and boundary element (BEM) formulations for the model solution.

2.1 Cost function

The CF, also called objective function, is usually defined (Eqn. 1) as a L-2 norm of the difference between the experimental measurements ψ^{EXP} and those simulated by the numerical procedure ψ^{NUM}, and integrated over the observed period of time T, in the case of dynamic measurements,

$$f = \frac{1}{2NT} \int_T \sum_{i=1}^{N} \left(\psi_i^{EXP} - \psi_i^{NUM} \right)^2 dt \tag{1}$$

Other alternatives are possible, like L-1 norm, or defining the residual between the frequency-domain counterpart of the measurement, or in any other domain.

When genetic algorithms or other heuristic search algorithms are used as for the minimization, an alternative form of the CF is defined, that improves the convergence of the algorithm, as argued by Gallego & Rus (2004),

$$f^L = -\lg (f + \varepsilon) \tag{2}$$

where ε is a small adimensional constant (typically $\varepsilon = 10^{-16}$) that ensures the existence of the logarithm when f vanishes.

2.2 Parametrization

In the context of inverse problems, parametrization of the model means to characterize the sought solution (the defect in this case) by a set of parameters p_i, which are the working variables and the output of the inverse problem. The choice of parametrization is not obvious, and it is a critical step in the problem setup, since the inverse problem is a badly conditioned one, in the sense that the solution may not be stable, exist or be unique, and the assumptions on the damage model that allow to represent it by a set of parameters can be understood as a strong regularization technique. In particular, a reduced set of parameters is chosen to facilitate the convergence of the search algorithm, and they are also defined to avoid coupling between them.

Typical parametrizations in damage location parameters are some geometric parameters that define the damage extent or size (Level-2 NDE, severity, while Level-1 NDE classification is attained by telling if there is or not damage), other parameters for damage location (Level-3 NDE, location). On the other hand, typical material constant parametrization contain the mechanical, electrical and piezoelectrical constants as coordinates of the parameter vector.

2.3 Search algorithm

The CF minimization can be performed by two alternative families of methods: gradient based methods (among which Gauss-Newton algorithms, BFGS or Simulated annealing are some of the most popular, see Dennis & Schnabel (1983, 1996)) and random search algorithms (Genetic Algorithms, See Goldberg (1989), particle sworm algotihms, simulated annealing, etc.). The latter family require significantly small amount of data in dealing with complex problems, while attaining global convergence as opposed to gradient-based methods, which in opposition are much less computationally expensive.

The IP can be mathematically formalized as a minimization problem, that starts from a measuring system from which the response is recorded, a computational idealization of the system that simulates the measurements (forward problem) depending on the unknown characterization of the defect (parametrization) and defining a CF as above. The solution of the IP is obtained by solving,

$$\min_{p_i} f(p_i) \tag{3}$$

3. Damage characterization

Piezoelectric ceramics are brittle and susceptible to fracture: the ultimate strength is less than 100 [MPa], while the fracture toughness is between 0.5 and 2.0 [MPa/m$^{1/2}$]. Furthermore, due to their ceramic nature, these materials are highly inhomogeneous, causing cracks and/or cavities in the manufacturing process or during their operation. When this occurs, high stress and electric fields concentrations appear around the defect, failing to serve the function for which their were designed. This has contributed to the emergence of many analytical and experimental works about fracture mechanics in the last two decades. However, there are few studies about damage detection, despite it is an interesting way to prevent the failure of these ceramics.

This section describes the damage effects from an analytical and an experimental point of view. Furthermore, inverse problems applied to damage characterization are supervised. Finally, the probability of detection is presented, which gives an idea of the probability that a defect is positively detected, is defined.

3.1 Damage effects: analytical

Analytical studies about piezoelectric fracture mechanics began in 1976 by Parton. Thereafter, traditional mathematical methods, as *Lekhnitskii* and *Stroh* formalisms, used in elastic fracture mechanics were applied to these ceramics. Note that fracture behavior of a piezoelectric ceramic under combined electro-mechanical loading is much more complex than that of the traditional ceramics.

3.1.1 Boundary conditions on defect faces

Several researchers have published theoretical studies on infinite piezoelectric ceramics containing an elliptic hole and have reported different results depending on the electric boundary

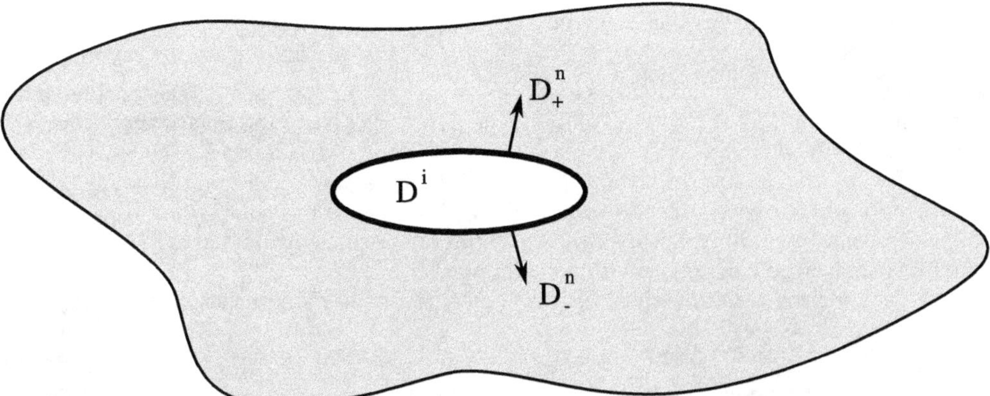

Fig. 3. Normal electric displacement on boundary faces

conditions chosen on defect faces. Therefore, one of the key problems in piezoelectric fracture mechanics is the selection of electric boundary conditions, as it it reported by Ou & Chen (2003).

Consider an infinite piezoelectric ceramic material with a elliptical hole inside, where $\mathbf{D}^n_+$ and $\mathbf{D}^n_-$ are the normal electric displacement to upper and lower defect faces, respectively, and $\mathbf{D}^i$ is the electric displacement inside the cavity, see Figure 3. There exist three (four for some authors) electric boundary conditions:

Impermeable The dielectric permittivity inside the cavity is assumed to be zero, because it is three order of magnitude less than the ceramic permittivity:

$$\mathbf{D}^n_+ = \mathbf{D}^n_- = 0 \tag{4}$$

Permeable The dielectric permittivity inside the cavity is assumed to be infinity or, alternatively, both sides of the defect is in electrical contact:

$$\mathbf{D}^n_+ = \mathbf{D}^n_- \tag{5}$$

Exact This boundary condition is also denominated semi–permeable. It is the only one with physical sense, since the cavity is assumed to behave like a capacitor, from an electrical point of view.

Ou & Chen (2003) reported that: i) impermeable boundary condition has no physical meaning and when it is applied, mathematical singularities appear and ii) permeable boundary condition can be applied when free traction on defect face is assumed. Therefore, the exact boundary condition is the only physically correct and should always be used.

3.1.2 Lekhnitskii formalism–based approach

A 3–D analytical work was developed by Sosa & Pak (1990) in order to study the fracture mechanics of a transversely isotropic ceramic with a crack inside, under electro–mechanics loads. Two important conclusions were obtained:

    i) Field variables at crack tip depend on the crack orientation

ii) Crack growth can be accelerated or delayed, depending on the direction of the electric field applied.

Sosa & Pak (1990) only computed the crack tip concentrations, so Sosa (1991) considered an infinity plain strain ceramic with an elliptic hole perpendicular to the polarization direction inside, in order to obtain the field variables around the cavity. Xu & Rajapakse (1999) extended this work considering an arbitrarily oriented elliptic hole and concluding that the highest concentrations occur when the elliptical hole is $33°$ respect to the polarization direction. In both works was concluded that the stress concentrations at crack tip are higher in piezoelectric than that in anisotropic materials without coupling.

Sosa (1992) used the result obtained by Sosa (1991) to define the intensity factors for piezoelectric ceramics:

$$K_I = \sqrt{a}\, T_{zz}^{\infty} \quad ; \quad K_{IV} = \sqrt{a}\, D_z^{\infty} \tag{6}$$

where a the length of the crack and T_{zz}^{∞} and D_z^{∞} are the stress and electric displacement (fracture mode I) applied, respectively. Note that an electric intensity factor K_{IV} is emerged in piezoelectric materials, in order to compute the electric contribution to the fracture.

In previous works, the impermeable boundary condition is assumed. Sosa & Khutoryansky (1996) applied the exact boundary conditions to expand the study developed by Sosa (1991), concluding that the impermeable boundary condition is unacceptable when the elliptic hole tends to a crack. The same conclusion was numerically obtained by Pérez-Aparicio et al. (2007), using the finite element method. Finally, Hao & Shen (1994) applied the exact boundary condition to obtain the fracture parameters, which were obtained by Sosa (1992) considering the impermeable boundary condition. Then, the fracture parameter of equation (6) are modified by:

$$K_I = \sqrt{\pi a}\, T_{zz}^{\infty} \quad ; \quad K_{IV} = \sqrt{\pi a}\, \left(D_z^{\infty} - D_z^i \right) \tag{7}$$

where D_z^i is the electric displacement inside the cavity. Stress intensity factors are not affected by the electric boundary condition. However, the new electric intensity factor K_{IV} depends on the mechanical load applied.

3.1.3 Stroh formalism–based approach

Suo et al. (1992) extend the *Stroh* formalism to piezoelectric problems, considering a semi–infinite piezoelectric ceramics with a crack inside. Plain strain assumption and impermeable boundary condition were assumed to calculate field concentrations and intensity factors. Furthermore, in this work was calculated the energy release rate G_I, which was divided into mechanical G_I^M and electrical parts by Park & Sun (1995a):

$$G_I^M = f\left[(T_{zz}^{\infty})^2 \right] \quad ; \quad G_I = f\left[(T_{zz}^{\infty})^2, T_{zz}^{\infty} D_z^{\infty}, (D_z^{\infty})^2 \right] \tag{8}$$

According to Park & Sun (1995a), the mechanical energy release rate G_I^M has to be used as fracture criteria. However, Gao et al. (1997) and McMeeking (2001) argued that electro–mechanical behaviors can not be studied by fracture mechanical parameters. Therefore, McMeeking (2004) and Wang & Sun (2004) obtained the energy release rate considering the exact boundary condition.

3.2 Damage effects: experimental

Experimental works about piezoelectric fracture mechanics started in 1980, in order to find a fracture criteria for these materials. Two experimental devices are generally used: *Compact Tension* (CT) and *Three Point Bending* (TPB).

Park & Sun (1995b) performed a CT fracture study, where the specimen was a PZT–4 ceramic with a crack inside, which was perpendicular to the poling direction and the applied electric field. It was published, an experimental curve that plots fracture loads F_c versus applied electric field E_a, observing the fracture dependence on applied electric field: F_c increases when $E_a < 0$ and decreases when $E_A > 0$. Similar results were obtained by Fu et al. (2000) and Soh et al. (2003) using PZT–4 and PZT–5 ceramics, respectively. However, Fu et al. (2000) observed that fracture load decreases for positive or negative electric fields applied for PZT–841. On the other hand, Park & Sun (1995b) and Soh et al. (2003) used the analytical results obtained by Park & Sun (1995a) and Gao et al. (1997), respectively, to conclude that the mechanical energy release rate fitted very well with the experimental results.

Jelitto, Kessler, Schneider & Balke (2005) and Jelitto, Felten, Hausler, Kessler, Balke & Schneider (2005) used a PZT–PIC 151 specimen and a TPB device to perform fracture experiments, publishing fracture criteria based on: intensity factors, mechanical and total energy release rates. From a numerical point of view, the intensity factors and the energy release rates ware calculated by the finite element method and considering the three boundary conditions. Finally, it was concluded that the mechanical energy release is more appropriate as fracture criteria, from an empirical point of view.

3.3 Numerical damage characterization approaches

After revising the existing literature (a full review on analytical and experimental piezoelectric fracture mechanics was performed by Chen & Hasebe (2005) and Schneider (2007), respectively), several conclusions are obtained:

- From an analytical point of view, there is no an electric boundary condition can reproduce the experimental results, which explains why current researching are based on non–linear fracture mechanics.

- From an experimental point of view, there is no unique trend for the fracture load as a function of the electric field. This implies that a general fracture criteria can not be developed, which explains the absence of work on piezoelectric reliability.

Therefore, there are many analytical and experimental works about piezoelectric fracture mechanics. However, there are few studies about damage detection, despite it is an interesting way to prevent the failure of these ceramics. Then, several questions appear to remain open: i) the use of the inverse problem to damage detection, ii) optimal experimental measurement setup for improving the inverse problem solutions, iii) probability of detection, iv) inverse problem sensitivity to system uncertainties, v) main variables responsible for the experimental noise and vi) how can the noise be effectively reduced?. All these questions were resolved by the Non–destructive Evaluation Laboratory of the University of Granada, see Rus (2010), by the publishing works: Rus et al. (2009), Palma et al. (2009) and Rus, García-Sánchez, Sáez & Gallego (2010a). Nowadays, these woks are being extended in Rus, Palma & Pérez-Aparicio (2010) and Rus, García-Sánchez, Sáez & Gallego (2010b).

Figure 4 shows a typical flow chart of the model–based inverse problem, where two inputs are introduced: i) parametrization is a number of parameters, which characterize the sought damage and are the inverse problem output and ii) experimental measurements. Finally, a

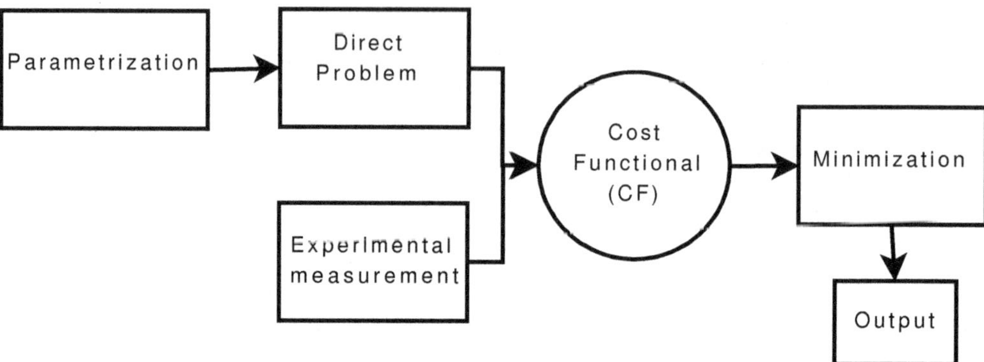

Fig. 4. Flow chart of the model-based inverse problem, see Palma et al. (2009)

cost function is defined and it is minimized by genetic algorithms, as it has been reported in previous sections.

3.3.1 Parametrization

The choice of parametrization is not obvious, and it is a critical step in the problem setup (see Palma et al. (2009)), since the inverse problem is a badly conditioned one, in the sense that the solution may not be stable, exist or be unique, and the assumptions on the damage model that allow to represent it by a set of parameters can be understood as a strong regularization technique. In particular, a reduced set of parameters is chosen to facilitate the convergence of the search algorithm, and they are also defined to avoid coupling between them.

Figure 5 shows a 2D piezoelectric plate of dimensions L_x and L_z with (a) a circular defect or (b) a crack inside, which has to be detected. Note that z is the poling direction. The damage location and size presented suggests the definition of the immediate parameters:

Circular defect $p_i = (x_0, z_0, r)$,

Crack $p_i = (x_0, z_0, L, \theta)$, where L is the length of the crack.

The true (and unknown) parameters are denoted by $\bar{p}_i$. In Rus et al. (2009), Palma et al. (2009) and Rus, Palma & Pérez-Aparicio (2010) a circular defect was considered as a first approximation, to simplify the meshing using the finite element method. On the other hand, in Rus, García-Sánchez, Sáez & Gallego (2010a) and Rus, García-Sánchez, Sáez & Gallego (2010b) a crack was considered, due to the simplicity of the mesh using the boundary element method.

3.3.2 Experimental measurements

Consider the specimen shown in Figure 5 with the circular defect or crack inside. This sample is excited by electrical or mechanical loads, and its response (displacements and/or voltage) is measured at N_i points along the lower boundary of the plate. Note that the piezoelectric coupling makes mechanical loads generate voltages (direct effect) and the electrical loads generate mechanical displacements (inverse effect). On the other hand, the boundary conditions are selected to avoid rigid solid motions.

In the literature there are no works about experimental measurements for damage characterization, so the Non–destructive Evaluation Laboratory of the University of Granada, Rus (2010), is undertaking experimental tests to verify all its numerical results. In Palma et al.

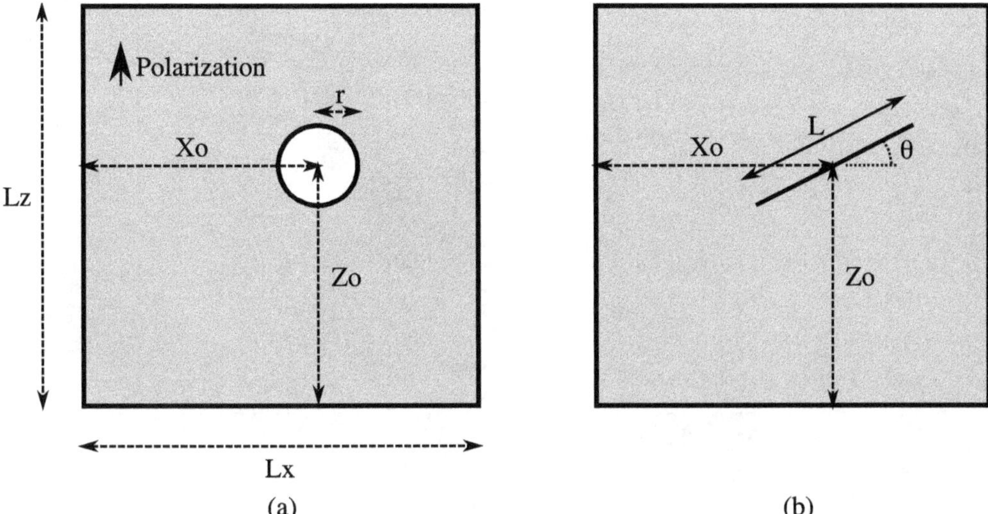

Fig. 5. Piezoelectric plate with (a) a circular defect or (b) a crack inside

(2009); Rus, García-Sánchez, Sáez & Gallego (2010a;b); Rus et al. (2009); Rus, Palma & Pérez-Aparicio (2010) the experimental measurements were generated solving the forward problem by finite or boundary element method and adding noise generated by a gaussian distribution:

$$\psi_i = \psi_i^{FWD} + \xi_i RMS(\psi_i^{FWD})\sigma \tag{9}$$

where ψ_i, ψ_i^{FWD}, ξ_i and σ are the simulated experimental measurements, measurements generated by the solution of the forward problem, random variables, and a parameter defined to study the influence of the noise in the inverse problem solution, respectively. On the other hand, RMS is the root mean square given in Rus et al. (2009).

3.3.3 Results

Figure 6 shows the dependency of the cost functional on the spatial location of the defect (fixing the size at the real value) for increasing noise levels, see Rus et al. (2009). If no noise is simulated, the cost function shows a clear optimum that the search algorithm is able to find. The shape of the cost function is distorted when the noise level increases.
Figure 7 shows the cost function and genetic algorithm convergence for different noise levels. For the case without noise the full convergence is obtained for less than 200 generations. A larger noise level is associated with slower convergence, probably due to the wavy and fuzzy shapes of the cost functional.

3.4 Probability of detection

The Probability of Detection (POD) gives an idea of the probability that a defect is positively detected, given a specimen, a defect size and some noise and system uncertainty conditions. The detection and characterization of defects is based on the interpretation of the alterations of the measurements due to the presence of the defect, however uncertainties and system noises also alter these measurements. For this reason, the POD is estimated by the probability that

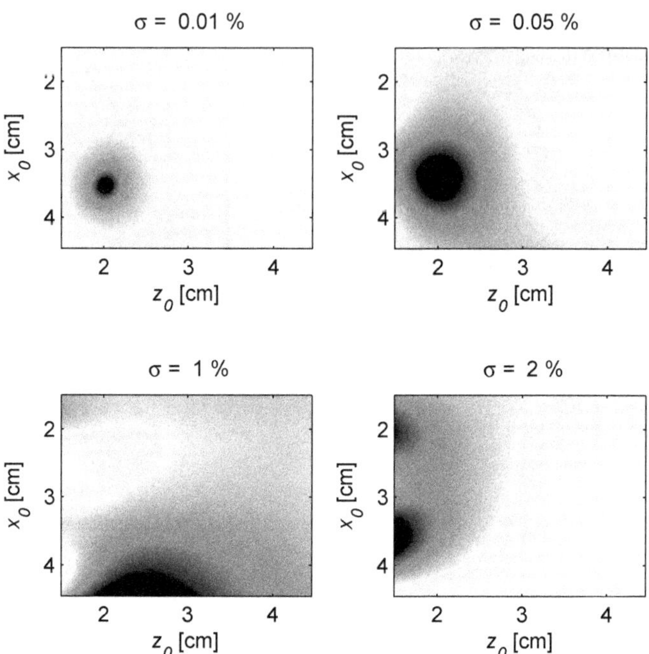

Fig. 6. Cost functional for increasing levels of experimental noise, Rus et al. (2009)

the alteration of the measurement caused by the defect is larger than that caused by the noise, see Rus et al. (2009):

$$\text{POD} = P\left(\frac{|\text{SIGNAL}|^2}{|\text{NOISE}|^2} > 1\right) \tag{10}$$

where SIGNAL and NOISE terms are obtained assuming: i) a linear dependence between the variation of the measurements with increasing noise, see 9 and ii) linear piezoelectric behaviors. Then, the experimental measurements depend on damage area A and noise σ, expanding in *Taylor* series and neglecting higher-order terms:

$$\psi_i(A,\sigma) = \psi_i(0,0) + \underbrace{A\frac{\partial\psi_i}{\partial A}(0,0)}_{\text{SIGNAL}} + \underbrace{\sigma\frac{\partial\psi_i}{\partial\sigma}(0,0)}_{\text{NOISE}} + hot \tag{11}$$

The first term of the right member is the measurement at point i without noise and defect. The second term is the alteration of that measurement due to the presence of the defect only SIGNAL. Finally, the third term is the alteration of the signal originated by the noise only NOISE.

Finally, in Rus et al. (2009) is obtained and validated, using *Monte Carlo* techniques, an analytical expression of the POD, which allows to estimate a priori the minimum defect findable given a specimen geometry and a noise level on measurements.

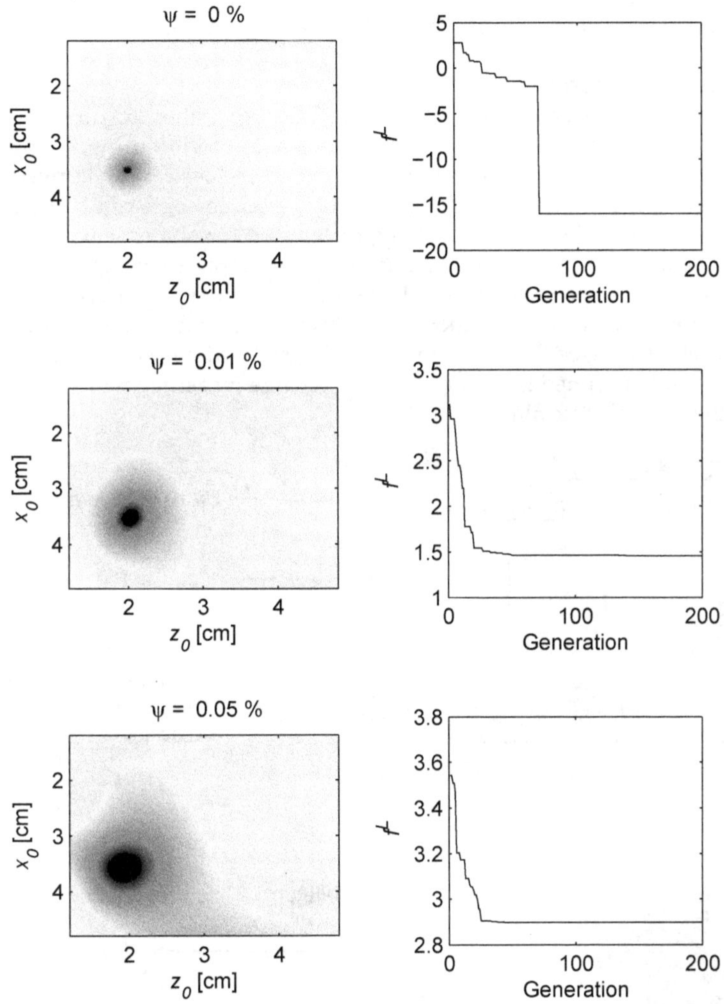

Fig. 7. Cost Function and Genetic Algorithm convergence for different noise levels, see Palma et al. (2009)

4. Property characterization

The performance as sensors or as actuators of piezoelectric elements, usually manufactured in the form of thin films, is defined by the piezoelectric properties, responsible of the electromechanical coupling, but also on the mechanical and electrical properties. These properties need to be established at the material design stage. However, when using piezoelectric crystals for simulation and active noise and vibration control applications, The properties obtained through manufacturer data are in most of the cases not enough to predict the structural behaviour and implement effcient control algorithms due to non-homogeneities in materials, and differing geometries and material properties in the regions of actuators and sensors.

4.1 Experimental procedures

Characterization of piezoelectric constants can be based on either the direct or the converse piezoelectric effect. In direct measurements the voltage produced by an applied stress is measured; in indirect measurements the displacement produced by an applied electric field is measured through different methods. The direct measurements can be performed either quasi-statically or dynamically. The first is typically used for nonresonant applications (sensors and actuators), while the second is typical for those applications where the piezoelectric thin film is used for the generation or detection of high-frequency bulk acoustic waves or surface acoustic waves. Various methods have been described for the measurement of the piezoelectric coefficients (see Fig. 8. For instance, the longitudinal piezoelectric coefficient d_{33} (see section 5.1) is determined from the generated charge measured when a calibrated stress is applied to the piezoelectric film.

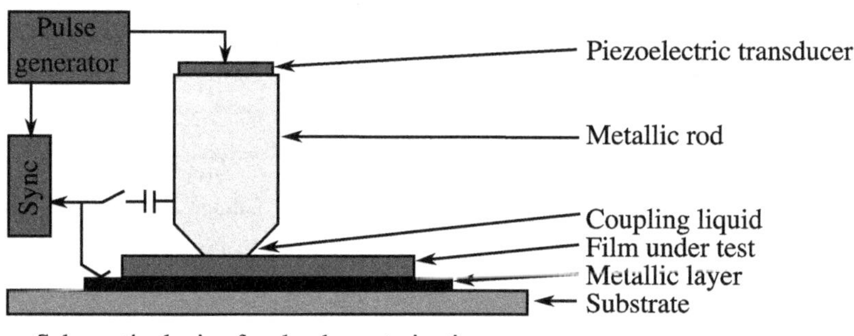

Schematic device for d$_{33}$characterization

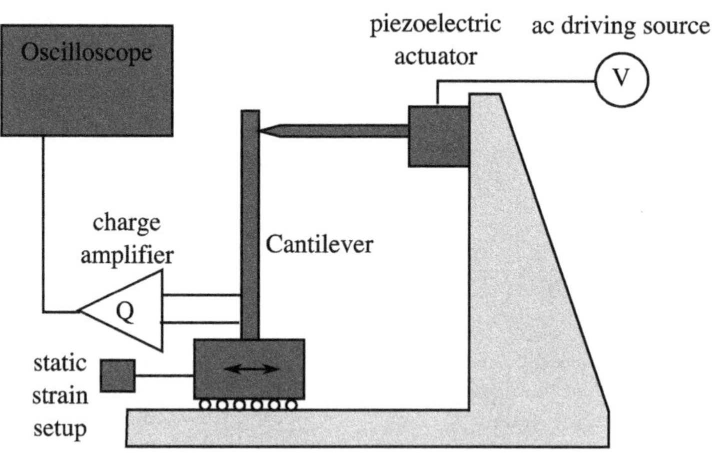

Schematic device for d$_{13}$characterization

Fig. 8. Experimental setups for constitutive property characterization

A recent and more sophisticated alternative method is based on the Atomic Force Microscopy (AFM), in which a variety of excitation-measurement configurations is applied in order to minimize the effect of the applied electric field with the microscope tip. The evaluation of the transverse e_{31} constant requires different techniques. Here the film is deposited on a cantilever beam and two electrodes are deposited on the free film surface. The free edge of the cantilever is first statically displaced, causing a bending of the cantilever, and then it is suddenly released. The consequent damped oscillation of the cantilever at its natural frequency can be revealed through the voltage appearing between the electrodes and displayed on an oscilloscope. It is proved that the first maximum of the damped oscillation is proportional to the e_{31} constant.

On the other hand, the measurement of the dielectric constants is usually performed with a standard impedance meter.

The complete characterization of piezoelectric constants requires the elastic stiffness coefficients. These can be determined from Surface Acoustic Wave (SAW) velocity dispersion curves that can be obtained in several ways. One method consists of measurement of the phase velocity of several vibration modes. Due to the small amplitude of the vibration, contact-less vibration measurements are usually employed, such as laser interferometry or reflection angle measurement.

4.2 Numerical characterization approaches

When a mathematical model of the piezoelectric system dynamics behaviour of the elements is needed, the physical constants that define the general constitutive laws (see section 5.1) need to be estimated, and this is done by through fit-to-data techniques. The inverse problem is the general theoretical and algorithmic framework to be applied, and in particular, the model-based inverse problem.

To address the limitations of current characterization methods in determining elastic and piezoelectric constants (Araújo et al. (2009); Araújo et al. (2002)) propose a finite element model-based, associated to gradient optimisation-based inverse problem algorithm using vibration data to carry out the identification of electromechanical properties in composite plate specimens with surface bonded piezoelectric patches or layers. This method has been later refined with neural networks to aid the inversion algorithms Araújo et al. (2006).

The problem of how to define which measurements and how to measure in order to minimize the uncertainties of material parameters given the unavoidable measurement noise has recently been addressed by Lahmer et al. (2010). An approach for a similar question based on the definition and maximization of the probability of detection was also addressed by Palma et al. (2009).

4.3 Experimental advances

A few experimental studies regarding the estimation of parameters in mathematical models for homogeneous beams (Banks et al. (1994)), plates and grid structures (Banks & Rebnord (1991)) have been reported. A primary concern when estimating parameters is that the model fits to data consistently across experiments after a time-domain inverse problem solution, for example in the presence of passive damping or for a range of eigenfrequencies (Banks et al. (1997)), to ensure the effectiveness of the model independently of the number of excited frequencies.

Using conventional methods to determine the elastic and piezoelectric constants faces some difficulties in engineered crystals with high coupling constants. The piezoelectric coeffcients

e_{33} and e_{31} (see section 5.1) become unstable using standard methods because they are highly sensitive to the elastic stiffness coefficients c_{12} and c_{13} due to the large d_{31} and d_{33}, yielding serious property fluctuations from sample to sample that often make it impossible to obtain self-consistent data. The derived values of s_{11}, s_{33}, s_{12} and s_{13} also become too sensitive to the variation of c_{13} when its value is close to those of c_{11} and c_{33}. A reversed strategy was suggested by Jiang et al. (2003) using detailed error analysis to determine the best strategy for each given material.

5. Numerical considerations

5.1 Governing equations

Consider a deformable electrically sensitive body, which occupies the region Ω of boundary Γ in a three–dimensional *Euclidean* space, in the absence of electro–mechanical loads.
From a mechanical point of view, the equilibrium and compatibility equations are given by, see Eringen (1980):

$$\nabla \cdot \mathbf{T} + \mathbf{f} = \rho \ddot{\mathbf{u}} \quad ; \quad \mathbf{S} = \nabla^s \mathbf{u} \tag{12}$$

where $\mathbf{T}$, $\mathbf{S}$, $\mathbf{f}$, $\mathbf{u}$ and ρ denote the stress tensor, strain tensor, body forces, displacement and density, respectively. Furthermore, the mechanical boundary conditions are:

$$\text{Dirichlet} \quad \rightarrow \quad \mathbf{u} = \bar{\mathbf{u}}$$

$$\tag{13}$$

$$\text{Newmann} \quad \rightarrow \quad \mathbf{T} \cdot \mathbf{n} = \mathbf{t}$$

where $\bar{\mathbf{u}}$, $\mathbf{n}$ and $\mathbf{t}$ denote the prescribed displacements, normal unit vector and tractions, respectively.
From an electrical point of view, the governing equations are obtained from the uncoupled *Maxwell* equations, see Eringen & Maugin (1990), and neglecting the free electric charges:

$$\nabla \cdot \mathbf{D} = 0 \quad ; \quad \nabla \times \mathbf{E} = 0 \tag{14}$$

where $\mathbf{D}$, $\mathbf{E}$ are the electric displacement or induction and electric field, respectively. According to the fields theory, the induction and the electric field can be obtained from vector $\mathbf{V}$ and scalar ϕ potentials:

$$\nabla \cdot \mathbf{D} = 0 \quad \rightarrow \quad \mathbf{D} = \nabla \times \mathbf{V}$$

$$\tag{15}$$

$$\nabla \times \mathbf{E} = 0 \quad \rightarrow \quad \mathbf{E} = -\nabla \phi$$

This duality allows to define two types of boundary condition, as working with vector or scalar potentials:

$$\text{Dirichlet} \quad \rightarrow \quad \mathbf{V} = \bar{\mathbf{V}} \quad ; \quad \phi = \bar{\phi}$$

$$\tag{16}$$

$$\text{Newman} \quad \rightarrow \quad \mathbf{n} \times \mathbf{E} = \mathbf{E}_s \quad ; \quad \mathbf{n} \cdot \mathbf{D} = D_n$$

where $\bar{\mathbf{V}}$, $\bar{\phi}$ and D_n are the prescribed electric vector potential, electric scalar potential or voltage and normal electric displacement, respectively, and $\bar{\mathbf{V}} = -\mathbf{n} \times \nabla \phi$.
Finally, the piezoelectric constitutive equations, which couple mechanical and electric fields are given by:

$$\mathbf{T} = \mathbf{c}^E : \mathbf{S} - \mathbf{e}^t \cdot \mathbf{E} \quad ; \quad \mathbf{D} = \mathbf{e} : \mathbf{S} + \epsilon^S \cdot \mathbf{E} \tag{17}$$

where $\mathbf{c}^E$, $\mathbf{e}$ and ϵ are the elastic, coupling and dielectric material properties, respectively.

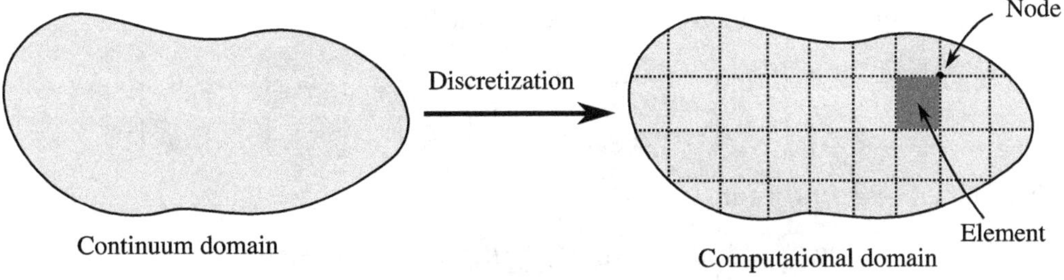

Fig. 9. FEM discretization

5.2 Finite element method

The Finite Element Method (FEM) is a numerical technique for the solution of linear and non–linear partial differential equations, which is used to model many problems in science and engineering. The FEM is the most advanced method for the solution of electro–mechanical field problems, but involves complex mathematical concepts, see Zienkiewicz et al. (2005) or Hughes (1987) for further study. In general, the FEM implies the following steps:

i) The continuum domain Ω is divided into subdomains or elements, which are interconnected at nodal points, see Figure 9.

ii) The nodal values, denominated degrees of freedom, are assumed to be the unknown parameters of the problem.

iii) A set of functions, denominated shape functions, is chose to interpolate the solution within each finite element in terms of its nodal values.

iv) The principle of virtual work is applied to the governing and constitutive equations to obtain the weak form of the problem.

iv) The solution is obtained by solving a set of linear (or non–linear) equations. In order to solve non–linear problems, algorithms such as *Newton–Raphson* has to be applied.

For the piezoelectric problem, two alternative FEM formulations can be developed, according to the equations 15: scalar and vector formulations, which are described in the following sections.

5.2.1 Scalar formulation

The Scalar Formulation (SF) developed in 1970 by Allik & Hughes (1970) has four degree of freedom per node: three displacement and the scalar potential or voltage: $\mathbf{U} = (\mathbf{u}, \phi)^t$, where $(.)^t$ denotes the transpose.

The weak form is obtained applying the principle of the virtual work to the governing equations and constitutive equations. Then, *Newmann* boundary conditions are satisfied automatically, while *Dirichlet* boundary conditions are applied to the nodal values. Subsequently, a discretization is performed using the shape functions $\mathbf{N}$ to interpolate the nodal values $\mathbf{a_u}$ and $\mathbf{a}_\phi$:

$$\mathbf{u} = \mathbf{N}\mathbf{a_u} \quad \rightarrow \quad \mathbf{S} = \mathbf{B}^S \mathbf{a_u}$$

$$\phi = \mathbf{N}\mathbf{a}_\phi \quad \rightarrow \quad \mathbf{E} = \mathbf{B}^E \mathbf{a}_\phi$$

(18)

where $\mathbf{B}^S = \nabla^S \mathbf{N}$ and $\mathbf{B} = \nabla \cdot \mathbf{N}$.

Finally, the set of linear equations to be solved is given by:

$$\begin{bmatrix} \mathbf{M} & 0 \\ 0 & 0 \end{bmatrix} \begin{Bmatrix} \ddot{u} \\ \ddot{\phi} \end{Bmatrix} + \begin{bmatrix} \mathbf{K_{uu}} & \mathbf{K_{u\phi}} \\ \mathbf{K_{\phi u}} & \mathbf{K_{\phi\phi}} \end{bmatrix} \begin{Bmatrix} u \\ \phi \end{Bmatrix} + \begin{Bmatrix} \mathbf{f_u} \\ \mathbf{f_\phi} \end{Bmatrix} = \begin{Bmatrix} 0 \\ 0 \end{Bmatrix} \tag{19}$$

where the stiffness matrices are:

$$\mathbf{K_{uu}} = \int_\Omega \mathbf{B}^S \mathbf{c}^E \mathbf{B}^S d\Omega \quad ; \quad \mathbf{K_{u\phi}} = \int_\Omega \mathbf{B}^S \mathbf{e}^t \mathbf{B}^E d\Omega$$

$$\mathbf{K_{\phi u}} = \int_\Omega \mathbf{B}^E \mathbf{e} \mathbf{B}^S d\Omega \quad ; \quad \mathbf{K_{\phi\phi}} = -\int_\Omega \mathbf{B}^E \mathbf{\epsilon}^S \mathbf{B}^E d\Omega \tag{20}$$

the kinematically consistent mass matrix is given by:

$$\mathbf{M} = \int_\Omega \mathbf{B}^S \rho \mathbf{B}^S d\Omega \tag{21}$$

and the body forces vector are:

$$\mathbf{f_u} = -\int_\Omega \mathbf{B}^S \mathbf{f} d\Omega - \int_\Gamma \mathbf{B}^S \mathbf{t} d\Gamma$$

$$\mathbf{f_\phi} = -\int_\Gamma \mathbf{B}^E \rho_\Gamma d\Gamma \tag{22}$$

Note that this formulation is monolithic, that is, the coupling is at the stiffness matrix. In Gaudenzi & Bathe (1995) an iterative (stagger) formulation was proposed, which iteratively solves the mechanical and the electric problems. On the other hand, elastic, coupling and dielectric properties have different orders of magnitude, making extremely ill-conditioned to the stiffness matrix. This drawback was resolved in Qi et al. (1997) by rescaling the dimensions of the properties.

Zeng & Rajapakse (2004) developed a FEM to include the remanent strain and remanent polarization induced by poling. This was the first step to model the hysteresis, which is a non–linear behavior inherent to ferroelectric materials. As it is reported in Kaltenbacher et al. (2010), three different approaches are developed to model the hysteresis, namely: i) thermodynamically consistent models (see, Kamlah & Bohle (2001)), ii) micromechanical models and iii) models with hysteresis operator, see Kaltenbacher et al. (2010).

5.2.2 Vector formulation

The Vector Formulation (VF) was developed by Landis (2002) in 2002. In this formulation there are six degrees of freedom per node (three displacements and three components of the vector potential): $\mathbf{U} = (\mathbf{u}, \mathbf{V})^t$. The discretization, weak form and set of equations are obtained in a similar way to that used in SF. However, the main difference arises choosing the constitutive equations, where the electric displacement has to be chosen such as independent variable for this formulation.

A problem of the VF is the loss of uniqueness. Thus, in Semenov et al. (2006) some gauging procedures were investigated and, finally, the *Coulomb* gauge was incorporated to the FEM formulation, using the penalty function method.

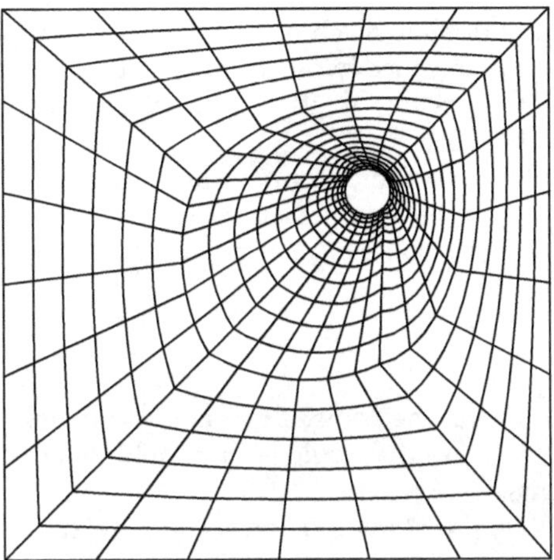

Fig. 10. Optimized mesh for damage detection using the FEM

The main drawback of the VF is the increase of CPU time, because there are six degrees freedom per node (four in the SF). However, this disadvantage can be offset by the rapid convergence in non–linear problems.
In terms of the inverse problem solutions the SF is better than VF, because in the laboratory it is easier to measure voltages than vector potentials, which are a mathematical artifice and not a physical observable.

5.2.3 Mesh generation for damage detection
The mesh generation is the main step to achieve accurate and reliable results using the finite element method. Therefore, Rus et al. (2009) developed an algorithm that automatically generates a high quality mesh. Given a square domain with a random (x_0, y_0, r) circle hole inside (see Figure 5), the fully automatic algorithm to construct multi–block structured 2D meshes consists of three steps:

i) The domain is subdivided into simple blocks by means of medial axis transform, which is the locus of centers of maximal empty circles inside the domain. Therefore, the domain skeleton is obtained in this step.

ii) Blocks are meshed by transfinite interpolation, which is a technique to draw meshes mapping the unit square onto the interior of the physical domain.

iii) Elements are concentrated close to the circular hole using a stretching function, which is a monotonously increasing function defined in the computational domain.

Finally, Figure 10 shows an optimized mesh generated with the previous algorithm, where the maximum element size is selected to provide an error in the measurements below an acceptable threshold, see Rus et al. (2009).

5.3 Boundary element method
The boundary element method is an alternative technique to model the response of piezoelectric materials in the case of linear constitutive laws particularly for three reasons: (1) the dimensionality of the problem is reduced by one in comparison with the FEM, requiring only

boundary meshing instead of domain meshing; (2) the cracks are easily modeled by a single boundary after introducing the so-called hypersingular formulation; and (3) infinite boundary conditions in the case of dynamics are satisfied automatically, without need for infinite finite elements, as is the case for the FEM.

However the formulation of the problem is complex for the case of piezoelectric materials with cracks and has only been accomplished in recent years, for the case of static piezoelectricity, by García-Sanchez et al. (2005); García-Sánchez et al. (2005; 2008) and for transient dynamics by Rojas-Díaz et al. (2009; 2010).

6. Future challenges

Several issues remain unsolved in the fields of damage and material properties characterization in piezoelectrics, which makes it a fertile researh field where concepts from other disciplines may give new insights.

In the case of damage, fracture growth criteria are a controversial issue, since the mechanical, electrical and total energy release rates have inconsistent relationships with crack opening at the theoretical level. The question of the identifiability, or minimal detectable damage, has neither been addressed, except for some preliminar approaches to the concept of probability of detection.

Regarding the property characterization, the established methods have a limited range of validity, and optimal measurement/excitation setups need to be determined and rationally justified. Furthermore, the classical characterization procedures extract some scalar parameters from the measurement, dumping large amounts of data that may contain useful information about the material constants. The complete measurements may be taken advantage of using model-based inverse problems to reconstruct the unknown coefficients of the constitutive model.

7. References

Allik, H. & Hughes, T. (1970). Finite elment method for piezoelectric vibration, *International Journal for Numerical Methods in Engineering* **2**: 151–157.

Araújo, A. L., Soares, C. M. M., Herskovits, J. & Pedersen, P. (2006). Parameter estimation in active plate structures using gradient optimisation and neural networks., *Inverse Prob. in Sci. and Eng.* **5**: 483–493.

Araújo, A., Soares, C. M., Herskovits, J. & Pedersen, P. (2009). Estimation of piezoelastic and viscoelastic properties in laminated structures, *Composite Structures* **87**(2): 168 – 174. US Air Force Workshop Structural Assessment of Composite Structures.

Araújo, A., Soares, C. M., Herskovits, J. & Pedersen, P. (2002). Development of a finite element model for the identification of mechanical and piezoelectric properties through gradient optimisation and experimental vibration data, *Composite Structures* **58**: 307–318.

Aster, R. C., Borchers, B. & Thurber, C. H. (2005). *Parameter Estimation and Inverse Problems*, Elsevier, USA.

Banks, H. T. & Rebnord, D. A. (1991). Estimation of material parameters for grid structures, *Journal of Mathematical Systems, Estimation and Control* **1**: 188–197.

Banks, H. T., Smith, R. C., Brown, D. E., Metcalf, V. L. & Silcox, R. J. (1997). The estimation of material and patch parameters in a pde-based circular plate model, *Journal of Sound and Vibration* **199**(5): 777 – 799.

Banks, H. T., Wang, Y. & Inman, D. J. (1994). Bending and shear damping in beams: frequency domain estimation techniques, *Journal of Vibration and Acoustics* **116**: 188–197.

Bonnet, M. & Constantinescu, A. (2005). Inverse problems in elasticity, *Inverse Problems* **21**: R1–R50.

Chen, Y. & Hasebe, N. (2005). Current understanding on fracture behavior of ferroelectric/piezoelectric materials, *J. Intell. Mater. Syst. Struct.* **16**: 673–687.

Chou, J. & Ghaboussi, J. (2001). Genetic algorithm in structural damage detection, *Comput. Strut.* **79**: 1335–1353.

Dennis, Jr., J. E. & Schnabel, R. B. (1983, 1996). *Numerical Methods for Unconstrained Optimization and Nonlinear Equations*, SIAM, Philadelphia.

Eringen, A. (1980). *Mechanics of Continua*, Robert E. Krieger Publishing Company.

Eringen, A. & Maugin, G. (1990). *Electrodynamics of Continua I*, Springer-Verlag New York, Inc.

Fu, R., Qian, C. & Zhang, T. (2000). Electric fracture toughness for conductive cracks driven by electric fields, *Applied Physics Letters* **76**(1): 126–128.

Gallego, R. & Rus, G. (2004). Identification of cracks and cavities using the topological sensitivity boundary integral equation, *Computational Mechanics* **33**.

Gao, H., Zhang, T. & Tong, P. (1997). Local and global energy release rates for an electrically yielded crack in a piezoelectric ceramic, *Journal of the Mechanics and Physics of Solids* **45**(4): 491–510.

García-Sanchez, F., Sáez, A. & Domínguez, J. (2005). Anisotropic and piezoelectric materials fracture analysis by bem, *Computers & Structures* **83**: 804–820.

García-Sánchez, F., Sáez, A. & Domínuez, J. (2005). Two-dimensional time-harmonic bem for cracked anisotropic solids, *Engineering Analysis with Boundary Elements* .

García-Sánchez, F., Zhang, C. & Sáez, A. (2008). 2-d transient dynamic analysis of cracked piezoelectric solids by a time-domain bem, *Computer Methods in Applied Mechanics and Engineering* **197**: 3108–3121.

Gaudenzi, P. & Bathe, K. (1995). An iterative finite element procedure for the analysis of piezoelectric continua, *Journal of Intelligent Material Systems and Structures* **6**: 266–273.

Goldberg, D. (1989). *Genetic algorithms in search, optimization and machine learning*, Addison-Wesley Publ, Reading, Massachussets, etc.

Hao, T. & Shen, Z. (1994). A new electric boundary condition of electric fracture mechanics and its applications, *Engineering Fracture Mechanics* **47**(6): 793–802.

Hughes, T. (1987). *Th Finite Element Method. Linear Static and Dynamic Finite Element Analysis*, Prentice-Hall, Inc.

Jelitto, H., Felten, F., Hausler, C., Kessler, H., Balke, H. & Schneider, G. (2005). Measurement of energy release rates for cracks in pzt under electromechanical loads, *Journal of the European Ceramic Society* **25**: 2817–2820.

Jelitto, H., Kessler, H., Schneider, G. & Balke, H. (2005). Fracture behavior of poled piezoelectric pzt under mechanical and electrical loads, *Journal of the European Ceramic Society* **25**: 749–757.

Jiang, W., Zhang, R., Jiang, B. & Cao, W. (2003). Characterization of piezoelectric materials with large piezoelectric and electromechanical coupling coefficients, *Ultrasonics* **41**(2): 55 – 63.

Kaltenbacher, B., Lahmer, T., Mohr, M. & Kaltenbacher, M. (2006). Pde based determination of piezoelectric material tensors, *Europ. J. App. Mathem.* **17**: 383–416.

Kaltenbacher, M., Kaltenbacher, B., Hegewald, T. & Lerch, R. (2010). Finite element formulation for ferroelectric hysteresis of piezoelectric materials, *Journal of Intelligent Material Systems and Structures* **21**: 773–785.

Kamlah, M. & Bohle, U. (2001). Finite element analysis of piezoceramic components taking into account ferroelectric hysteresis behavior, *International Journal of Solids and Structures* **38**: 605–633.

Lahmer, T., Kaltenbacher, B. & Schulz, V. (2010). Optimal measurement selection for piezoelectric material tensor identification, *Inverse Problems in Science and Engineering* **16**: 369–387.

Landis, C. (2002). A new finite-element formulation for electromechanical boundary value problems, *International Journal for Numerical Methods in Engineering* **55**: 613–628.

Lee, S. Y., Rus, G. & Park, T. (2007). Detection of stiffness degradation in laminated composite plates by filtered noisy impact testing, *Comp. Mech.* pp. 1–15.

Lee, S. Y., Rus, G. & Park, T. H. (2008). Quantitative nondestructive evaluation of thin plate structures using the complete frequency from impact testing, *Struct. Eng. Mech.* **28**: 525–548.

Liu, P. & Chen, C. (1996). Parametric identification of truss structures by using transient response, *J. Sound Vib.* **191**: 273–287.

Mares, C. & Surace, C. (1996). On application of genetic algorithms to identify damage in elastic structures, *J. Sound Vib.* **195**: 195–215.

McMeeking, R. (2001). Towards a fracture mechanics for brittle piezoelectric and dielectric materials, *International Journal of Fracture* **108**: 25–41.

McMeeking, R. (2004). The energy release rate for a griffith crack in a piezoelectric material, *Engineering Fracture Mechanics* **71**: 1149–1163.

Menke, W. (1984). *Geophysical data analysis, Discrete Inverse Theory*, Academic Press INC.

Oh, J., Cho, M. & Kim, J. (2005). Dynamic analysis of composite plate with multiple delaminations based on higher–order zigzag theory, *Int. J. Solids & Struct.* **42**: 6122–6140.

Ou, Z. & Chen, Y. (2003). Dicussion of the crack face electric boundary condition in piezoelectric fracture mechanics, *International Journal of Fracture* **123**: L151–L155.

Pagano, N. (1970). Exact solution for rectangular bidirectional composites and sandwich plate, *J. Comp. Mater.* **4**: 20–34.

Palma, R., Rus, G. & Gallego, R. (2009). Probabilistic inverse problem and system uncertainties for damage detection in piezoelectrics, *Mechanics of Materials* **41**: 1000–1016.

Park, S. & Sun, C. (1995a). Effect of electric field on fracture of piezoelectric ceramics, *International Journal of Fracture* **70**: 203–216.

Park, S. & Sun, C. (1995b). Fracture criteria for piezoelectric ceramics, *J. Am. Ceram. Soc.* **78**(6): 1475–1480.

Pérez-Aparicio, J., Sosa, H. & Palma, R. (2007). Numerical investigation of field-defect interactions in piezoelectric ceramics, *International Journal of Solids and Structures* **44**: 4892–4908.

Qi, H., Fang, D. & Yao, Z. (1997). Fem analysis of electro-mechanical coupling effect of piezoelectric materials, *Computational Materials Science* **8**: 283–290.

Rojas-Díaz, R., García-Sánchez, F. & Sáez, A. (2009). Dynamic crack interactions in magneto-electroelastic composite materials, *International Journal of Fracture* **47**: 119–130.

Rojas-Díaz, R., García-Sánchez, F. & Sáez, A. (2010). Analysis of cracked magnetoelectroelastic composites under time-harmonic loading, *International Journal of Solids and Structures* **47**: 71–80.

Ruíz, A., Ramos, A. & San-Emeterio, J. (2004a). Estimation of some transducer parameters in a broadband piezoelectric transmitter by using an artificial intelligence technique, *Ultrasonics* **42**: 459–463.

Ruíz, A., Ramos, A. & San-Emeterio, J. (2004b). Evaluation of piezoelectric resonator parameters using an artificial intelligence technique, *Integrated Ferroelectric* **63**: 137–141.

Rus, G. (2010). Nondestructive evaluation laboratory.
URL: *http://www.ugr.es/local/grus/research.htm*

Rus, G. & Gallego, R. (2002). Optimization algorithms for identification inverse problems with the boundary element method, *Engineering Analysis with Boundary Elements* **26**: 315–327.

Rus, G. & Gallego, R. (2005). Boundary integral equation for inclusion and cavity shape sensitivity in harmonic elastodynamics, *Engineering Analysis with Boundary Elements* **25**: 77–91.

Rus, G. & Gallego, R. (2006). Solution of identification inverse problems in elastodynamics using semi-analytical sensitivity computation, *Engineering Analysis with Boundary Elements* p. in press.

Rus, G. & García-Martínez, J. (2007). Ultrasonic tissue characterization for monitoring nanostructured tio2-induced bone growth, *Phys. Med. Biol.* **52**: 3531–3547.

Rus, G., García-Sánchez, F., Sáez, A. & Gallego, R. (2010a). Damage detection in piezoceramics via bem, *Key Engineering Materials* **417**: 381–384.

Rus, G., García-Sánchez, F., Sáez, A. & Gallego, R. (2010b). Feasibility of model-based damage identification in piezoelectric ceramics, *In preparation -: –*.

Rus, G., Lee, S. & Gallego, R. (2005). Defect identification in laminated composite structures by bem from incomplete static data, *Int. J. Solids and Structures* **42**: 1743–1758.

Rus, G., Lee, S.-Y. & Gallego, R. (2004). Defect indentification in laminated composite structures by BEM from incomplete static data, *ijss* **4**(5-6): 1743–1758.

Rus, G., Palma, R. & Pérez-Aparicio, J. (2009). Optimal measurement setup for damage detection in piezoelectric plates, *International Journal of Engineering Science* **47**: 554–572.

Rus, G., Palma, R. & Pérez-Aparicio, J. (2010). Experimental design of dynamic model-based damage identification in piezoelectric ceramics, *In preparation -: –*.

Rus, G., Wooh, S. C. & Gallego, R. (2007a). Processing of ultrasonic array signals for characterizing defects. part i: Signal synthesis, *IEEE T. Ultrason. Ferr.* **october**: 2129–2138.

Rus, G., Wooh, S. C. & Gallego, R. (2007b). Processing of ultrasonic array signals for characterizing defects. part ii: Experimental work, *IEEE T. Ultrason. Ferr.* **october**: 2139–2145.

Russell, R. & Ghomshei, M. (1997). Inverting piezoelectric measurements, *Tectonophysics* **271**: 21–35.

Schneider, G. (2007). Influence of electric field an mechanical stresses on the fracture of ferroelectrics, *Annu. Rev. Mater. Res.* **37**: 491–538.

Semenov, A., Kessler, H., Liskowsky, A. & Balke, H. (2006). On a vector potential formulation for 3d electromechanical finite element analysis, *Communications in Numerical Methods in Engineering* **22**: 357–375.

Soh, A., Lee, K. & Fang, D. (2003). On the effect of an electric field on the fracture toughness of poled piezoelectric ceramics, *Materials Science and Engineering A* **360**: 306–314.

Sosa, H. (1991). Plane problems in piezoelectric media with defects, *International Journal of Solids and Structures* **28**(4): 491–505.

Sosa, H. (1992). On the fracture mechanics of piezoelectric solids, *International Journal of Solids and Structures* **29**(21): 2613–2622.

Sosa, H. & Khutoryansky, N. (1996). New developments concerning piezoelectric materials with defects, *International Journal of Solids and Structures* **33**(23): 3399–3414.

Sosa, H. & Pak, Y. (1990). Three-dimensional eigenfuction analysis of a crack in a piezoelectric material, *International Journal of Solids and Structures* **26**(1): 1–15.

Suo, Z., Kuo, C., Barnett, D. & Willis, J. (1992). Fracture mechanics for piezoelectric ceramics, *Journal of the Mechanics and Physics of Solids* **40**(4): 739–765.

Tarantola, A. & Valette, B. (1982). Inverse problems = quest for information, *J. Geophys.* **50**: 159–170.

Tardieu, N. & Constantinescu, A. (2000). On the determination of elastic coefficients from indentation experiments, *Inverse Problems* **16**: 577–588.

Tikhonov, A. & Arsenin, V. (1979). *Methods for solving ill-posed problems*, Nauka, Moscow.

Tikhonov, A. & Arsénine, V. (1974). *Méthodes de résolution de problémes mal posés*, Editions Mir, Moscou.

Wang, B. & Sun, Y. (2004). Intensity factors for some common piezoelectric fracture mechanics specimens with conducting cracks or electrodes, *International Journal of Engineering Science* **42**: 1129–1153.

Xu, X. & Rajapakse, R. (1999). Analytical solution for an arbitrarily oriented void/crack and fracture of piezoceramics, *Acta Mater.* **47**(6): 1735–1747.

Zeng, X. & Rajapakse, R. (2004). Effects of remanent field on an elliptical flaw and a crack in a poled piezoelectric ceramic, *Computational Materials Science* **30**: 433–440.

Zienkiewicz, O., Taylor, R. & Zhu, J. (2005). *The Finite Element Method: The Basis*, Elsevier Butterworth-Heinemann.

5

Analysis of mechanical and electrical damages in piezoelectric ceramics

Xinhua Yang, Guowei Zeng and Weizhong Cao
Huazhong University of Science and Technology,
China

1. Introduction

For recent decades, due to the intrinsic electromechanical coupling, piezoelectric ceramics have been used widely in the modern aeronautical and astronautic engineering, the medical apparatus and the instruments, etc. However, their brittleness might result in damage and failure. Therefore, it is very important theoretically and practically for fracture analysis, reliability design and lifetime prediction of piezoelectric ceramics to investigate the fracture mechanism.

On the basis of linear piezoelectric fracture mechanics, some solutions were obtained by Pak (1990, 1992), Sosa and Pak (1990), Sosa (1991, 1992), Suo et al (1992), Suo (1993) and others, but they contradict the piezoelectric fracture experiments (Park and Sun, 1995; Heyer et al, 1998). The linear piezoelectric fracture mechanics predicted that the stress intensity factors were independent of the applied electric field, but the experimental results (Park and Sun, 1995; Heyer et al, 1998) showed that an existing electric field might impede or accelerate crack propagation, which was related to the applied electric field direction.

Regarding piezoelectric ceramics as a class of mechanically brittle and electrically ductile solids, Gao et al (1997) put forward a strip saturation model analogous with the classical Dugdale model for plastic yielding, in which electrical displacement is assumed to yield when electric field is up to its linear limit. Yang and Zhu (1998) thought that piezoelectric nonlinearity comes from domain switching, so they proposed a small-scale domain-switching model to explain the toughening mechanism. Fulton and Gao (2001) proposed an electrical nonlinearity model based on piezoelectric microstructure. They simulated polarization switching and saturation in ferroelectrics by using a collection of discrete electric dipoles superimposed on a medium satisfying the linear piezoelectric constitutive law, and then derived a local crack driving force which generates a qualitative match to experimental observations. Based on the field limiting space charge model, Zhang et al (2003) proposed a charge-free zone model, and estimated the electric field at the tip of an electrically conductive crack according to the electric field intensity factor. Assuming that the electric field in a strip ahead of the crack tip is equal to the dielectric breakdown strength, Zhang et al (2005) developed a strip dielectric breakdown model similar to the strip saturation model.

In the above nonlinear models, only electrical nonlinearity is considered. However, the experiments (Zhang and Gao, 2004) demonstrated that in addition to the nonlinearities in the relationships of polarization versus electric field and strain versus electric field, the stress-strain curves are nonlinear. From the viewpoint of continuum damage theory, material nonlinearity is caused by damage, so that the continuum damage mechanics can be extended to describe nonlinear behaviors of piezoelectric ceramics. Mizuno (2002) introduced a single damage variable into the constitutive equation of piezoelectric ceramics to take the effect of damage on material properties into account, and after that, Mizuno and Honda (2005) analyzed the steady-state crack growth in a double cantilever piezoelectric beam.

In our work, an anisotropic damage constitutive model involving mechanical and electrical damages was proposed (Yang et al, 2003 and 2005a), and the phenomenological damage tensors were qualitatively connected with microstructure parameters by the unit cell method (Yang et al, 2007 and 2008; Zeng et al, 2008). The linear and nonlinear piezoelectric fracture criteria were established after the damage analysis of the well-known piezoelectric fracture experiments of Park and Sun's (Yang et al, 2005b). Finally, the damage constitutive model and the nonlinear fracture criterion were used to investigate the quasi-static propagation behavior of the midspan crack in a three-point bending PZT-4 beam, and effects of mechanical and electric loads on crack growth were observed.

2. Piezoelectric damage constitutive model

Damage in a piezoelectric solid can come from a number of sources, including fabrication, connection, service, etc. After homogenization, a damaged material element from the solid is regarded as a homogeneous continuum, so that its constitutive behavior can be characterized by new three-dimensional anisotropic constitutive equations with $\tilde{\mathbf{c}}$, $\tilde{\mathbf{e}}$, and $\tilde{\boldsymbol{\lambda}}$ as the elastic, piezoelectric and dielectric constants.

$$\sigma_{ij} = \tilde{c}_{ijkl}\varepsilon_{kl} - \tilde{e}_{kij}E_k, \quad D_i = \tilde{e}_{ikl}\varepsilon_{kl} + \tilde{\lambda}_{ik}E_k. \tag{1}$$

These constants are called effective elastic, piezoelectric, and dielectric constants, respectively. They are related to the undamaged constants $\mathbf{c}$, $\mathbf{e}$, and $\boldsymbol{\lambda}$ by

$$\tilde{c}_{ijmn} = \tfrac{1}{4}(\omega_{ki}\delta_{lj} + \delta_{ki}\omega_{lj})(\omega_{om}\delta_{pn} + \delta_{om}\omega_{pn})c_{klop},$$

$$\tilde{e}_{imn} = \tfrac{1}{2}(\omega_{km}\delta_{ln} + \delta_{km}\omega_{ln})G_{ji}e_{jkl}, \quad \tilde{\lambda}_{il} = G_{ji}G_{kl}\lambda_{jk}. \tag{2}$$

where $\boldsymbol{\omega}$ and $\mathbf{G}$ are two second-order continuity tensors, and $\boldsymbol{\delta}$ is the Kronecker Delta. The mechanical and electrical damage variables $\mathbf{D}^{\mathrm{M}}$ and $\mathbf{D}^{\mathrm{E}}$, which characterize damage degree of the material element, can be expressed in terms of the continuity tensors as

$$D_{ij}^{\mathrm{M}} = \delta_{ij} - \omega_{ik}\omega_{kj}, \quad D_{ij}^{\mathrm{E}} = \delta_{ij} - G_{ik}G_{kj}. \tag{3}$$

Eq.(1) combined with Eqs.(2) and (3) are called piezoelectric damage constitutive model. It can be rewritten in matrix form as follows.

$$\{\sigma\} = [\tilde{c}]\{\varepsilon\} - [\tilde{e}]^{\mathrm{T}}\{E\}, \quad \{D\} = [\tilde{e}]\{\varepsilon\} + [\tilde{\lambda}]\{E\}. \tag{4}$$

where

$$\begin{aligned}
\{\sigma\} &= [\sigma_{11}\ \sigma_{22}\ \sigma_{33}\ \tau_{23}\ \tau_{13}\ \tau_{12}]^{\mathrm{T}}, \\
\{\varepsilon\} &= [\varepsilon_{11}\ \varepsilon_{22}\ \varepsilon_{33}\ \gamma_{23}\ \gamma_{13}\ \gamma_{12}]^{\mathrm{T}}, \\
\{E\} &= [E_1\ E_2\ E_3]^{\mathrm{T}}, \quad \{D\} = [D_1\ D_2\ D_3]^{\mathrm{T}}.
\end{aligned} \tag{5}$$

And according to Eq.(2), we have

$$[\tilde{c}] = [N]^{\mathrm{T}}[c][N], \quad [\tilde{e}] = [G]^{\mathrm{T}}e[N], \quad [\tilde{\lambda}] = [G]^{\mathrm{T}}[\lambda][G]. \tag{6}$$

where

$$[N] = \begin{bmatrix}
\omega_{11} & 0 & 0 & 0 & \frac{1}{2}\omega_{13} & \frac{1}{2}\omega_{12} \\
0 & \omega_{22} & 0 & \frac{1}{2}\omega_{23} & 0 & \frac{1}{2}\omega_{12} \\
0 & 0 & \omega_{33} & \frac{1}{2}\omega_{23} & \frac{1}{2}\omega_{13} & 0 \\
0 & \omega_{23} & \omega_{23} & \frac{1}{2}(\omega_{22}+\omega_{33}) & \frac{1}{2}\omega_{12} & \frac{1}{2}\omega_{13} \\
\omega_{13} & 0 & \omega_{13} & \frac{1}{2}\omega_{12} & \frac{1}{2}(\omega_{11}+\omega_{33}) & \frac{1}{2}\omega_{23} \\
\omega_{12} & \omega_{12} & 0 & \frac{1}{2}\omega_{13} & \frac{1}{2}\omega_{23} & \frac{1}{2}(\omega_{11}+\omega_{22})
\end{bmatrix},$$

$$[G] = \begin{bmatrix}
G_{11} & G_{12} & G_{13} \\
G_{12} & G_{22} & G_{23} \\
G_{13} & G_{23} & G_{33}
\end{bmatrix}. \tag{7}$$

The transversely isotropic constant matrices can be expressed as follows

$$[c] = \begin{bmatrix}
c_{11} & c_{12} & c_{13} & 0 & 0 & 0 \\
c_{12} & c_{11} & c_{13} & 0 & 0 & 0 \\
c_{13} & c_{13} & c_{33} & 0 & 0 & 0 \\
0 & 0 & 0 & c_{44} & 0 & 0 \\
0 & 0 & 0 & 0 & c_{44} & 0 \\
0 & 0 & 0 & 0 & 0 & \frac{1}{2}(c_{11}-c_{12})
\end{bmatrix}, \tag{8}$$

$$[e] = \begin{bmatrix}
0 & 0 & 0 & 0 & e_{15} & 0 \\
0 & 0 & 0 & e_{15} & 0 & 0 \\
e_{31} & e_{31} & e_{33} & 0 & 0 & 0
\end{bmatrix}, \quad [\lambda] = \begin{bmatrix}
\lambda_{11} & 0 & 0 \\
0 & \lambda_{11} & 0 \\
0 & 0 & \lambda_{33}
\end{bmatrix}.$$

If damage is transversely isotropic and has the same principal axes as the material, Eq.(7) can be reduced to

$$[N] = \begin{bmatrix} \omega_{11} & 0 & 0 & 0 & 0 & 0 \\ 0 & \omega_{11} & 0 & 0 & 0 & 0 \\ 0 & 0 & \omega_{33} & 0 & 0 & 0 \\ 0 & 0 & 0 & \frac{1}{2}(\omega_{11}+\omega_{33}) & 0 & 0 \\ 0 & 0 & 0 & 0 & \frac{1}{2}(\omega_{11}+\omega_{33}) & 0 \\ 0 & 0 & 0 & 0 & 0 & \omega_{11} \end{bmatrix}, \tag{9}$$

$$[G] = \begin{bmatrix} G_{11} & 0 & 0 \\ 0 & G_{11} & 0 \\ 0 & 0 & G_{33} \end{bmatrix}.$$

Accordingly, only four scalar quantities are needed for transversely isotropic piezoelectric damage, namely, ω_{11}, ω_{33}, G_{11}, and G_{33}, and the correponding four damage variables are

$$D_{11}^M = 1 - \omega_{11}^2 = 1 - \frac{\tilde{c}_{11}}{c_{11}}, \quad D_{33}^M = 1 - \omega_{33}^2 = 1 - \frac{\tilde{c}_{33}}{c_{33}}, \tag{10}$$

$$D_{11}^E = 1 - G_{11}^2 = 1 - \frac{\tilde{\lambda}_{11}}{\lambda_{11}}, \quad D_{33}^E = 1 - G_{33}^2 = 1 - \frac{\tilde{\lambda}_{33}}{\lambda_{33}}.$$

3. Relation between damages and microstructure parameters

It is well known that damage degree depends on microstructure geometry, density, interactions, etc. As the first step, it needs to be solved how the phenomenological damage tensors is quantitatively connected with microstructure parameters. Eq.(10) reveals a way to determine quantitative connections between the mechanical and electrical damages and microstructure parameters, because the effective properties of the damaged material element can be calculated by means of micromechanical mechanics methods, such as the unit cell method.

3.1 Insulating microvoids

PZT-7A piezoelectric ceramics with the material constants listed in Table 1 is assumed to contain many periodically distributed ellipsoidal insulating microvoids, shown in Fig.1(a). As the smallest periodical unit which contains sufficient microstructure information, a unit cell is taken out from the material, and only quarter of the cell model needs to be considered because of symmetry. See Fig.1(b). Accordingly, the displacement boundary conditions of the model can be described as

$$\begin{aligned} u_1 &= 0 \quad \text{for} \quad x_1 = 0; \\ u_2 &= 0 \quad \text{for} \quad x_2 = 0; \\ u_2 &= 0 \quad \text{for} \quad x_3 = 0. \end{aligned} \tag{11}$$

Materials	c_{11}	c_{12}	c_{13}	c_{33}	c_{44}	e_{31}	e_{33}	e_{15}	λ_{11}	λ_{33}
	GPa					C/m²			10⁻⁹F/m	
PZT-7A	148	76.2	74.2	131	25.4	-2.1	9.5	9.2	4.07	2.08
PZT-4	139	77.8	74.3	113	25.6	-6.98	13.84	13.44	6	5.47

Table 1. Material properties

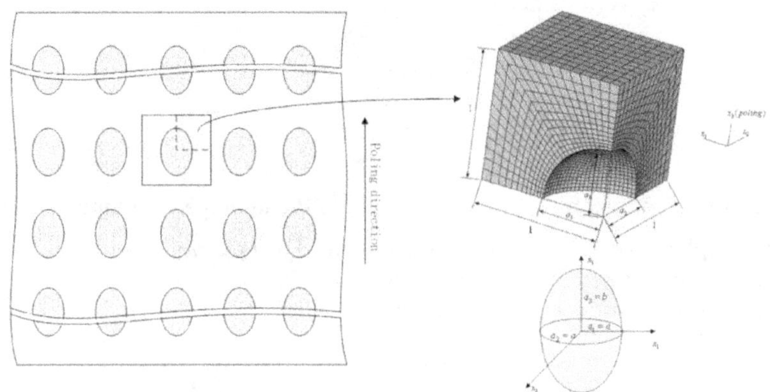

(a) PZT-7A with periodically distributed microvoids   (b) Quarter of the cell model
Fig. 1. Computational model

Different loading and boundary conditions are imposed on the cell model in order to evaluate its effective properties. For example, when the displacement degree of freedom $U_{x_1=1} = 1$ is applied in the boundary planes $x_1 = 1$ or $U_{x_3=1} = 1$ applied in the planes $x_3 = 1$, the effective elastic constant $\tilde{c}_{11}$ can be evaluated from Eq. (4) by

$$E_1 = E_2 = E_3 = 0, \quad \varepsilon_{22} = \varepsilon_{33} = 0, \quad \varepsilon_{11} \neq 0,$$

$$\tilde{c}_{11} = \frac{\sigma_{11}}{\varepsilon_{11}} \approx \frac{Q_{x_1=1}}{U_{x_1=1}}. \tag{12}$$

And $\tilde{c}_{33}$ can also be evaluated by

$$E_1 = E_2 = E_3 = 0, \quad \varepsilon_{11} = \varepsilon_{22} = 0, \quad \varepsilon_{33} \neq 0,$$

$$\tilde{c}_{33} = \frac{\sigma_{33}}{\varepsilon_{33}} \approx \frac{Q_{x_3=1}}{U_{x_3=1}}. \tag{13}$$

In the above two expressions, $Q_{x_1=1}$ and $Q_{x_3=1}$ are the active force in the planes $x_1 = 1$ and $x_3 = 1$, respectively. When only the electric field E_2 is imposed along the x_2-direction through applying the voltage $V_{x_2=1}$ in the plane $x_2=1$ and $V_{x_2=0}$ in the plane $x_2=0$ of the cell, we have $D_2 = \tilde{e}_{15}\varepsilon_{32} + \tilde{\lambda}_{33}E_2$, so that the dielectric constant $\tilde{\lambda}_{11}$ can be evaluated by

$$E_1 = E_3 = 0, E_2 \neq 0, \quad \varepsilon_{11} = \varepsilon_{22} = \varepsilon_{33} = \varepsilon_{32} = 0,$$

$$\tilde{\lambda}_{11} = \frac{D_2}{E_2} \approx \frac{D_{x_2=1}}{V_{x_2=1} - V_{x_2=0}}$$

(14)

Applying the voltage $V_{x_3=1}$ in the plane $x_3=1$ and $V_{x_3=0}$ in the plane $x_3=0$ of the cell, the dielectric constant $\tilde{\lambda}_{33}$ can be obtained by

$$E_1 = E_2 = 0, E_3 \neq 0, \quad \varepsilon_{11} = \varepsilon_{22} = \varepsilon_{33} = 0,$$

$$\tilde{\lambda}_{33} = \frac{D_3}{E_3} \approx \frac{D_{x_3=1}}{V_{x_3=1} - V_{x_3=0}}$$

(15)

In the above two expressions, $D_{x_2=1}$ and $D_{x_3=1}$ are the average components of electric flux density in the planes $x_2=1$ and $x_3=1$, respectively. Further, the mechanical and electrical damages can be determined in accordance with Eq.(10).
Similarly, the piezoelectric coefficients can be evaluated by

$$E_3 \neq 0, \quad \varepsilon_{11} = \varepsilon_{22} = \varepsilon_{33} = 0,$$

$$\tilde{e}_{13} = -\frac{\sigma_{11}}{E_3} \approx -\frac{F_{x_1=1}}{V_{x_3=1} - V_{x_3=0}}, \quad \tilde{e}_{33} = -\frac{\sigma_{11}}{E_3} \approx -\frac{F_{x_3=1}}{V_{x_3=1} - V_{x_3=0}}$$

(16)

and

$$E_2 \neq 0, \quad \varepsilon_{11} = \varepsilon_{22} = \varepsilon_{33} = \varepsilon_{32} = 0,$$

$$\tilde{e}_{15} = -\frac{\sigma_{32}}{E_2} \approx \frac{F_{x_2=1}}{V_{x_2=1} - V_{x_2=0}}$$

(17)

In the above two expressions, $F_{x_1=1}$, $F_{x_2=1}$ and $F_{x_3=1}$ are the reactive forces on the planes $x_1=1$, $x_2=1$ and $x_3=1$, respectively. On the other hand, according to the obtained mechanical and electrical damage components, the piezoelectric coefficients can be also obtained from Eq.(6). By comparison of these results from the above two kinds of method, it can be checked if the hypothesis of transversely isotropic damage is rational.
In the unit cell, the ellipsoidal void geometry is characterized by its volume fraction $f = V_v / V_m$, where V_v and V_m are respectively the void and matrix volumes, and aspect ratio $S = b/a$, where a and b are respectively the minor and major semi-axes of the void. The longitudinal cross-section area fraction of ellipsoidal void can be expressed by $A_3 = (3f\sqrt{\pi}/4S)^{\frac{2}{3}}$ and the transverse cross-section area fraction by $A_1 = A_2 = (3f\sqrt{\pi S}/4)^{\frac{2}{3}}$.
The void aspect ratio S is fixed at 1.0 and the void volume fraction f is changed from 0 to 32.1%, in order to evaluate the influences of the void volume fraction f on the damages. The variation curves of damages with the void volume fraction f are drawn in Fig.2(a). It is apparent that both the mechanical and electrical damages increase with the increasing void volume fraction f and the curves of D_{11}^M and D_{33}^M superpose each other, which reveals that the mechanical damage induced by spherical voids is isotropic. This is in good agreement

with our knowledge. And then, the void volume fraction f is fixed at 6.54% and the void aspect ratio S is changed from 0.3 to 2.0, in order to observe influences of the void aspect ratio S on the damages. The variation curves of the damages with the void aspect ratio S are plotted in Fig.2 (b). While D_{33}^M goes down rapidly in forepart but slowly afterward with the increasing void aspect ratio S, D_{11}^M always goes up slowly. This is understandable because A_3 decreases but A_1 or A_2 increases with the increasing aspect ratio S for a fixed volume fraction f. Nevertheless, D_{11}^E and D_{33}^E hardly change, which shows that the electrical damage components are independent of the void aspect ratio S.

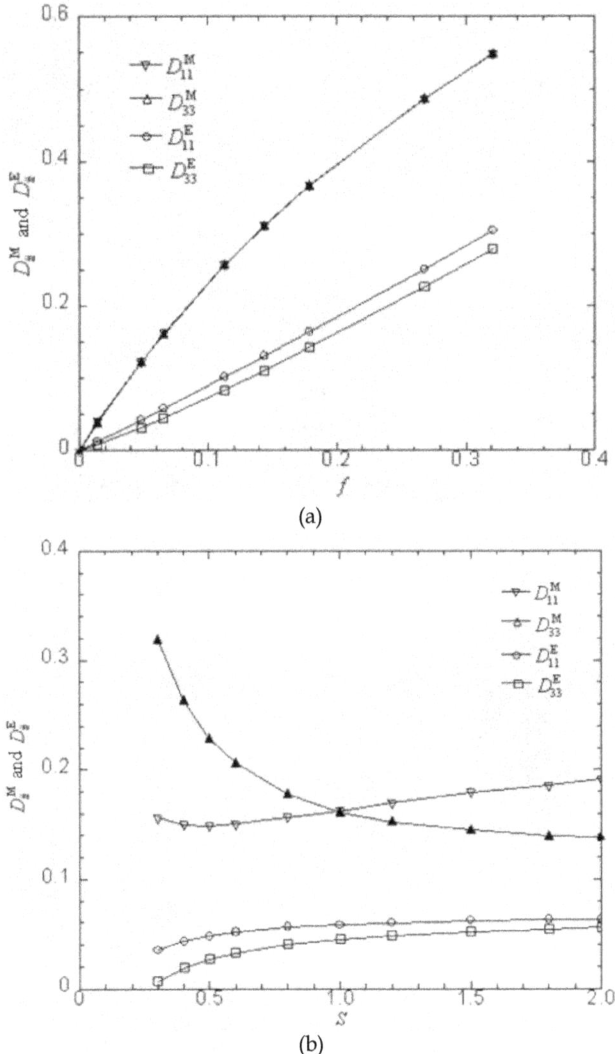

(a)

(b)

Fig. 2. Variation curves of damages with (a) volume fraction and (b) aspect ratio of an insulating void

The effective piezoelectric coefficients are calculated from Eq.(6) and from Eq.(16) or (17), respectively, and the variation curves of the effective piezoelectric coefficients with the void volume fraction and the void aspect ratio are drawn in Fig.3 (a) and (b), respectively. Apparently, the results from the two methods are very approximate, which reflects that the hypothesis of transversely isotropic damage is reasonable for the analyzed piezoelectric solid with voids.

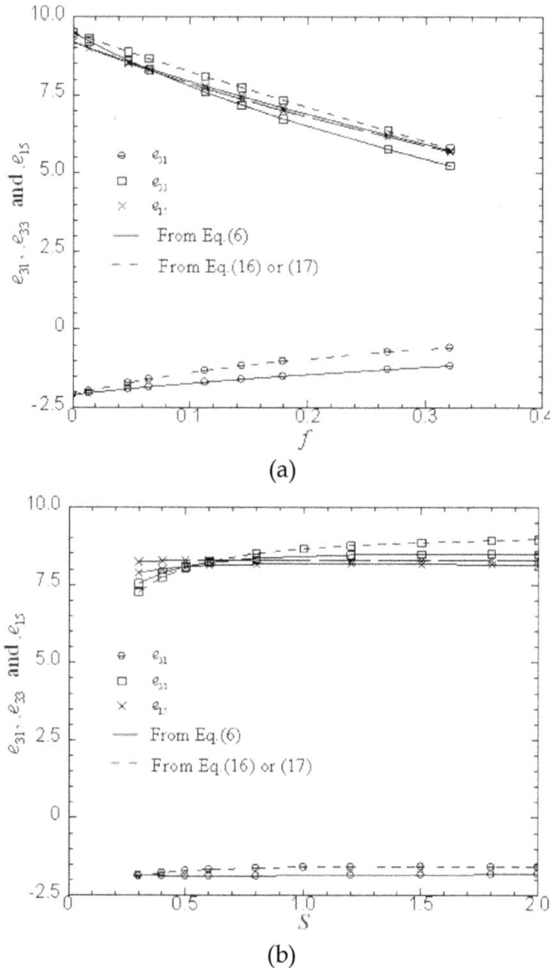

(a)

(b)

Fig. 3. Variation curves of effective piezoelectric coefficients with (a) the void volume fraction and (b) the void aspect ratio

3.2 Conductive microvoids

For conductive microvoids, the shape ratio S is fixed at 1.0 (namely the microvoids are spherical) and the volume fraction f is changed from 0 to 32.1%, in order to investigate the effect of the volume fraction on damages. The variation curves of mechanical and electrical damages with the volume fraction are plotted in Fig.4(a). It can be found that both the mechanical and electrical damage components increase with the increasing volume fraction and the curves of D_{11}^{M} and D_{33}^{M} overlap each other. After that, the volume fraction f is fixed at 6.54% and the shape ratio S is changed from 0.4 to 2 in order to investigate the effect of the shape ratio on damages. The variation curves of mechanical and electrical damages with the shape ratio are plotted in Fig.4(b). For the mechanical damage components, D_{33}^{M} rapidly decreases but D_{11}^{M} slowly increases with the increasing shape ratio. For the electrical damage components, however, D_{11}^{E} slowly increases but D_{33}^{E} slowly decreases with the increasing shape ratio.

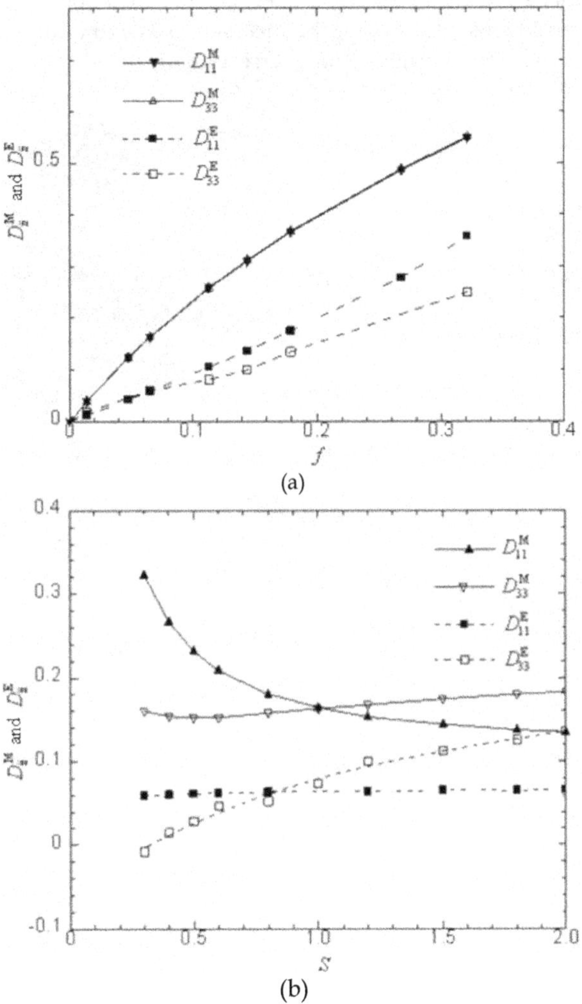

(a)

(b)

Fig. 4. Variation curves of damages with (a) the volume fraction and (b) the aspect ratio for conducting voids

3.3 Conductive inclusions

Consider a PZT-7A piezoelectric solid with periodically distributed ellipsoidal conductive inclusions, shown in Fig.5. The solid is polarized in the z direction. The inclusions have Poisson's ratio of 0.25 and elastic modulus of $E = \dfrac{5}{6}\alpha c_{11}$. In order to evaluate the effect of inconlusion elastic modulus on damages, we fix f in 6.54% and S in 1, and change α from 0 to 1. The curves of mechanical and electrical damages with α are plotted in Fig.6. It can be seen that the mechanical damage rapidly decreases as α increases, and goes to zero when α tends toward 1. This is because inclusions transform into microvoids when $\alpha = 0$ and approximates to the matrix material when $\alpha = 1$. However, the electric damage slowly increases when α increases, because the variation of α does not change the conductivity of inclusions. Obviously, the inclusion elastic modulus has great influence on the mechanical damage but a little on the electric damage.

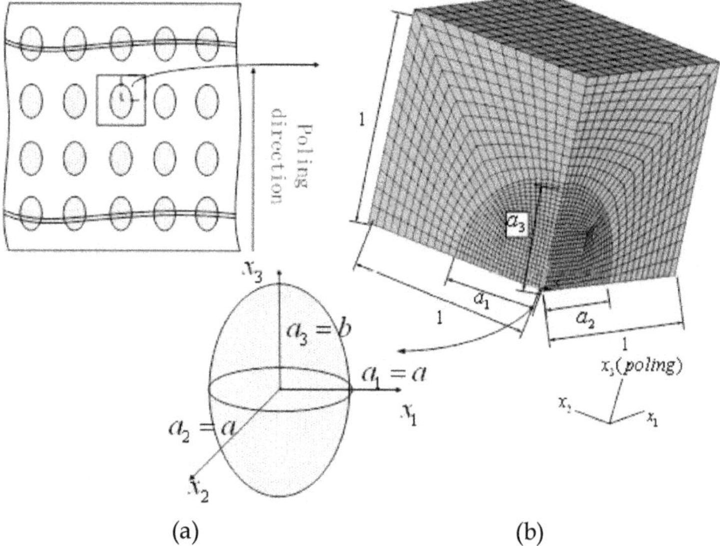

Fig. 5. Unit cell model and its coordinate system: (a) PZT-7A with periodically distributed inclusions; (b) 1/8 model of the cell

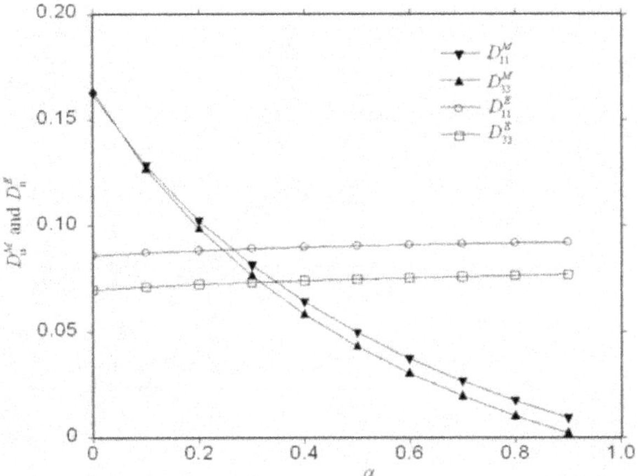

Fig. 6. Variation curves of mechanical and electrical damages with elastic coefficient ratio

In succession, we fix S in 1 and set α as 0.4 and 0.6, respectively, to investigate variation law of mechanical and electrical damages with inclusion volume fraction. The curves are drawn in Fig.7. It can be found that mechanical and electrical damages monotonously increase with inclusion volume fraction for different elastic moduli. This is possibly because increasing inclusion volume has greater influence on electromechanical properties of the materials.

Finally, the effects of inclusion shape on damages are discussed, when f is fixed in 6.54% and S is set as 0.4 and 0.6, respectively. The computational results are shown in Fig.8. It is easily found that inclusion shape has very complex influence on mechanical and electrical damages, but the variation curves of damages with inclusion shape ratio have similar shape for different elastic moduli. Moreover, the inclusion shape variation has a little influence on the mechanical damage, but great influence on the electrical damage. For a constant inclusion volume, when the shape ratio increases, A_3 decreases, but A_1 and A_2 increase, so that D_{11}^E and D_{33}^E have opposite variation laws.

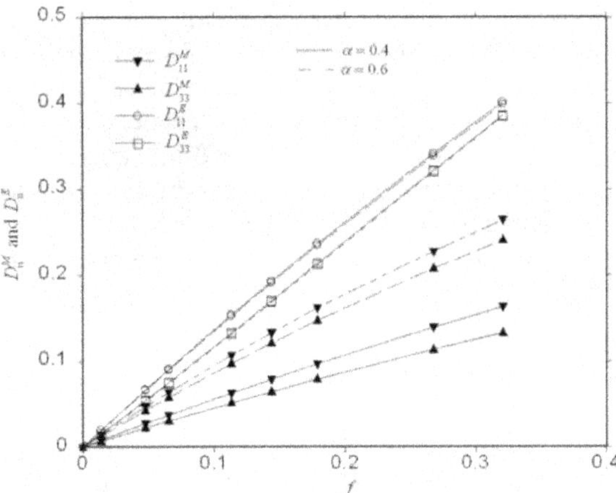

Fig. 7. Variation curves of mechanical and electrical damages with the volume fraction

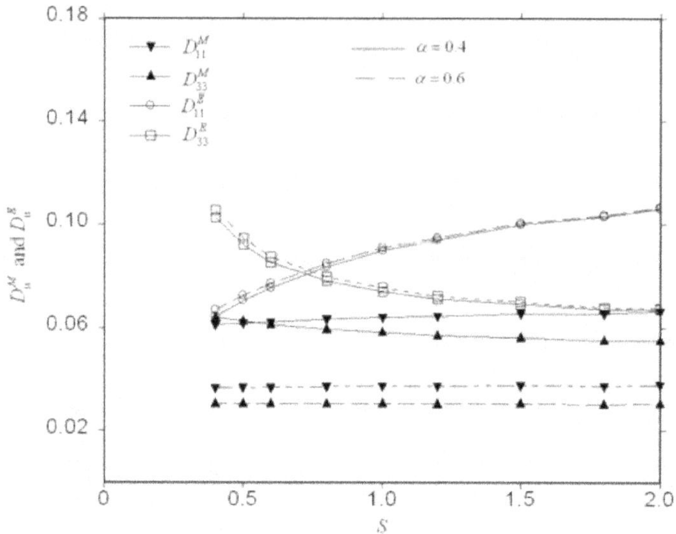

Fig. 8. Variation curves of mechanical and electrical damages with the aspect ratio

4. Fracture criteria

For the brittle piezoelectric ceramics, the onset of crack growth almost inevitably results in failure, so it is very important for understanding piezoelectric fracture behaviors to establish a useful fracture criterion. While the natural extension of linear elastic fracture mechanics provided an agreeably simple and general criterion (Suo et al, 1992), it was not supported by experimental studies (Park and Sun, 1995; Heyer et al, 1998). Arguing that fracture was a mechanical process and should be controlled only by the mechanical part of the energy, Park and Sun (1995) used the maximum mechanical strain energy release rate for predicting fracture loads. However, there was no fundamental reason to separate a physical process into two parts: mechanical and electrical. Shen and Nishioka (2000), Sih and Zuo (2000), and Zuo and Sih (2000) proposed an energy density criterion.

In order to establish a damage-based fracture criterion, the damage constitutive equations in Section 2 are used to characterize the constitutive behavior of PZT-4 piezoelectric ceramics. The undamaged properties are given in Table 1. The specimens are respectively subjected to different critical mechanical loads and their corresponding electric fields given by Park and Sun (1995). The mechanical and electrical damages near the crack tips in the compact tension specimens and the three-point bending specimens are computed.

The damage components D^M_{33} and D^E_{33} are dominant in the problems under consideration, so only these two components at crack-tips are observed. Their relationship is imitated by means of linear and nonlinear methods, and the parameters in the equations are calculated according to the method of least squares.

$$D^M_{33} = 0.5238 + 0.5651 D^E_{33},$$
$$D^M_{33} = 0.5029 + 0.8878 D^E_{33} + 3.5384 (D^E_{33})^2. \tag{18}$$

The curves are plotted in Fig.9. It is apparent that the nonlinear imitation is better than the linear. Eq.(18) can be rewritten as

$$1.9092 D_{33}^{M} - 1.0789 D_{33}^{E} = 1,$$
$$1.9885 D_{33}^{M} - 1.7653 D_{33}^{E} - 7.0361 (D_{33}^{E})^{2} = 1.$$

(19)

Define (1) linearly combined damage:

$$D = a_1 D_{33}^{M} + a_2 D_{33}^{E},$$

(20)

and (2) nonlinearly combined damage:

$$D = b_0 D_{33}^{M} + b_1 D_{33}^{E} + b_3 (D_{33}^{E})^{2}.$$

(21)

In the above two formulations, coefficients a_1, a_2, b_0, b_1, and b_2 are material dependent, but independent of geometries, loading conditions and boundary conditions of specimens. For PZT-4, a_1 =1.9092, a_2 =-1.0789; b_0 =1.9885, b_1 =-1.7653, b_2 =-7.0361. Accordingly, a piezoelectric fracture criterion can be given as follows.

$$D = D_{c},$$

(22)

where, $D_{c} = 1$. This means that piezoelectric fracture occurs when the combined damage D reaches its critical value D_{c}.

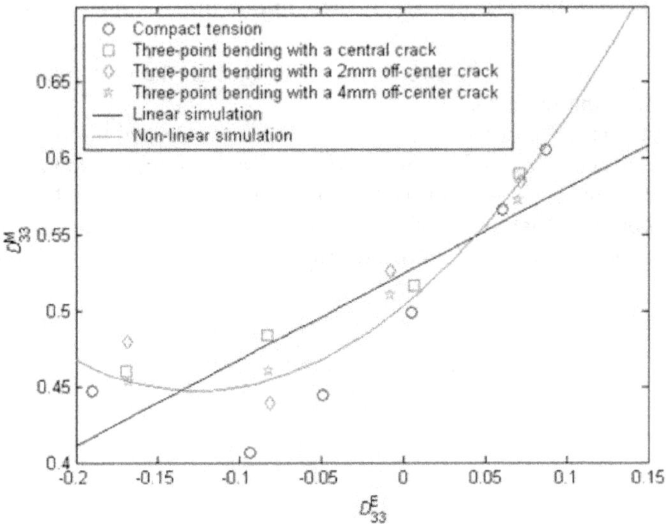

Fig. 9. Linear and nonlinear imitations of mechanical and electrical damages at the crack-tips

5. Quasi-static crack propagation in piezoelectric beam

As an important application of the aformentioned piezoelectric damage constitutive model and nonlinear fracture criteria, the quasi-static propagation behaviors of the midspan crack in a three-point bending PZT-4 beam with sizes of 19.1mm×9mm×5.1mm are simulated in different mechanically and electrically loading conditions. The beam is homogeneously poled along the x_3-direction and subjected to a downward concentrated force F in the midspan and an electric voltage U excitateding an electric field parallel to its length direction. A midspan insulating edge crack with a depth of 4mm is perpendicular to the poling direction. The combined-damage determined by the dominant mechanical and electrical damage components is regarded as the fracture criterion and its gradient is assumed to control cracking direction. The influences of mechanical and electrical loads on the crack growth rate are evaluated.

5.1 Effect of mechanical load on crack propagation

Referring to the experimental data by Park and Sun (1995), we fix the electrical load at U=3KV and change the mechanical load F from 400N to 500 N, 600N, 700N and 800N in order to evaluate the effect of mechanical load on crack propagation. The crack growth increment Δa at different computational steps is observed, and its variation and fitting curves with the computational step T are drawn in Fig.10. It can be seen that the crack propagation has an increasingly accelerating trend and the mechanical load plays a very important role in crack growth. As an example, within the computational step range given in this paper, the crack depth increment Δa has the minimum lower limit $(0.5 \pm 0.01) \times 10^{-4}$ mm at the start and the minimum upper limit $(0.96 \pm 0.01) \times 10^{-4}$ mm at the final computational step for the external concentrated force F=400N, but the maximum lower limits $(1.1 \pm 0.01) \times 10^{-4}$ mm and the maximum upper limit $(1.79 \pm 0.01) \times 10^{-4}$ mm for F=800N. Accordingly, it is very obvious that mechanical load enhances crack propagation. Fig.11 exhibits the effect of mechanical load on the variation curve of the crack depth with the computational step.

Mechanical loading is a dominant factor controlling piezoelectric fracture. This is why the conventional piezoelectric fracture theories using the stress intensity factor (Pak, 1992; Sosa, 1992; Suo, 1992), the energy release rate (Park and Sun, 1995; Gao et al, 1997) and the energy density factor (Shen and Nishioka, 2000; Sih and Zuo, 2000; Zuo and Sih, 2000) as fracture criteria can give good fracture predictions for high stress cases.

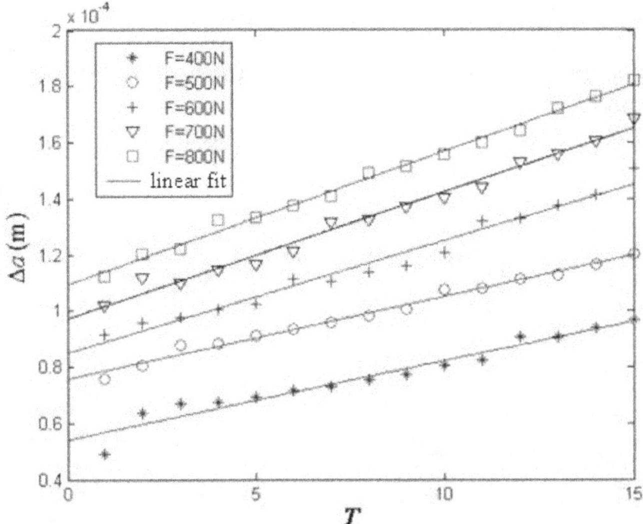

Fig. 10. Crack growth rate curves at different mechanical loads

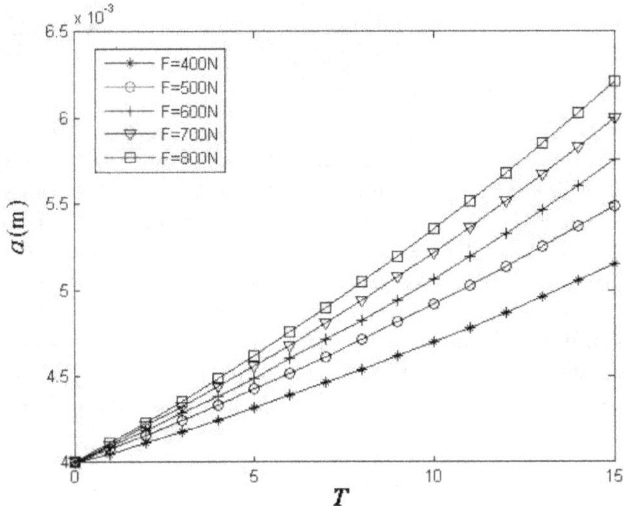

Fig. 11. Crack depth curves at different mechanical loads

5.2 Effect of electric field on crack propagation

Electric field is another important influencing factor. To observe the electromechanical coupling fracture behavior of piezoelectric media, Tobin and Pak (1993) early performed a series of indentation tests on PZT-8 ceramics under a given static electric field and an indentation load of 4.9N on different locations. The results showed that the cracks perpendicular to the poling direction became longer or shorter under a positive or negative electric field. It was concluded that the positive field assisted the crack propagation, whereas the negative field retarded it. The analogous conclusion was also given by Park and Sun (1995) after completing the compact tension and three-point bending tests for PZT-4 piezoelectric ceramics.

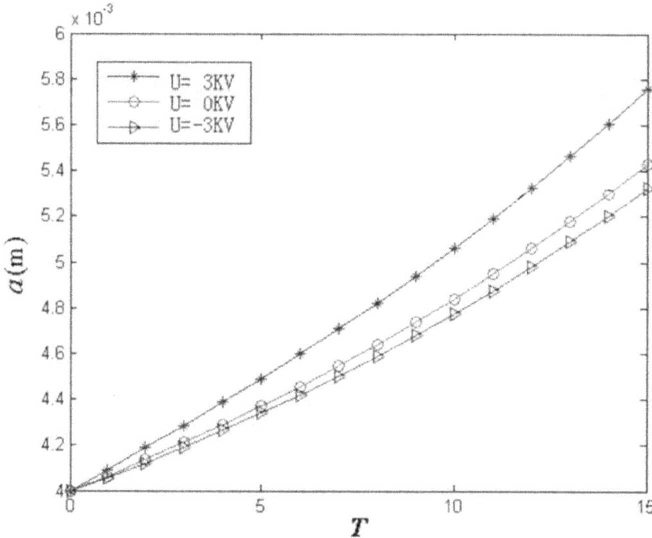

Fig. 12. Crack depth curves at different electrical loads

In order to evaluate the influence of electric field on crack propagation, an invariable mechanical load of F=600N is imposed on the beam, and the electrical voltage U is changed from 3KV to 0KV and -3KV. Some results are plotted in Fig.12. It is apparent that the applied electric field has great effect on the crack propagation. The crack growth rate comes to its maximum for the electrical voltage of 3KV. Along with the reducing voltage, however, the crack growth rate decreases rapidly, and reaches the minimum value at the electrical voltage of -3KV. It can be concluded that a positive electric field enhances crack propagation, whereas a negative electric field inhibits it. This is qualitatively consistent with the experimental conclusions given by Park and Sun (1995), and Tobin and Pak (1993).

6. Conclusions

In this chapter, a three-dimensional anisotropic piezoelectric damage constitutive model is developed by extension of continuum damage mechanics into piezoelectric materials. Using the unit cell method, the quantitative connections between the phenomenological damage tensors and microstructure parameters are established. These discussed microstructures include: (1) periodically distributed insulating microvoids, (2) conductive microvoids, and (3) conductive inclusions. Based on the compact tension tests and the three-point bending tests performed by Park and Sun, damage-based linear and nonlinear piezoelectric fracture criteria are obtained by using the method of least squares combined with numerical damage analysis for the fracture critical cases. As an application example, both the proposed piezoelectric damage constitutive model and the nonlinear fracture criterion are used to investigate the quasi-static propagation behavior of the midspan crack in a three-point bending PZT-4 beam, and the influences of mechanical and electrical loads on cracking behavior are evaluated.

7. References

Fulton, C.C. & Gao, H. (2001). Effect of local polarization switching on piezoelectric fracture. Journal of the Mechanics and Physics of Solids, Vol. 49, pp. 927 – 952.

Gao, H.; Zhang, T.Y., & Tong, P. (1997). Local and global energy release rates for an electrically yielded crack in a piezoelectric ceramic. J. Mech. Phys. Solids. Vol. 45, pp. 491–510.

Heyer, V.; Schneider, G.A.; Balke, H.; Drescher, J. & Bahr, H.A. (1998). A fracture criterion for conducting cracks in homogeneously poled piezoelectric PZT-PIC 151 ceramics. Acta Mater. Vol. 46, pp. 6615–6622.

Mizuno, M. (2002). Constitutive equation of piezoelectric ceramics taking into account damage development. Key Eng. Mater. Vol. 233–236, pp. 89–94.

Mizuno, M. & Honda, Y. (2005). Simplified analysis of steady-state crack growth of piezoelectric ceramics based on the continuum damage mechanics. Acta Mech. Vol. 179, pp. 157-168.

Pak, Y.E. (1990). Crack extension force in a piezoelectric material. Trans. ASME J. Appl. Mech. Vol. 57, pp. 647–653.

Pak, Y.E. (1992). Linear electro-elastic fracture mechanics of piezoelectric materials. Int. J. Fracture Vol. 54, pp. 79–100.

Park, S.B. & Sun, C.T. (1995). Fracture criteria of piezoelectric ceramics. J. Am. Ceram. Soc. Vol. 78, pp. 1475–1480.

Shen, S. & Nishioka, T. (2000). Fracture of piezoelectric materials: energy density criterion. Theor. Appl. Fract. Mech. Vol. 33, pp. 57–65.

Sih, G.C. & Zuo, J.Z. (2000). Multiscale behavior of crack initiation and growth in piezoelectric ceramics. Theor. Appl. Fract. Mech. Vol. 34, pp. 123–141.

Sosa, H. (1991). Plane problems in piezoelectric media with defects. Int. J. Solids Struct. Vol. 28, pp. 491–505.

Sosa, H. (1992). On the fracture mechanics of piezoelectric solids. Int. J. Solids Struct. Vol. 29, pp. 2613-2622.

Sosa, H. & Pak, Y.E. (1990). Three-dimensional eigenfunction analysis of a crack in a piezoelectric material. Int. J. Solids Struct. Vol. 26, pp. 1–15.

Suo, Z. (1993). Models for breakdown-resistant dielectric and ferroelectric ceramics. J. Mech. Phys. Solids. Vol. 41, pp. 1155-1176.

Suo, Z.; Kuo, C.M.; Barnett, D.M. & Willis, J.R. (1992). Fracture mechanics for piezoelectric ceramics. J. Mech. Phys. Solids. Vol. 40, pp. 739–765.

Tobin, A.G. & Pak, Y.E. (1993). Effect of electric fields on fracture behavior of PZT ceramics. Proc. SPIE--Int. Soc. Opt. Eng. Vol. 1916, pp. 78–86.

Yang, W. & Zhu, T. (1998). Switch-toughening of ferroelectrics subjected to electric field. Journal of the Mechanics and Physics of Solids, Vol. 46, pp. 291-311.

Yang, X.H.; Chen, C.Y. & Hu, Y.T. (2003). Analysis of damage near a conducting crack in a piezoelectric ceramic. Acta Mechanica Solida Sinica, Vol. 16, pp. 147-154.

Yang, X.H.; Chen, C.Y.; Hu, Y.T. & Wang, C. (2005). Damage analysis and fracture criteria for piezoelectric ceramics. International Journal of Non-Linear Mechanics, Vol. 40, pp. 1204-1213.

Yang, X.H.; Zeng, G.W. & Chen, C.Y. (2007). Determination of Mechanical and Electrical Damages of Piezoelectric Material with Periodically Distributed Microvoids. Journal of Mechanics, Vol. 23, pp. 107-112.

Yang, X.H.; Zeng, G.W. & Chen, C.Y. (2008). Analysis of Mechanical and Electrical Damages of Piezoelectric Ceramic with Periodically Distributed Conducting Inclusions. Journal of Mechanical Strength, Vol. 30, pp. 844-847. (In Chinese)

Yang, X.H.; Chen, C.Y.; Hu, Y.T. & Wang, C. (2005). Combined damage fracture criteria for piezoelectric ceramics. Acta Mechanica Solida Sinica, Vol. 18, pp. 21-27.

Zeng, G.W.; Yang, X.H. & Chen, C.Y. (2008). Mechanical and electrical damages of piezoelectric ceramics with periodically distributed conductive voids. Journal of Wuhan University of Science and Technology, Vol. 31, pp. 424-426. (In Chinese)

Zhang, T.Y.; Wang, T.H. & Zhao, M.H. (2003). Failure behavior and failure criterion of conductive cracks (deep notches) in thermally depoled PZT-4 ceramics. Acta Materialia, Vol. 51, pp. 4881-4895

Zhang, T.Y.; Zhao, M.H. & Gao, C.F. (2005). The strip dielectric breakdown model. International Journal of Fracture, Vol. 132, pp. 311-327.

Zhang, T.Y. & Gao, C.F. (2004). Fracture behaviors of piezoelectric materials. Theoretical and Applied Fracture Mechanics, Vol. 41, pp. 339-379.

Zuo, J.Z. & Sih, G.C. (2000). Energy density theory formulation and interpretation of cracking behavior for piezoelectric ceramics. Theor. Appl. Fract. Mech. Vol. 34, pp. 17-33.

6

Porous piezoelectric ceramics

Elisa Mercadelli, Alessandra Sanson and Carmen Galassi
Institute of Science and Technology for Ceramics, National Research Council,
CNR-ISTEC, Via Granarolo 64, I-48018 Faenza,
Italy

1. Introduction

Lead Zirconate Titanate (PZT) is the best performing, cost effective class of piezoelectric materials known to date (Kumamoto et al., 1991; Martin et al., 1993; Perez et al. 2005). Its success is strongly related to the flexibility in terms of composition (Zr/Ti ratio, use of different dopants) and microstructure. In particular, when PZT is doped with Nb and coupled with a controlled porous microstructure, it becomes a promising candidate for ultrasonic transducer applications.

To obtain high piezoelectric responses it is important to produce dense ceramics: in this respect, the porosity is generally considered a defect that causes the decreasing of the mechanical and piezoelectric properties. On the other hand, the introduction of tailored porosity can considerably improve the performances of ultrasonic devices, such as hydrophones (Geis et al., 2000) or medical diagnostic devices (Smith, 1989). Dense PZT-type piezoceramics show low hydrostatic figure of merit (FOM) and poor acoustic coupling to the media with which it is in contact and are therefore not suitable for these applications. On the contrary, in porous piezoelectric materials, a partial decoupling between transverse and longitudinal effects leads to an increase of the FOM while the transfer of acoustical energy to water or biological tissues is improved as a consequence of a lower acoustical impedance (Z) (Li et al., 2003; Okazaki & Nagata, 1973). A low Z value, in fact, reduces the mismatch between the device and the media through which the signal is transmitted or received, leading to a more efficient acoustic wave transfer (Roncari et al., 2001).

Porous piezoceramic can be designed as a functionally graded material (FGM) (Mercadelli et al., 2010). This allows to match the need of an high response, typical of dense piezoceramic, to a good compatibility with the investigated media given by a porous material.

In this chapter the main processing routes necessary to obtain PZT-based materials with tailored porosity will be thoroughly analyzed and discussed. The electrical and acoustic properties of porous piezoelectric ceramics will be evaluated considering both porosity amount and pore morphology.

2. Porous Ceramics Processing

Materials containing tailored porosity exhibit special properties and features that usually cannot be achieved by their conventional dense counterparts. Therefore, porous materials find nowadays many applications in several technological processes.

Such applications include the filtration of molten metals, high-temperature thermal insulation, support for catalytic reactions, filtration of particulates, and filtration of hot corrosive gases in various industrial processes. The advantages of using porous ceramics in these applications are usually their high melting point, specific electronic properties, low thermal mass, low thermal conductivity, controlled permeability, high surface area, low density, high specific strength, and low dielectric constant.

Porous ceramics are classified (IUPAC) according to their pore size (macroporous ceramics, pore size > 50 nm, microporous, d < 2 nm and mesoporous ceramics, 2 nm < d < 50 nm) and in terms of pore geometry (Araki & Halloran, 2005): foam, interconnected, pore spaces between particles, plates, and fibres, large or small pore networks.

The main processes to produce porous ceramics are well detailed in a review proposed by Studart et al., 2006. The most straightforward processing route for the preparation of porous ceramics is the partial sintering of cold-compacted powders. In this method many experiments have to be conducted in order to develop an appropriate sintering procedure to achieve the desirable porosity. The degree of porosity is in fact controlled by the degree of sintering, which, in turn, is controlled by temperature and/or soaking time. Many novel methods have been developed for the preparation of porous ceramics with controlled microstructure. The most widely used for macroporous ceramics could be classified into replica, sacrificial template and direct foaming methods (Fig. 1).

2.1 Replica Technique

The replica method is based on the impregnation of a cellular structure (sponge) with a ceramic suspension. The organic structure is than removed by controlled thermal treatments in order to produce a macroporous ceramic exhibiting the same morphology of the original porous material. Many synthetic and natural sponges can be used as templates for this method.

Porous ceramics obtained with the sponge replica method can generally reach total open porosity levels within the range 40%–95% and are characterized by a reticulated structure of highly interconnected pores with sizes between 200 µm and 3 mm. The minimum cell size of replica-derived porous ceramics is however limited to approximately 200 µm, for the difficulty of impregnating polymeric sponges with excessively narrow cells.

2.2 Sacrificial Template Method

The sacrificial template technique usually consists on the preparation of a biphasic composite comprising a continuous matrix of ceramic particles and a dispersed sacrificial phase. The latter is initially homogeneously distributed throughout the matrix and is ultimately extracted to generate pores within the microstructure. Predominantly open pores of various different morphologies can be produced with this method. The most crucial step in this technique is the removal of the sacrificial phase that can be done by pyrolysis, evaporation, or sublimation. These processes might involve the release of an excessive amount of gases and must be carried out at sufficiently slow rates in order to avoid cracking of the cellular structure. The most widely used sacrificial templates can be summarized as follow:

- <u>Synthetic organics</u>: polyvinyl chloride (PVC), polystyrene (PS), polyethylenoxide (PEO) or Polyviniylbutiral (PVB), polymethylmethacrylate (PMMA) or polymethylmethacrylate-ethyleneglycole (PMMA-PEG) beads, methylhydroxyethyl cellulose (MHEC), phenolic resin, nylon, cellulose acetate, polymeric gels, naphtalene.
- <u>Natural organics</u>: gelatine, peas and seeds, cellulose/cotton, glucide, sucrose, dextrin, wax, alginate, starch.
- <u>Liquids</u>: water, camphene, emulsion-oils.
- <u>Salts</u>: NaCl, $BaSO_4$, K_2SO_4.
- <u>Metals/ceramics</u>: nickel, carbon (graphite, fiber, nanotubes), SiO_2 (particles, fibers), ZnO.

The natural or synthetic organic components are essentially removed by decomposition/combustion thermal treatments, while inorganic pore formers (salts, ceramic or metallic composites) are generally eliminated though chemical processes (aqueous or acidic leaching).

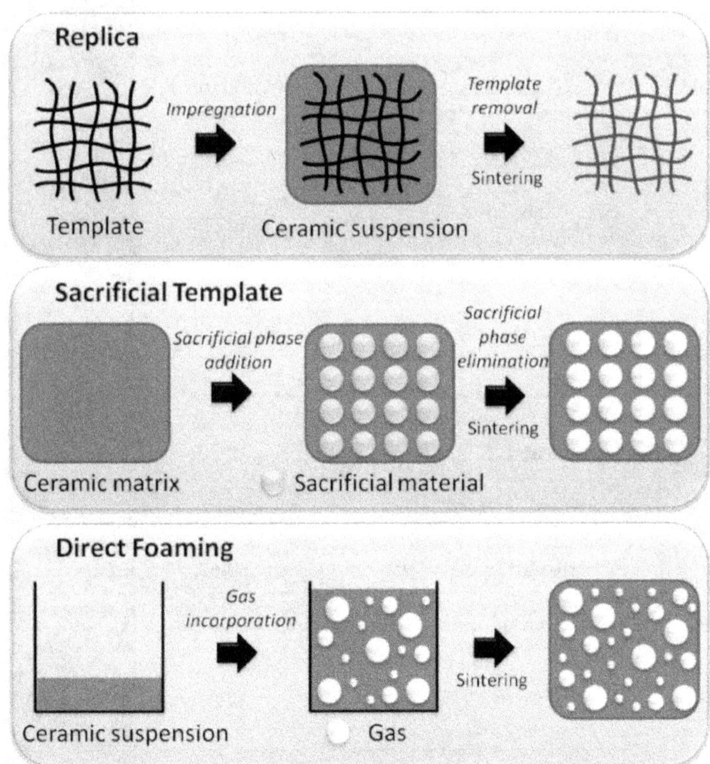

Fig. 1. Scheme of the main processing routes used for the production of porous ceramics

Pore former, Process used	Pore former shape and size (μm)	Ceramic pore size (μm)	Porosity (%)	Microstructure	Composition
Synthetic organic					
PMMA, die pressing	Spherical, 220	120-125 (3-3 connectivity)	20-50		**PZT** (Kumar et al., 2005; Praveenkumar et al., 2006))
	Spherical, 15	10-15	4-18		**PZT** (Zeng et al., 2007) **PMN-PZT** (Zeng et al., 2006)
	Irregular, 100-120	100-120	4-18		**PZT** (Zeng et al., 2007)
	Spherical, 5-6	1-7	30-50		ZrO_2, $LaGaO_3$ (Kaleva et al., 2006)
	120-170	> 100	55-35		**PZT** (Zeng et al., 2006)
	34-76	10-80	Gradient 10-40		**PZT** (Zhang et al., 2007)
	Spherical, 150-200	120-170	10-55		Y_2O_3 (Gain et al. 2006)
PEO, die pressing	Spherical, 150	100 (3-3 connectivity)	25-50		**PZT** (Kumar et al., 2005)
PVC, die pressing	Spherical, 125	6 (0-3 connectivity)	40-50		**PZT** (Kumar et al., 2005)
MHEC, die pressing	10-140	01-10	30-60		**PZT** (Roncari et al., 2001)
PVB, tape casting	-	01-1	45		**PZT** (Roncari et al., 2001)
	-	0.8-1.6	60		**PZT** (Craciun et al., 1998)

	Pore former, Process used	Pore former shape and size (µm)	Ceramic pore size (µm)	Porosity (%)	Microstructure	Composition
Synthetic organic	Stearic acid, die pressing	Irregular	10-100 (3-3 connectivity)	Gradient 10-40		**PZT** (Zhang et al., 2007)
		-	10-30 (long axis connectivity)	Gradient 2.4-21.4		**PZT** (Li et al., 2003)
Natural organics	Dextrin, die pressing	Irregular, ~ 15	-	4-12		**PZT** (Zeng et al., 2007)
	Glucide, die pressing	~ 20	10-1000	60-90		**PZT** (Montanaro et al., 1998)
	Potato starch, starch consolidation	45-50	>50	25-50		Al$_2$O$_3$ (Gregorovà & Pabst, 2007; Gregorovà et al. 2006)
		-	30-40	30-50		**PZT** (Galassi et al., 2005)
	Potato starch, die pressing	10-20	-	-		ZrO$_2$, LaGaO$_3$ (Kaleva, 2006)
	Potato starch, tape casting	-	~ 30-40	25		**PZT** (Palmqvist et al., 2007)
	Wheat starch, starch consolidation	~ 20	~ 20	25-50		Al$_2$O$_3$ (Gregorovà & Pabst, 2007; Gregorovà et al. 2006)
	Tapioca starch	12-14	-	-		(Gregorovà et al. 2006)
	Corn starch, starch consolidation	12-14	~ 14	25-50		Al$_2$O$_3$ (Gregorovà & Pabst, 2007; Gregorovà et al. 2006)
	Corn starch, tape casting	~ 15	~ 5	5-10		**PZT** (Shaw et al., 2007)
	Corn starch, die pressing	12	10-40 (3-3 connectivity)	35		**PZT** (Galassi et al., 2005; Roncari et al., 1999)

	Pore former, Process used	Pore former shape and size (µm)	Ceramic pore size (µm)	Porosity (%)	Microstructure	Composition
Natural organics	Rice starch	4-5	-	-		(Gregorovà et al. 2006)
	Rice starch, die pressing	6	10-20 (3-3 connectivity)	35		**PZT** (Galassi et al., 2005)
	Ammonium citrate monohydrate, die pressing	10-50	-	-		ZrO₂, LaGaO₃ (Kaleva et al., 2006)
	Ammonium oxalate monohydrate, die pressing	-	5-30 (3-3 connectivity)	24-40		**PZT** (Eremkin et al., 2004)
	K₂CO₃ (solid state mtayhesis with PbSO₄ and TiO₂), and leaching	-	5-30 (3-3 connectivity)	-		**PbTiO₃** (Toberer & Seshadri, 2006; Toberer et al., 2004)
Inorganics	Graphite, die pressing	Elongated, 200	5-30 (3-3 connectivity)	Gradient1 0-40		**PZT** (Piazza et al., 2005), **YSZ** (Corbin & Apte, 1999)
	Carbone	Spherical, 5-50	1.4-3.5	Gradient 10-40		**Ni-YSZ** (Holtappels et al., 2006)
		Spherical, 1-30	≤ 1	48		**Ni-YSZ** (Sanson et al., 2008)
Liquids	Canfene, sublimazione	-	(3-3 connectivity)	90		**PZT** (Lee et al., 2007; Lee et al., 2008)

Table 1. Pore formers and processes used to produce porous ceramics: state of the art.

2.3 Direct Foaming Method

The pores produced with this approach result from the direct incorporation of air bubbles into a ceramic suspension, therefore there is no need for extensive pyrolysis steps before sintering of the green body. The stabilization and setting of the wet foams is the decisive step in direct foaming methods. The total porosity of directly foamed ceramics is proportional to the amount of gas incorporated into the suspension or liquid medium during the foaming process and can reach values ranging between 45 and 95%. The pore size, on the other hand, is determined by the stability of the wet foam before setting. Wet foams are thermodynamically unstable: these destabilization processes significantly increase the size of incorporated bubbles, resulting in large pores in the final cellular microstructure. The use of surface modified particles to stabilize the wet foam has decreased the lower limit of pore size achievable via direct foaming to 10 µm, with the upper one being around 1.2 mm.

3. Porous ceramics via the sacrificial template methods: state of the art

The three techniques above mentioned allow a wide range of pore size and porosity amount. The foaming method is generally used to produce a high level of interconnected porosity, while the replica technique, (depending on the template) leads to high porosity amount and tailored pore size ranging between 10 µm and 1.2 mm. The sacrificial template method not only allows a strict control of the amount, mean size and morphology of the porosity produced but, tailoring the nature and size of the pore former agent, guarantees a wider range of porosity level and dimension. Moreover this method is the only one able to produce small pores (< 10 µm).

For these reasons the sacrificial template method must be considered for those applications that require micrometric pore size. A panoramic of the pore forming agents and of the processes involved in the fabrication of porous piezoelectric materials is reported in Table 1.

4. Porous piezoelectric ceramics: characteristics and applications

As already mentioned, the porosity is usually unwanted for its detrimental effect on the piezoelectric and mechanical properties. However the introduction of a controlled porosity into a piezoelectric ceramic could strongly improve its acoustic performances and therefore its ultrasonic responce.

4.1 Porous ceramics as composites

Composites are materials that are receiving a growing attention by the scientific as well as the industrial word. They can in fact show specific properties (electronic, magnetic, mechanical, etc.) that cannot be otherwise achieved by the single phase materials. This approach has been applied for the first time to the piezoelectric materials at the end of the 1970s combing them with polymeric and/or metallic phases to obtain actuators or transducers (Akdogan et al., 2005).

In the production and design of a piezocomposite the right choice of the spatial distribution between the two phases determines the effective improvement of the piezoelectric properties (Levassort et al., 2007). For this reason, the concept of connectivity has been defined to describe the way in which the individual phases are self-connected (that is, continuous) (Skinner et al., 1978). There are 10 connectivity patterns for a two-phase (diphasic) system, in which each phase can be continuous in zero, one, two, or three dimensions. The internationally accepted nomenclature to describe such composites is (0-0), (0-1), (0-2), (0-3), (1-1), (1-2), (1-3), (2-2), (2-3) and (3-3). The first digit refers to the number of dimensions of connectivity for the piezoelectrically active phase, and the second digit is

used for the electromechanically inactive phase. A porous ceramic can be as well considered a composite where the main phase is the active ceramic and the second phase is the porosity. The development of such piezoelectric composites aims to combine the specific properties of each single phase to maximize either the electromechanical and ultrasonic response of a particular device. A single material, or a unique phase, cannot in fact satisfy the need of maximize the piezoelectric response and at the same time minimize the material density to acoustically match the transducer with the water or the media to which it is in contact.
In these cases, when two opposite requirements have to coexist, the production of a composite is the only way to produce an efficient device (Akdogan et al., 2005).

4.2 Properties

Nowadays, many types of piezoelectric composites are used as active materials for transducers. These composites are fabricated by incorporating different second phases, e.g. dielectric ceramics (Jin et al., 2003), metals (Li et al., 2001), polymers (Klicker et al., 1981) or pores (Li et al., 2003), to modulate the electrical properties. The porosity has several advantages as non-piezoelectric second phase. Firstly, porous piezoelectric ceramics are composed of ceramics only, thus there is no possibility of detrimental chemical reactions between the piezoelectric ceramic and the second phase during production. Secondly, their piezoelectric properties are easily tailored by changing the porosity level and the pore morphology. Finally, they are cheaper to produce and lighter than other possible piezoelectric composites (Jin et al., 2003; Li et al., 2001).
Since porous piezoelectric ceramics have lower acoustic impedances than dense ceramics, they could be used to improve the mismatch of acoustic impedances at the interfaces of medical ultrasonic imaging devices or underwater sonar detectors (Kumar et al., 2006). Therefore, the electrical and acoustic properties of porous piezoelectric ceramics need to be thoroughly evaluated in the light of either porosity level and pore morphology. Three are the characteristics that a porous piezoceramic has to satisfy to be used for ultrasonic applications:

1. high hydrostatic figure of merit;
2. acoustic impedance similar to the one of the investigated media;
3. low mechanical quality factor.

<u>Hydrostatic figure of merit</u>

The efficiency of a piezoceramic for ultrasonic applications and for hydrophone devices in particular is measured by the hydrostatic figure of merit $d_h g_h$. In hydrostatic condition, the transducer behaviour is proportional to this figure of merit through the equation:

$$d_h g_h = (d_{33} - 2\,|d_{31}|) \cdot (g_{33} + 2\,|g_{31}|) \tag{1}$$

where for PZT ceramics $d_{33} \approx 2\,|d_{31}|$ and $g_{33} \approx 2\,|g_{31}|$. From this equation, the figure of merit for dense PZT tends to zero. On the other hand in the porous piezoceramics a partial decoupling of the piezoelectric response between the longitudinal and transversal direction occurs. As a consequence, the more the porosity increases the more the $|d_{31}|$ value decreases in respect to the d_{33} one. In this way, the electrical response is maximized along the most useful direction (i.e. in the thickness, 33) and, at the same time, minimized in the other ones.

The figure of merit is in this way improved by more than three orders of magnitude, decreasing the effect in the 31 direction while keeping constant the one in 33.

<u>Acoustic impedance</u>

The acoustic impedance (Za) of a material is a measure of the propagation of the acoustic wave throughout the interfaces of different media. The wave transfer is maximized when the two media have the same acoustic impedance. The higher is the difference of impedance between the two media, the higher would be the acoustic wave's fraction reflected at the interface. In the case of a transducer, its efficiency is linked to the acoustic matching between the piezoceramic (high Za ~ 30-33 [10^6 Kg / m^2s]) and the investigated media (water, biological tissues) to which is in contact (low Za ~ 1-1.5 [10^6 Kg / m^2s]) (Bowen et al., 2004). This large impedance difference leads to poor acoustic matching and low axial resolution. This problem is generally overcome reducing the ceramic density. It has been shown (Levassort et al., 2007) that the introduction of 40% vol of isotropic porosity into a piezoceramic leads to a reduction of 60% the acoustic impedance thus dramatically increasing the device efficiency.

<u>Mechanical quality factor (Qm)</u>

The increasing of mechanical losses (and the consequent reduction of the mechanical quality factor) is another effect induced by the presence of porosity into a piezoceramic. This property is useful to increase the transducers bandwidth, i.e. to reduce the electrical signal losses in operating conditions.

4.3 Piezoelectric ceramics as ultrasonic transducers

Piezoceramics devices are widely used as ultrasonic transducers because they can generate powerful ultrasonic waves useful for cleaning, drilling and welding, as well as to stimulate chemical processes. Moreover, they act either as transmitters and receivers of ultrasonic waves in medical diagnostic equipments and non-destructive material testing apparatus for locating defects within a structure. Ultrasonic waves are mechanical vibrations that propagate in a material as a consequence of a series of very small continuous displacements of atoms and chain segments around their equilibrium positions. Displacements are induced at neighbouring zones by the forces within a chain segment and between adjacent chain segments, propagating in this way a stress-strain wave. Several kinds of ultrasonic waves may propagate in solid materials: longitudinal waves, shear waves, Rayleigh waves (or surface acoustic waves), Lamb waves (or plate waves). The most common method of ultrasonic wave generation and detection uses longitudinal waves in which the piezoceramic transducers are required to move like pistons at very high frequencies (from 20 kHz to hundreds of MHz) (Lionetto et al., 2004).
Recently the research interest in this field is focused on the development of suitable transducers for the medical diagnosis.

4.4 Transducers for ultrasonic medical diagnostics

A classical ultrasonic transducer (Fig. 2) is composed by:

- an active piezoelectric element (that works as either transmitter and receiver);
- a backing layer (that absorbs the acoustic emission in the back side of the actuator)
- a matching layer (that matches the acoustic impedance of the piezoelectric material with the one of the investigated surface).

With this architecture, the energy emission is maximized and therefore the axial resolution (that is the capability to distinguish two points along the axis of an ultrasonic beam) increased thus improving the image resolution (Levassort et al., 2004).

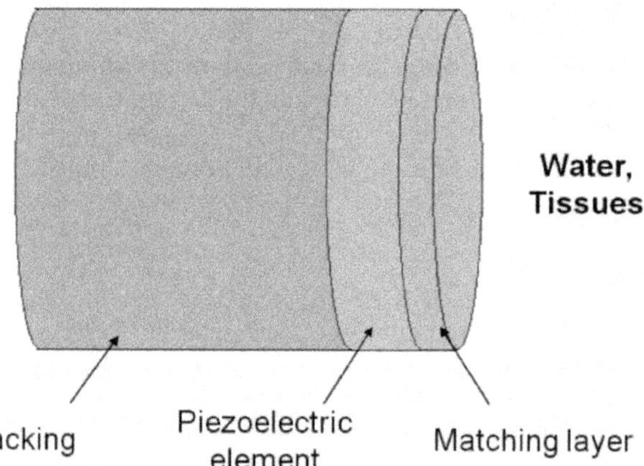

Fig. 2. Scheme of a classical ultrasonic transducer.

<u>Active piezoelectric element</u>

The active piezoelectric element is typically a piezoelectric disk, 1-2 cm thick, with two electroded parallel faces. The application of an alternating current between the two electrodes generates a synchronous thickness variation of the transducer (inverse piezoelectric effect); on the other hand a mechanical perturbation (ultrasonic pulse) induces an electrical signal as a consequence of the direct piezoelectric effect.

The piezoelectric element thickness is a crucial parameter that defines the resonance frequency of the device. High-frequency ultrasonic devices are, for example, needed for the medical diagnosis in organs such as skin, eye and blood vessels, where high resolution is requested at low depth of field (Levassort et al., 2004). The depth of penetration of the ultrasounds is in fact inversely proportional to the frequency. High-frequency ultrasounds are therefore only used to study relatively "superficial" structures.

In the case of a transducer with a fixed thickness d, the resonance condition occurs at a wave length λ equal to:

$$\lambda = 2d \tag{2}$$

The corresponding frequency is linked to the ultrasound speed (c) into the transducer material through the equation:

$$f = \frac{c}{\lambda} \tag{3}$$

It is clear that the frequency and the piezoceramic thickness are inversely proportional. Therefore it is necessary to reduce the thickness of the piezoelectric element to work at high resonance frequencies.

Backing layer

The backing layer has two functions: on one hand it works as mechanical support for the active element, on the other hand it has to attenuate the acoustic energy lost from the back face of the transducer. In this way no energy is radiated back to the active layer hindering therefore the production of parasitic echoes (Levassort et al., 2004). To optimize the transmission of the ultrasounds to the back surface of the transducer, the backing and active layer materials must have similar impedance values. For this reason epoxy resins filled with tungsten particles are generally used as materials for backing layers.

Matching layer

On the front of the transducer (i.e. between the piezoceramic and the propagation medium) a matching layer is generally used. This layer must mediate the impedance values of the ceramic with those of the biologic tissues thus maximizing the acoustic energy amount transmitted by the transducer and improving consequently its efficiency.
The optimal impedance value for a matching layer is defined as:

$$Z_m = \sqrt{Z_t \cdot Z_{st}} \tag{4}$$

where Z_m, Z_t and Z_{st} are the acoustic impedance values of respectively the matching layer, the transducer and the biological tissue. The matching layer thickness must be $\lambda/4$ to produce destructive interference with the waves reflected back. For example, in the case of a single porous matching layer, suitable for a PZT transducer working at a frequency of 50 MHz, with a speed of sonar wave propagation of v = 1882 m/s and acoustic impedance of Zm = 9.64 [10^6 Kg / m²s], its thickness t would be:

$$t = \frac{v}{f} = \frac{1882 \; ms^{-1}}{50 \times 10^6 s^{-1}} = 38 \; \mu m \tag{5}$$

Whereas the optimal impedance value for the matching layer would be:

$$Z_m = \sqrt{Z_t \cdot Z_{st}} = \sqrt{25 \cdot 1} = 5 \; [10^6 Kg/m^2 s] \tag{6}$$

Therefore the matching layer must be optimized in terms of either the material, to match the acoustic impedance, and the thickness to maximize the device performances. The porous piezoceramics could be successfully used as matching layer, allowing an optimal chemical-physical compatibility with the active piezoelectric element (dense ceramic) and offering at the same time the possibility to tailor the microstructural properties to improve the efficiency.

5. Porosity Graded Piezoelectric Ceramics (PGPC)

The ultrasonic properties of the piezoelectric materials could be improved producing a transducer (the active layer) with a functionally-graded structure (Fig. 3). The porosity-graded structure could match a dense active layer with high piezoelectric constant able to efficiently transmit/receive the acoustic/electrical signal with a porous layer with a lower piezoelectric response but with an acoustic impedance appropriate for the porous matching layer and the investigated media.

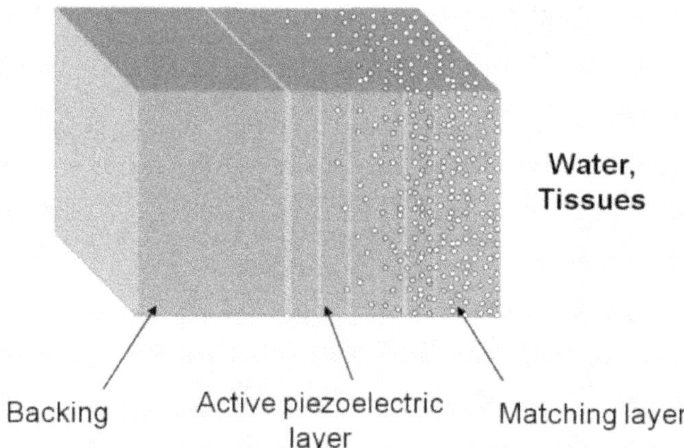

Fig. 3. Scheme of a transducer with a porosity-graded active layer and a porous matching layer.

The transducer performances could be improved even further by adding more matching layers (Fig. 4): in fact a gradient of the acoustic impedance values could minimize the differences between the ceramic and the media, maximizing, as a consequence, the acoustic energy transmitted to the biological tissues. The ideal would be the production of a continuous porosity-gradient material able to assure a uniform properties-variation from the transducer to the tissues.

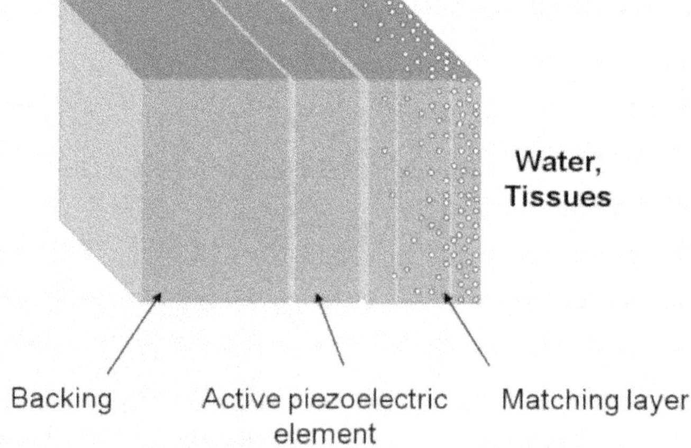

Fig. 4. Scheme of a transducer with a porosity-graded matching layer.

5.1 Processes for the production of (PGPC): state of the art

The production of functionally graded materials has to face and overcome the technological difficulties related to the fabrication of layers with different microstructural characteristics (different shrinkage, various organics content to be eliminated, etc.). Few works are reported till now regarding this kind of architectures.

Porous graded piezoelectric materials more than 1 mm thick and with porosity ranging from 10 to 40 % have been already produced by die-pressing layers of powders with different pore former amounts (see Table 1). With this technique it is however difficult to produce graded materials with thickness less than 1 mm. As mentioned above, for the design of an ultrasonic medical device it is very important to cover a wide range of thicknesses to assure resonance frequency from 20 kHz to hundreds of MHz (Opielinski & Gudra, 2002). Working at high frequency (f > 20 MHz) means to have a suitable resolution at low depth of field, requirement needed for the ultrasound probes for the skin, eye and blood vessels diagnostic. In this case it is particularly important to have a transducer with thicknesses below 1 mm.

In the next paragraph we will consider the tape casting process to obtain porous piezoelectric thick layers with these characteristics. Tape casting is in fact the most used technique for the production of ceramic thin layers. These can be than stacked to obtain graded materials less than 1 mm thick. With this process a functionally-graded material can be obtained by sintering layer-stacked green tapes with stepwise varied compositions.

This process has been already used to produce porous piezoelectric multilayers with a sandwich-like structure (dense/porous/dense layers and vice versa) for piroelectric applications (Palmqvist et al., 2007; Shaw et al., 2007).

6. Tape cast porosity-graded piezoelectric ceramics

We have recently reported the optimization needed to produce a flat, 400 μm thick, functionally-graded porous Nb-doped PZT material ($Pb_{0.988}(Zr_{0.52}Ti_{0.48})_{0.976}Nb_{0.024}O_3$, PZTN) by tape casting (Mercadelli et al., 2010). Casting, lamination, debinding and sintering are necessary steps to obtain multilayer structures with this technique. Any variation of each of these steps strongly affects the final product. Therefore each of them (slurry formulation, lamination and thermal treatments) has to be thoroughly investigated.

Combining the sacrificial template method with tape casting, an engineered porosity could be produced. The choice of the right pore former deeply influences the size and morphology of the pores and, as a consequence, the final electrical properties of the piezoelectric product. A preliminary study was conducted to identify the most suitable pore former to obtain micrometric and isotropic porosity. A literature analysis indicated rice starch (RS) and carbon black (CB) as the more appropriate pore formers to obtain fine porosity. Sintered tapes with 30%vol of CB and RS lead to about the same amount of total porosity (35%). However the tapes fracture sections in figure 5 clearly show that CB produces uniform micrometric pores while RS leads to 5-6 μm size pores. The CB was therefore chosen as pore former for the fabrication of porous piezoelectric layers and multilayers.

Fig. 5. SEM micrographs of the fracture section: sintered tape with CB (a) and RS (b). The insertions show the pore former morphology of a) CB and b) RS.

The porosity gradient was formed in situ by sintering layer-stacked green tapes with stepwise varied contents of pore former.

Graded porous piezoceramics were fabricated by gradually increasing CB content, adjusting the binder burnout procedure and tailoring the multilayer thickness. In this way cracks and delaminations were avoided, leading to a well developed and controlled microstructure with porosity ranging from 10 to 30 vol% (Fig. 6).

The preliminary tests done on PZTN-FGM multilayer showed a mean acoustic impedance of 15 10^6 Kg/(m^{2}s) and piezoelectric properties of $S_{11}{}^E$ = 26 10^{-12} m/N, k_p=0.38, d_{31} = -106 10^{-12} m/V. Those data confirmed the potentiality of this material for ultrasonic devices and tape casting as suitable technique to produce porosity-graded piezoelectric ceramics.

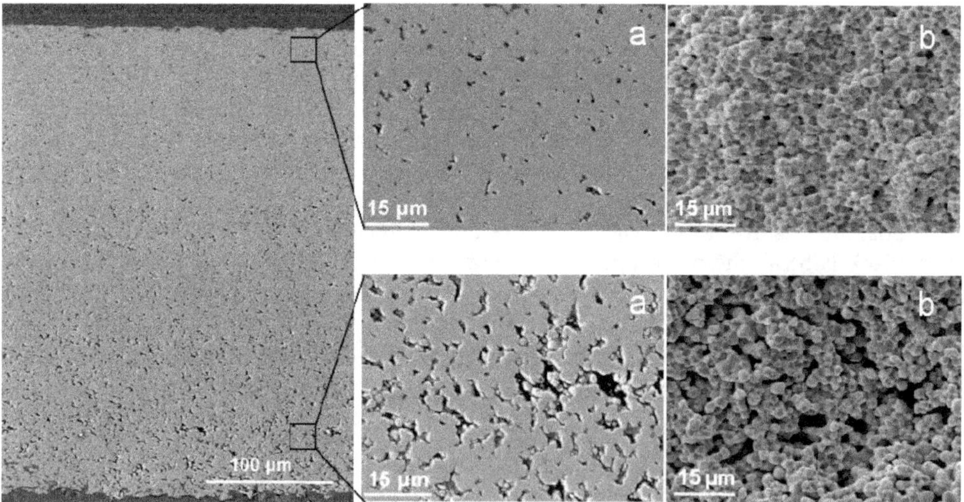

Fig. 6. SEM micrographs of the 6 layers porous graded PZTN material: a) polished surface, b) fracture surface (Mercadelli et al., 2010).

7. Conclusions and feature trends

Porous piezoceramics find nowadays many applications as ultrasonic transducers. The introduction of a controlled porosity into a piezoelectric ceramic could in fact strongly improve the acoustic performances for this kind of applications.

In this chapter the more recent literature on the processing of porous ceramics has been reviewed. The decisive influence of the processing method on the material's microstructure and properties was pointed out as well as the role of the application needed for the choice of the more suitable processing route for the production of porous ceramics. The sacrificial templating methods provide a straightforward way for the fabrication of macroporous ceramics with porosities and average pore sizes ranging from 20% to 90% and 1–700 µm respectively. The possibility to easily tailor the morphology and amount of porosity makes this method effective for the production of porous piezoceramics. The combination of this technique with the tape-casting process allows the production of sub-millimetre porous and porous-graded piezoelectric structure.

Reducing the piezoceramic thickness is a key point to reach higher resonance frequency and, as a consequence, high resolutions medical transducers application. In this respect, the drive toward device miniaturization has created a strong interest in PZT thick-film technology and as a consequence into the screen printing process. With thicknesses in the range 5–80 µm, screen-printed PZT thick films fill an important technological gap between thin-film and bulk ceramics offering the advantage of miniature scale and direct integration into hybrid electronic packages. This technique therefore could be very promising for the production of porosity-graded structure for high-frequency (from 20 to 50 MHz) ultrasonic transducers.

8. References

Akdogan, E.K.; Allahverdi M. & Safari, A. (2005). Piezoelectric composites for sensor and actuator applications, *IEEE Trans. Ultrason., Ferroelect., Freq. Contr.*, 52, 746-775. Republished with permission of Elsevier B.V., Copyright 2010

Araki K. & Halloran J.W. (2005). Porous ceramic bodies with interconnected pore channels by a novel freeze casting technique, *J. Am. Ceram. Soc.*, 88, 1108-1014.

Bowen, C.R.; Perry, A.; Lewis A.C.F. & Kara, H. (2004). Processing and properties of porous piezoelectric materials with high hydrostatic figures of merit, *J. Eur. Ceram. Soc.*, 24, 541–545.

Corbin S.F. & Apte, P.S. (1999). Engineered porosity via tape casting, lamination and the percolation of pyrolyzable particulates, *J. Am. Ceram. Soc.*, 82, 1693-1701.

Craciun, F.; Guidarelli, G.; Galassi C. & Roncari, E. (1998). Elastic wave propagation in porous piezoelectric ceramics, *Ultrasonics*, 36, 427-430.

Eremkin, V.V.; Smotrakov, V.G.; Aleshin V.A. & Tsikhotskii, E.S. (2004). Microstructure of porous piezoceramics for medical diagnostics, *Inorg. Mater.*, 40, 775-779.

Gain, A.K.; Song H. & Lee, B. (2006). Microstructure and mechanical properties of porous yttria stabilized zirconia ceramic using poly methyl methacrylate powder, *Scripta Materialia*, 54, 2081-2085.

Geis, S.; Lobmann, P.; Seifert, S. & Fricke, J. (2000). Dielectric properties of PZT aerogels. *Ferroelectrics*, 241, 1719-1726.

Galassi, C.; Capiani, C.; Craciun F. & Roncari, E. (2005). Water-based technique to produce porous PZT materials, *J.Phys. IV*, 128, 27-31.

Galassi, C.; Snijkers, F.; Cooymans, J.; Piazza, D.; Capiani C. & Luyten, J. (2005). Influence of the pore size and morphology on the piezoelectric properties of PZT material, *Proc. of the Conference PCM*, pp.20-21, October 2005, Bruge.

Gregorová, E.; Pabst W. & Bohacenko, I. (2006). Characterization of different starch types for their application in ceramic processing, *J. Eur. Ceram. Soc.*, 26, 1301-1309.

Gregorová E. & Pabst, W. (2007). Porosity and pore size control in starch consolidation casting of oxide ceramics-Achievements and problems, *J. Eur. Ceram. Soc.*, 27, 669-672.

Gregorová, E.; Zivcová Z. & Pabst, W. (2006). Porosity and pore space characteristics of starch-processed porous ceramics, *J. Mater. Sci.*, 41, 6119-6122.

Holtappels, P.; Sorof, C.; Verbraeken, M.C.; Rambert, S. & Vogt, U. (2006). Preparation of porosity–graded SOFC anode substrates, *FUEL CELLS*, 06, 113–116.

Jin, D.R.; Meng Z.Y. & Zhou, F. (2003). Mechanism of resistivity gradient in monolithic PZT ceramics, *Mater. Sci. Eng. B*, 99, 83-87.

Kaleva, G.M.; Golubko N.V. & Suvorkin, S.V. (2006). Preparation and microstructure of ZrO_2- and $LaGaO_3$-based high-porosity ceramics, *Inorg. Mater.*, 42, 799-805.

Klicker, K.A.; Biggers J.V. & Newnham, R.E. (1981). Composites of PZT and epoxy for hydrostatic transducer applications, *J. Am. Ceram. Soc.*, 64, 5-9.

Kumamoto, S.; Mizumura, K.; Kurihara, Y.; Ohhashi, H. & Okuno, K. (1991). Experimental evaluation cylindrical ceramic tubes composed of porous $Pb(ZrTi)O_3$ ceramics. *Jpn. J. Appl. Phys.*, 30, 2292–2294.

Kumar, B.P.; Kumar H.H. & Kharat D.K. (2005). Study on pore-forming agents in processing of porous piezoceramics, *J. Mater. Sci. Mater. Electr.*, 16, 681-686.

Kumar, B.P.; Kumar H.H. & Kharat, D.K. (2006). Effect of porosity on dielectric properties and microstructure of porous PZT ceramics, *Mater. Sci. Eng. B*, 127, 130–133.

Lee, S.H.; Jun, S.H.; Kim H.E. & Koh, Y.H. (2008). Piezoelectric properties of PZT-based ceramic with highly aligned pores, *J. Am. Ceram. Soc.*, 91, 1912–1915.

Lee, S.H.; Jun, S.H.; Kim H.E. & Koh, Y.H. (2007). Fabrication of porous PZT–PZN piezoelectric ceramics with high hydrostatic figure of merits using camphene-based freeze casting, *J. Am. Ceram. Soc.*, 90, 2807–2813.

Levassort, F.; Holc, J.; Ringgaard, E.; Bove, T.; Kosec M. & Lethiecq, M. (2007). Fabrication, modelling and use of porous ceramics for ultrasonic transducer applications, *J. Electroceram.*, 19, 125–137.

Levassort, F.; Tran-huu-hue, L.P.; Gregoire J.M. & Lethiecq, M. (2004). High frequency ultrasonic transducer, *Material technology and design of integrated piezoelectric devices proceedings*, pp. 53-69, by Polecer, European Thematic Network on Polar Electroceramics, Courmayeur, Italy.

Li, J.F.; Takagi, K.; Ono, M.; Pan, W.; Watanabe, R. & Almajid, A. (2003). Fabrication and evaluation of porous piezoelectric ceramics and porosity-graded piezoelectric actuators. *J. Am. Ceram. Soc.*, 86, 1094-1098.

Li, J.F.; Takagi, K.; Terakubo N. & Watanabe, R. (2001). Electrical and mechanical properties of piezoelectric ceramic/metal composites in the $Pb(Zr,Ti)O_3$/Pt system, *Appl. Phys. Lett.*, 79, 2441-2443.

Lionetto, F.; Licciulli, A.; Montagna F. & Maffezzoli, A. (2004). Piezoceramics: an introductive guide to their practical applications, *Materials & Processes*, 3-4, 107-127.

Martin, L.D. & Minoru, T. (1993). Electromechanical properties of porous piezoelectric ceramics. *J. Am. Ceram. Soc.*, 76, 1697–1706.

Mercadelli E.; Sanson A.; Pinasco P.; Roncari E. & Galassi C. (2010). Tape cast porosity-graded piezoelectric ceramics. *J. Eur. Ceram. Soc.*, 30, 1461-1467.

Montanaro, L.; Jorand, Y.; Fantozzi G. & Negro, A. (1998). Ceramic foams by powder processing, *J. Eur. Ceram. Soc.*, 18, 1339-1350.

Okazaki, K. & Nagata, K. (1973). Effects of grain size and porosity on electrical and optical properties of PLZT ceramics. *J. Am. Ceram. Soc.*, 56, 82-86.

Opielinski K.J. & Gudra, T. (2002). Influence of the thickness of multilayer matching systems on the transfer function of ultrasonic airborne transducer, *Ultrasonics*, 40, 465–469.

Palmqvist, L.; Lindqvist K. & Shaw, C. (2007). Porous multilayer PZT materials made by aqueous tape casting, *Key Eng. Mater.*, 333, 215-218.

Perez, J.A.; Soares, M.R.; Mantas, P.Q. & Senos, A.M.R. (2005). Microstructural design of PZT ceramics. *J. Eur. Ceram. Soc.*, 25, 2207–2210.

Piazza, D.; Capiani C. & Galassi, C. (2005). Piezoceramic material with anisotropic graded porosity, *J. Eur. Ceram. Soc.*, 25, 3075-3078.

Praveenkumar, B.; Kumar H.H. & Kharat, D.K. (2006). Study on microstructure, piezoelectric and dielectric properties of 3-3 porous PZT composites, *J. Mater. Sci. Mater. Electr.*, 17, 515-518.

Roncari, E.; Galassi, C.; Craciun, F.; Capiani C. & Piancastelli A. (2001). A microstructural study of porous piezoelectric ceramics obtained by different methods, *J. Eur. Ceram. Soc.*, 21, 409-417.

Roncari, E.; Galassi, C.; Craciun, F.; Guidarelli G.; Marselli S. & Pavia, V. (1999). Ferroelectric ceramics with included porosity for hydrophone applications, ISAF 1998 in *Proc. of the Eleventh IEEE International Symposium on Applications of ferroelectrics*, pp. 373-376, Ed. E. Colla, D. Damjanovic and N. Setter, IEEE catalog n. 98CH36245 The Institute of Electrical and Electronic Engineers, Ultrasonic, Ferroelectrics and frequency Control Society.

Sanson, A.; Pinasco P. & Roncari, E. (2008). Influence of pore formers on slurry composition and microstructure of tape cast supporting anodes for SOFCs, *J. Eur. Ceram. Soc.*, 28, 1221–1226.

Shaw, C.P.; Whatmore R.W. & Alcock, J.R. (2007). Porous, functionally gradient pyroelectric materials, *J. Am. Ceram. Soc.*, 90, 137-142.

Skinner, D.P.; Newnham R. E. & L. E. Cross, L. E. (1978). Connectivity and piezoelectric-pyroelectric composites, *Mater. Res. Bull.*, 13, 599–607.

Smith, W.A. (1989). Role of piezocomposites in ultrasonic transducers. *Ultrasonics Symposium Proceedings*, pp. 755-766, by IEEE, Piscataway, NJ, United States.

Studart, A.R.; Gonzenbach, U.T.; Tervoort E. & Gauckler L.J. (2006). Processing routes to macroporous ceramics: A review, *J. Am. Ceram. Soc.*, 89, 1771-1789.

Toberer E.S. & Seshadri, R. (2006). Template-free routes to porous inorganic materials, *Chemical Communications*, 30, 3159-3165.

Toberer, E.S. Weaver, J.C. Ramesha K. & Seshadri, R. (2004). Macroporous monoliths of functional perovskite materials through assisted metathesis, *Chem. Mater.*, 16, 2194-2200.

Zeng, T.; Dong, X.; Mao, C.; Zhou, Z. & Yang, H. (2007). Effects of pore shape and porosity on the properties of porous PZT 95/5 ceramics, *J. Eur. Ceram. Soc., 27*, 2025-2029.

Zeng, T.; Dong, X.; Mao, C.; Chen S. & Chen, H. (2006). Preparation and properties of porous PMN-PZT ceramics doped with strontium, *Mater. Sci. Eng. B, 135*, 50-54.

Zeng, T.; Dong, M.L.; Chen H. & Wang, Y.L. (2006). The effects of sintering behaviour on piezoelectric properties of porous PZT ceramics for hydrophone application, *Mater. Sci. Eng. B, 131*, 181-185.

Zhang, H.L.; Li J. & Zhang, B. (2007). Microstructure and electrical properties of porous PZT ceramics derived from different pore-forming agents, *Acta Materialia, 55*, 171-181.

7

Genetic identification of parameters the piezoelectric transducers for digital model of power converter in ultrasonic systems

Pawel Fabijanski and Ryszard Lagoda
Warsaw University of Technology, Institute of Control and Industrial Electronics
Poland

1. Introduction

This chapter is dedicated to ultrasonic piezoelectric ceramic power transducers. These elements are now the most popular source of high power ultrasound and is used in many industrial applications. High power ultrasonic waves are generally used in such industrial processes as welding, acceleration of chemical reactions, scavenging in gas medium, echo sounding and underwater communication (sonar systems), picture transmission, and, above all, ultrasonic cleaning. In practice is now the most widely used the sandwich type power transducers.

Stage design power converters high power ultrasonic devices usually preceded by computer analysis of currents and voltages waveforms the elements of the system, particularly in semiconductor instruments of power. Competent representation requires the use of these waveforms of electrical models of piezoelectric ceramic transducers under the parameters of line with reality and allows to calculate the electrical operating parameters used in the layout of semiconductor switches, capacitors and reactors. The proposed method in the chapter to identify the parameters using a genetic algorithm is one of many possible methods to use here, but has the advantage that it can be implemented automatically.

For example the standard ultrasonic system for cleaning technology (Fig.1.1) includes:

1. ultrasonic generator,
2. transducer or set of transducers,
3. cleaning tank.

Piezoelectric ceramic transducers placed in the tub generate ultrasonic waves that pass through the liquid and reach the element immersed in the tank. As a result, created in the liquid, with very high frequency, alternating areas of high and low pressure. In areas, where low pressure is forming millions of bubbles of vacuum. When the pressure in the alveoli increases and is high enough, bubbles implode, releasing enormous energy at the same time. This phenomenon is called cavitation. Emerging implosions work as a whole series of small cleaning brush. The phenomenon is spreading in all directions and causes intense but controlled detachment of particles of pollutants on the entire surface of cleaning detail. Washed away dirt particles collect on the surface of the cleaning solution from where they are blown into a nearby basin, and then be filtered and recycled.

Ultrasonic cleaning is more effective in cleaning hard materials, than the cleaning of soft or porous materials. It was found that, the harder the surface, including the operation of ultrasound is more efficient. Hence, metals, glass, hard plastics well led by ultrasound and are ideally suited for ultrasonic cleaning.

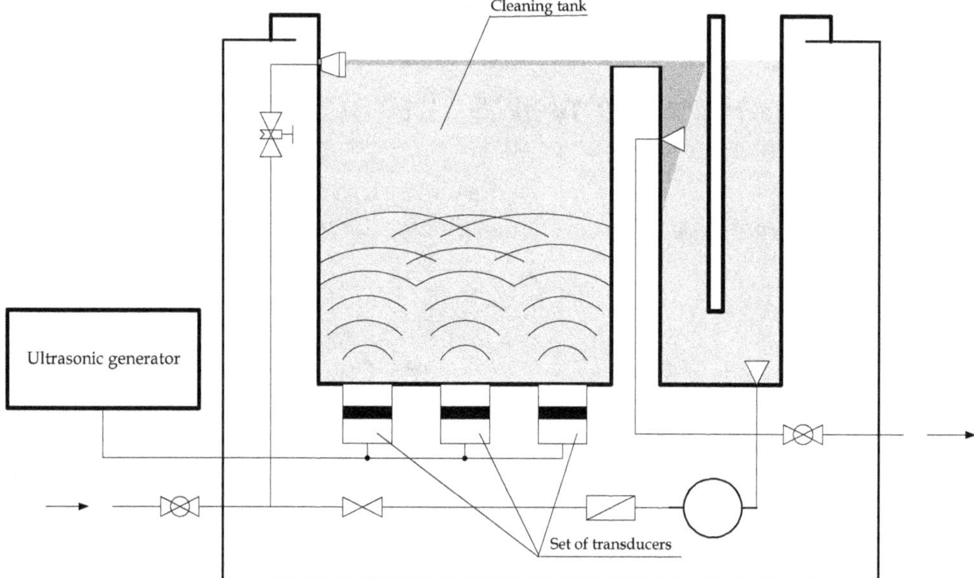

Fig. 1.1 Ultrasonic cleaning system

The quality of ultrasonic cleaning can be improved:
-increasing the ultrasonic power density (cavitation stronger, better cleaning effect),
-reducing the frequency of the generated ultrasound (a stronger phenomenon of cavitations),
-maintaining optimal cleaning bath temperature, the temperature range from 40 °C to 50 °C (cavitation effect is strongest).
By increasing the temperature of the cleaning bath can get more cleaning, but for temperatures ranging from 70 °C to 80 °C, the cleaning effect will be weaker now.

Block diagram and the main circuit of the ultrasonic generator (Fig. 1.2) consists of:
- converter AC/DC,
- full-bridge inverter FBI (T_1-T_4, D_1-D_4),
- isolating transformer T,
- special filter F (L_w, C_w),
- sandwich type transducer PT,
- sensor of vibrations S (optional),
- control unit CU.
The control unit consists of two parts. The first part, in FBI inverter, is the frequency feed-back control loop and the second part in AC/DC converter is the amplitude feed-back control loop.
Two coupling loop works independently. In the case of overload inverter system of current limiter disconnect supply of AC/DC converter. Signal f_{set} make possible to set up manually frequency switching inverter FBI and signal A_{set} establish amplitude ultrasonic oscillation.

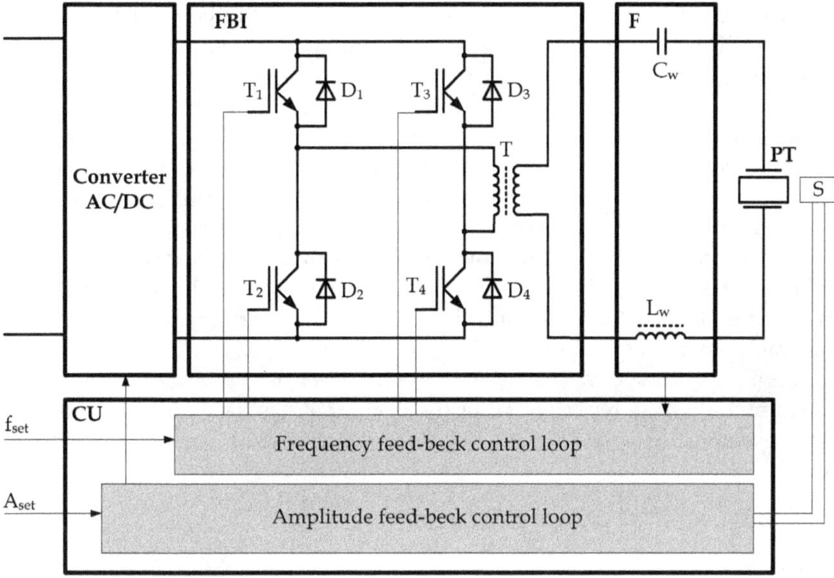

Fig. 1.2 Ultrasonic generator

In most applications, the stabilization of the amplitude of vibration is carried out by other means and do not apply vibration sensor S and the feedback loop. In simple systems, without the exact amplitude of vibration stability, instead of inverter-controlled AC/DC system used free-fall small bridge rectifier.

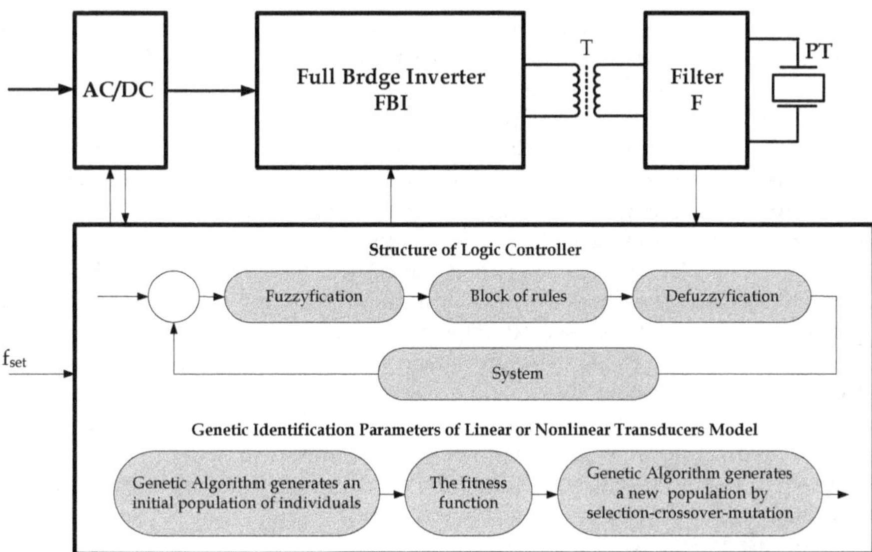

Fig. 1.3 Example of adaptive control system with an ultrasonic generator parameter identification of the transducer member

Tuning the frequency of the inverter output voltage to the FBI, the mechanical resonance of the transducer PT is also implemented in different ways. In large power systems apply adaptive systems, such as fuzzy logic systems, as shown in Figure 1.3.

In the case of untuning, the frequency feed-back control system of the converter change the output voltage frequency and untuning the generator to the mechanical resonance frequency of transducers. The output voltage frequency of converter most by equals the mechanical resonance frequency of the transducers and most by tuning with high precise.

In real circuit the mechanical resonant frequency is function of many parameters of piezoelectric material, among others, the most important are temperature, time, and for industrial cleaning systems the column of cleaning factor, and the surface of the cleaned elements. The frequency feed-back control system for the short power most by very simple, and for the high power very complicated.

The precise microprocessor control systems with an observer often requires knowledge of the electrical model of the replacement of the converter, especially for frequencies near the mechanical resonance. To identification this parameters is used genetic algorithm.

2. Sandwich type ultrasonic transducer

2.1 Construction

The view and construction of ultrasonic transducer by sandwich type is shown in Figure 2.1. This element is made of piezoelectric ceramics PZT and are for example a combination of steel-ceramics-aluminum blocks connected by one or several screws. The transmitter is constructed of two metal blocks between which are clamped piezoelectric ceramic plate (Fig.2.1.b).

This transducer design has much less to its resonant frequency compared to the same frequency of the vibration plate and more importantly allows you to generate ultrasound with high power ratings.

Change torsional screw adjusts the ratio of mechanical stiffness (elasticity), and reduces the weight of the mechanical resonance frequency of the entire transducer. Minor changes to the degree twist allows fine tuning of the resonance frequency transducer to the required value.

Piezoelectric ceramic plates squeeze between two metal blocks also has practical significance. Piezoelectric ceramic material from which the rings are made of a material that is very hard, but unfortunately also fragile. By increasing the stiffness of the mechanical transducer is thus protects the vibrating piezoelectric ceramic rings before cracking and crushing.

As mentioned above, the elements likely to attract the vibrations of sandwich transducers are piezoelectric ceramic plate. On the surface of the contact blocks and metal plates shall be a partial reflection of ultrasonic wave generated in the transducer. The energy associated with this wave is lost as heat and causes a reduction in efficiency of the converter and increasing its temperature. Any change in temperature alters the mechanical stresses occurring in the piezoelectric ceramic plates and changing the mechanical resonance frequency. Losses can be minimized if the acoustic impedance of the metal from which the blocks are made of has a value close to the acoustic impedance of piezoelectric ceramic material.

The sandwich transducers employed environmental overlays hydroacustic metal blocks should be small as possible acoustic impedance, acoustic impedance similar to water. Types of materials used and construction of the transmitter determines, therefore, two contradictory requirements, forcing the designer to compromise. In addition to the tiered sandwich transducers are often used pickups, flat and tubular bimorph.

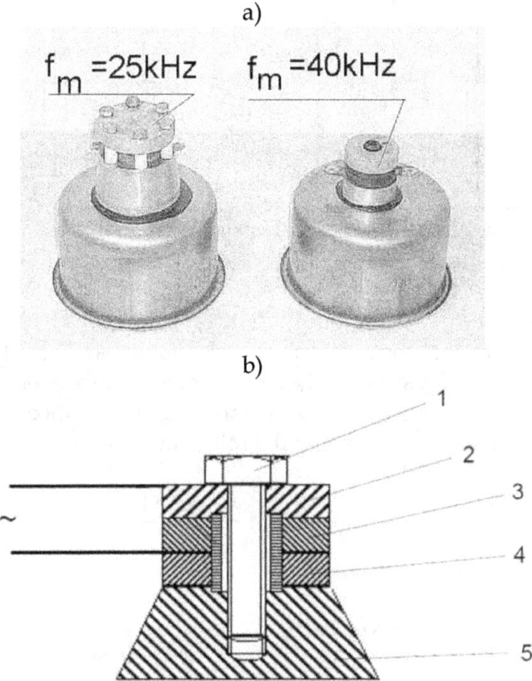

Fig. 2.1 Sandwich type piezoelectric ceramic transducers,
a) view (the resonant frequency about 25 kHz, 40 kHz)
b) construction   1 – screw or pin settings (gripping), transmitter
2.5 – blocks of metal (eg. aluminum, iron, brass)
3.4 – piezoelectric ceramic plates (cylindrical, annular)

2.2 Electric Model

Most assumed that the equivalent electrical circuit of piezoelectric ceramic transducers about the resonant frequency of mechanical vibration consist of connection in parallel resistance R_e, capacity C_e and dynamic RLC elements.

This model is shown in Figure 2.2., where
R_e = resistance of the piezoelectric ceramic plates,
C_e – static, electric capacity of the transducer,
C - equivalent mechanical capacity,
L - equivalent mechanical inductance,
R - equivalent resistance,
$R = R_m + R_a$
where: R_m - equivalent mechanical loss resistance, R_a - equivalent acoustic resistance.

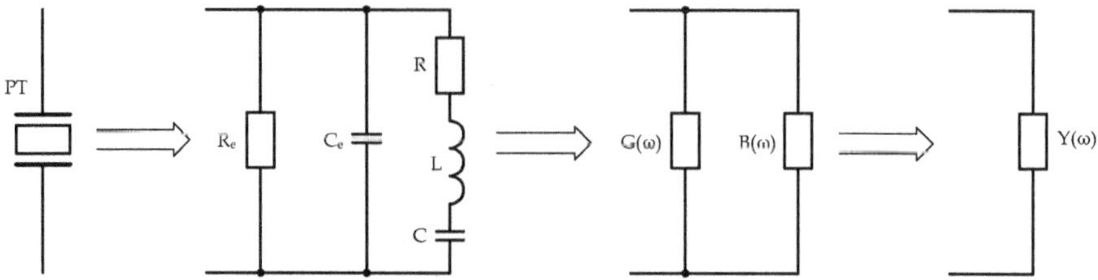

Fig. 2.2 The equivalent circuit of piezoelectric ceramic transducer about the resonant frequency of mechanical vibration

This model must be linear and nonlinear. If the resistance R_m changes as a function of pulsation untuning resonant converter model is nonlinear. Mostly for small untuning assumed that at constant load resistance R_m is constant and the model of transducer is linear.

For linear model the susceptance B, conductance G, admittance Y of the transducers in frequency ω function may be expressed by the following equations:

$$B(\omega) = \text{Im } Y(\omega) = \omega C_e + \frac{\omega C(1 - \omega^2 LC)}{\omega^2 C^2 R^2 + \left(1 - \omega^2 LC\right)^2} \tag{2.1}$$

$$G(\omega) = \text{Re } Y(\omega) = G_e + \frac{\omega^2 C^2 R}{\omega^2 C^2 R^2 + \left(1 - \omega^2 LC\right)^2} \tag{2.2}$$

$$Y(\omega) = \sqrt{G^2 + B^2} \tag{2.3}$$

where $G_e = \dfrac{1}{R_e}$, $\omega = 2\pi f$.

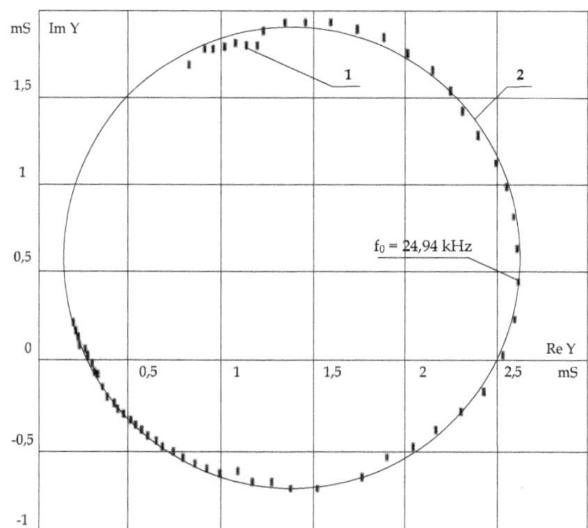

Fig. 2.3 Exemplary admittance characteristics of sandwich type transducers
1. Real characteristic,
2. Equivalent characteristic

The frequency f_0 of mechanical vibration may be calculated by known equation:

$$f_0 = \frac{1}{2\pi}\sqrt{LC} \qquad (2.4)$$

Using this model we can simulate and monitor the operation of the sandwich type transducer. The building is distinguished by its two components: a electrical branch, which includes R_e and C_e, and dynamic branch, which consists of R, C, L elements.
The values R_e, C_e, C, L, R most be calculated from admittance characteristic of transducers (Fig. 2.3), for the off-load and on-load conditions (R_a = var).

3. Application of genetic algorithm
to identify the parameters of the transducer

The electrical characteristics of ceramic transducer resistance R_e and electrical capacitance C_e can be determined with high accuracy on the basis of electrical measurement. Changes in these parameters while the transducer can be considered to be negligible because of the transmitter power is assumed that the C_e and R_e are fixed.
Parameters of dynamic branch R, L, C of power piezoelectric ceramic transducer model electric are not physically measurable. We can only designate empirically. Identification algorithm is therefore based on numerical calculations. Input parameter for these calculations is the image of the actual characteristics of the transducer admittance.

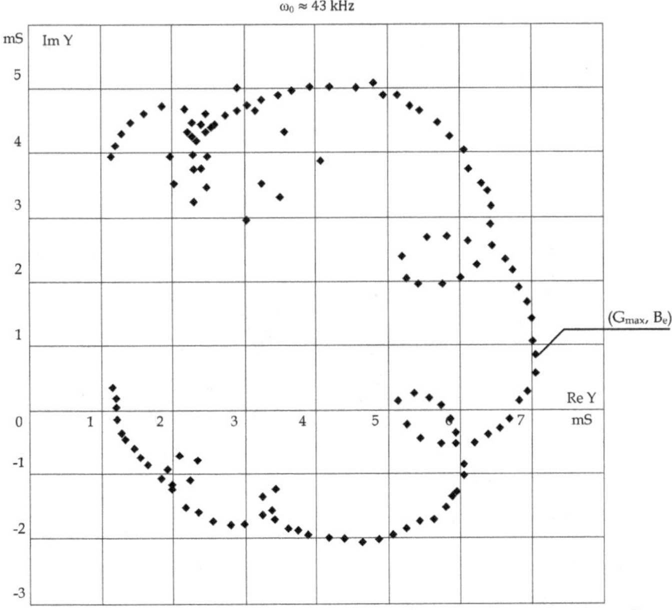

Fig. 3.1. Real admittance characteristics of the piezoelectric ceramic transducer

An example measurements of results for the transducer mechanical resonance frequency equal to about 43 kHz is locate in appendix in Table 1 and one-to-one correspondence real admittance characteristics of this transducer is shown in Fig.3.1. The graph consists of points whose coordinates correspond to the conductance G = Re(Y) and susceptance B = Im(Y) of the converter, measured at a certain frequency.

Analysing the shape of this characteristic can be clearly observed that with increasing frequency (Figure accordance with the frequency increasing clockwise), outlines the main loop of the graph. In the general case, the image of the curve is an ellipse. This ellipse is interpolated electrical admittance characteristics of an ideal replacement transducer schedule shown in Fig.3.1. It may be noted that the actual characteristics, in addition to the main loop also includes many smaller "loops" that testify to the presence of additional resonances in the transducer side. These resonances, however, will not occur in the adopted system replacement transmitter.

After eliminating these "loops" with the actual picture of the actual characteristics of the admittance characteristics of the transducer will be the figure presented in Fig. 3.2.

In Table 1 are marked in bold points, which missed the chart.

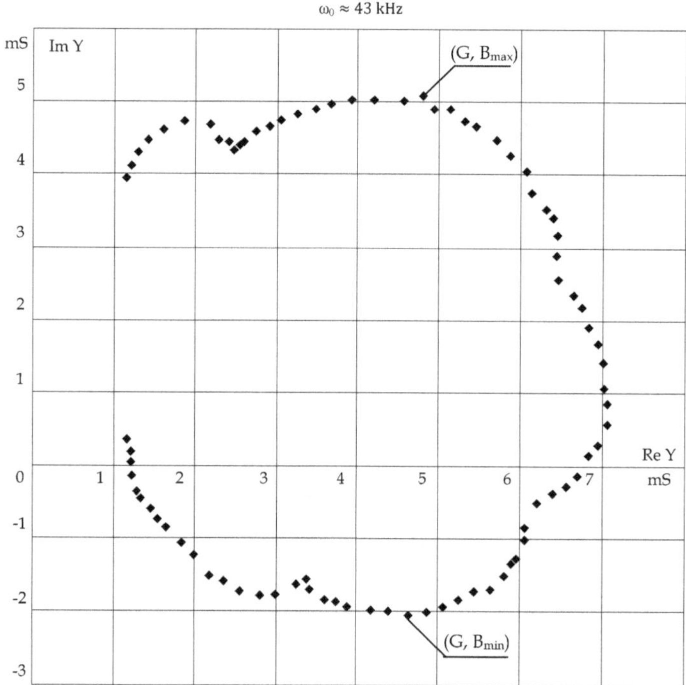

Fig. 3.2 The actual characteristics of the transducer after eliminating resonances fringe

The set of coordinates of the actual approximation characteristics transducer (Fig. 3.2) is a database for further calculations.

3.1 Calculation algorithm

Designation of alternative modes of dynamic parameters of the transmitter requires the implementation of numerical calculations and find such a set of R, L, C parameters, which in a given error will allow mapping of the main loop, the actual characteristics of the transducer admittance presented Fig.3.2.

One element of the identification of these parameters is a genetic algorithm.. Generally one can say that the genetic algorithm is a regula - system learner who makes a certain number of iteration steps. It is a group of strict security procedures that are based on the fundamental mechanisms of biological evolution such as natural selection and inheritance. It works interactively with the environment in discrete time. The algorithm of this type of reproduction may take place subject to the diversity of the population:
- model with preload,
- measures niche

A special feature of this type is that the algorithm is not seeking a single optimal solution, but a group of cooperating the best solutions. At any time, the algorithm works evolutionary principle of survival, which is always available some of the best solutions at the moment. As the proceedings algorithm solution to optimize and adapt to the conditions in which the algorithm works. For further calculations is always the best solutions are selected and rejected solutions are worse. Here there is a process of succession. In order to obtain optimal solutions group for further reproduction of the best solutions are selected at the time of the algorithm. It is a natural selection process occurs in nature. It is known that the probability of obtaining better result is greater if we use it to generate the best available solution than the use of inferior results. According to the law of nature and genetics "survive" the best and strongest.

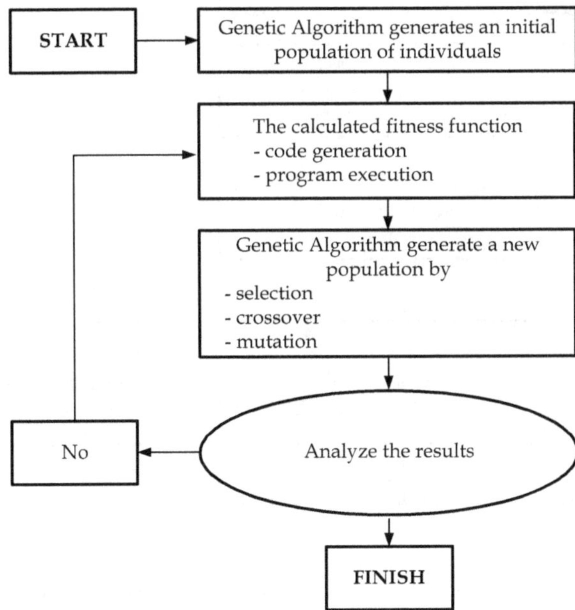

Fig. 3.3 Simply genetic algorithm

The genetic algorithms is a natural process of mating genes. It is no different genetic algorithm. There also are choosing the best "genes" of each solution and verify their combinations. Alongside the cross as a natural evolution is mutation, a random change in the gene. Both these processes are the values of the genetic operator.

Using the genetic algorithm, remember to keep the best balance between the transfer of genes to the next generation, and a draw solution space. Too broad conditions imposed solutions may give erroneous results in spite of every generation the best available solution at the moment. Genetic algorithm is an excellent tool to monitor and maintain the balance between these two dependencies.

The overall pattern of genetic algorithm is illustrated in Fig. 3.3.

Using a genetic algorithm to identify the parameters of dynamic model of the electrical branch of the transducer sandwich working near mechanical resonance is shown in Fig. 3.4.

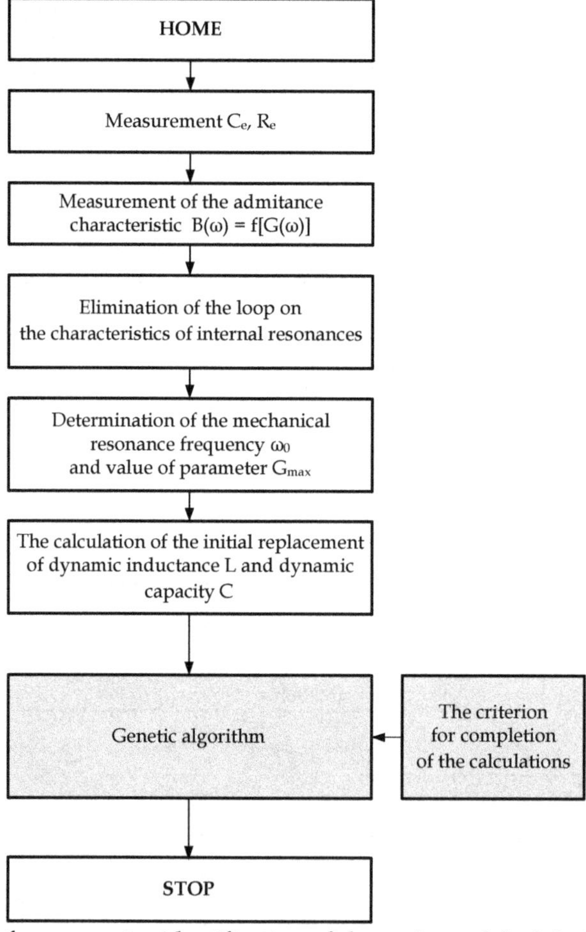

Fig. 3.4 Algorithm for parameter identification of dynamic model of the branch of electrical power ultrasonic transducer

The initial value of the dynamic resistance R is calculated by selecting the characteristics shown in Fig.3.1 point with coordinates (G_{max}, B_e). In this point there is a mechanical resonance of the transducer and is a good approximation condition.

$$\omega_0^2 LC - 1 = 0 \tag{3.1}$$

The dependence (2.2) shows that the resonance:

$$G_{max}(\omega_0) = G_e + \frac{1}{R} \tag{3.2}$$

Therefore:

$$R = \frac{1}{G_{max}(\omega_0) - G_e} \tag{3.3}$$

Choosing the actual characteristics of the admittance of any two points with coordinates (G_1, B_1), (G_2, B_2), which lie beyond the point of mechanical resonance $(\omega_1 \neq \omega_0, \omega_2 \neq \omega_0)$ can calculate the initial value of the replacement of mechanical capacity C.
Transforming the system of equations:

$$G_1 - G_e = \frac{\omega_1^2 C^2 R}{(1 - \omega_1^2 LC)^2 + \omega_1^2 R^2 C^2}$$
$$G_2 - G_e = \frac{\omega_2^2 C^2 R}{(1 - \omega_2^2 LC)^2 + \omega_2^2 R^2 C^2} \tag{3.4}$$

and taking into account that for G_{max} is equal to the pulsation vibrations ω_0 approximate value of C is described by the formula:

$$C' = \frac{1 - \frac{\omega_1^2}{\omega_0^2}}{\omega_1} \sqrt{\frac{(G_1 - G_e)[(G_{max}(\omega_0) - G_e]}{1 - \frac{(G_1 - G_e)}{G_{max}(\omega_0) - G_e}}} \tag{3.5}$$

$$C'' = \frac{1 - \frac{\omega_2^2}{\omega_0^2}}{\omega_2} \sqrt{\frac{(G_2 - G_e)[(G_{max}(\omega_0) - G_e]}{1 - \frac{(G_2 - G_e)}{G_{max}(\omega_0) - G_e}}} \tag{3.6}$$

$$C = \frac{C' + C''}{2} \tag{3.7}$$

The initial value of the replacement of mechanical inductance L determined from the relationship:

$$L = \frac{1}{\omega_0^2 C} \tag{3.8}$$

After substituting the calculated value of the R, L, C and the measurement R_e, C_e the relationship (2.1) and (2.2) is determined by numerical coordinates of the points of the electrical characteristics of the admittance model for the same pulse, for which measurements of actual performance. If the error resulting from a comparison of the approximation characteristic with real characteristic (Fig.3.2) is greater than the accepted values, generate a new population of R_e, C_e, R, L, C parameters, repeat the calculation of coordinates of the electrical characteristics of the model and then analyze the resulting error. Calculations should be continued until the error resulting from a comparison of the characteristics shown in Fig.3.2 with the characteristics of the accepted model of the electrical transducer is smaller than the set value. Measurement error should be performed in the pulsation ω changing between points B_{max} and B_{min} (Fig.3.2).

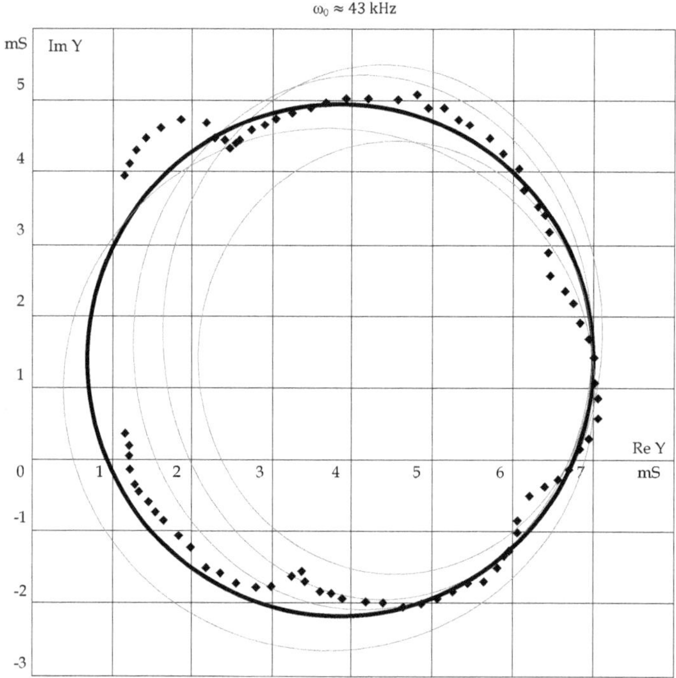

Fig. 3.5 Iterative improvement of the results obtained in numerical calculations and a computer simulation electric model of an sandwich type transducers

Procedure described above can be used and introduces the appropriate algorithm in the DSP program simulation.

//Program piezoelectric ceramic transducer
input {Circuit parameters [R_e,C_e,R,L,C,u(t)];

       Simulation parameters (t_p , t_k ,Δ,υ(0));

       Evolution parameters (Num,Size, NumM,Initial,NumG,stop);)

Output (t, u; t_p; u_o =u(o);};

for (t, t_k, t++){New function form;

start initial population;}
for (integer i=1,Num,i++){New population,
Select the best}

where:
Genotype dimension: Num,
Size of the generation created by mutation: Size,
Number of the best selected genotypes for mutation: NumM,
Number of the best selected genotypes for crossing: NumC,
initial parameter mutation range: Initial

Examples of results obtained in subsequent iterative steps for the transducer mechanical resonance frequency of 42.9 kHz is shown in Fig.3.5.
When the next iterative step will be the condition for the completion of the calculations should be considered that the characteristics of the model shown in Fig. 2.2 parameters R_e, C_e, R, L, C generated in the genetic algorithm in the last population of values, coincides with the actual characteristics of the interpolated transducer. The shape of this characteristic is shown in Fig. 3.6.

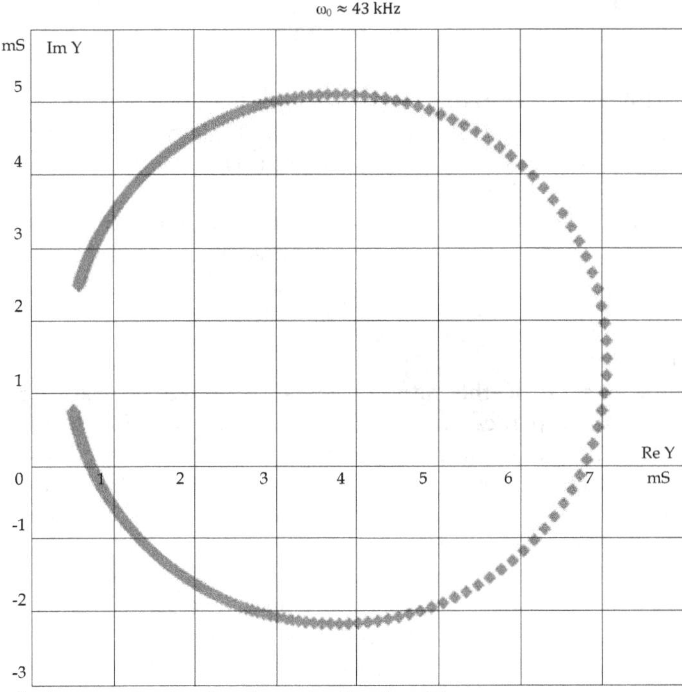

Fig. 3.6 Circle admittance obtained using genetic algorithm

3.2 Effect of the electrical equivalent circuit parameters
on the shape characteristics of the transducer admittance

Location of figures which creates a characteristic admittance of the transducer coordinate system $B(\omega) = f[G(\omega)]$ can be determined by changing the value of capacity C_e and the loss of electrical resistance of ceramics R_e. The higher value of Re the greater displacement figures to the right people along the conductance $G(\omega)$. If you forget in the pattern of replacement sensor resistance R_e ($R_e \rightarrow \infty$) will move to the left approximation of the characteristics. In this case, the graph will be almost tangential to the axis $B(\omega)$. Reduction in the capacity of C_e will move down the sheet, along the axis $B(\omega)$. Conductance G_{max} value determines the width of the main loop of the graph. From relation (3.2) shows that the value of G_{max} decreases when the resistance R increases.

The impact of dynamic capacity C can analyze the shape characteristics of saving the relationship (2.1) in the form

$$B(\omega) \approx \frac{\omega C(1 - \frac{\omega^2}{\omega_0^2})}{\omega^2 C^2 R^2 + (1 - \frac{\omega^2}{\omega_0^2})^2} \tag{3.9}$$

For small, $\pm 2\%$ of the pulsation untuning mechanical resonance, it can be assumed that:

$$B(\omega) \approx \frac{1 - \frac{\omega^2}{\omega_0^2}}{\omega C R^2} \tag{3.10}$$

From relation (3.10) shows that reducing the value of the dynamic capacity C will increase susceptance system. This also applies to the maximum B_{max} and minimum B_{min}. In this case, it means that the values of B_{max} and $|B_{min}|$ increased. The difference $B_{max} - B_{min}$ is equal to the amount of main loop of the graph.

These rule changes in the electrical equivalent circuit parameters of the transducer should be used during the generation of successive populations of a set of values of R_e, C_e, R, L, C genetic algorithm.

4. Conclusion

The actual characteristics of the admittance is a source of transparent, condensed information about the properties and parameters of the piezoelectric ceramic transducer stimulated by the mechanical vibrations. Used to identify the dynamic parameters of R, L, C, and optimization of electrical parameters R_e, C_e occurring electric transducer model, genetic algorithm uses these data and represents one of many possible methods to use here. It has the advantage that it can be realized in automatic cycle.

Based on preliminary data obtained experimentally from measurements of electrical parameters (R_e, C_e) and measurements of voltage and current waveforms created image of the actual transducer admittance characteristics of $B(\omega) = f[G(\omega)]$. Genetic algorithm can find the optimal approximation of the characteristics, correction of the value of R_e, and C_e and calculation of parameters R, L, C model of the electrical industry dynamic.

An important part of the described method is the proper preparation of input data. They should be given in the form of an ordered table of coordinates of points lying on the actual characteristics of the transducer according to those points that do not belong to the main loop of the graph. The process of eliminating points of the loop parasitic resonances can be implemented step by step analysis of the data included in the table of measurement results or by analyzing the image created by a set of characteristic points of $B(\omega) = f\,[G(\omega)]$.

Attempts to identify the parameters of a single transducer mechanical resonance frequency of 23.8 kHz and 43.2 kHz have confirmed the effectiveness of the genetic algorithm. Identification algorithm described above can be particularly useful in studies of larger groups of ultrasound transducers in high-power washing.

By introducing the necessary modifications can also be used to identify the parameters of transducers working in real conditions and to implement control systems that track the frequency of mechanical resonance.

5. References

Davis, L. (1991). *Handbook of Genetic Algorithms*, Van Nostrad Reinhold, New York

Gen, M. & Cheng, R. (1997). *Genetic Algorithms & Engineering Design,* John Wiley & Sons, New York

Wnuk, P. (2004). Genetic optimization of structure and parameters of TSK fuzzy models, *Elektronika 5th International Conference MECHATRONICS 2004*, No. 8-9, (September 2004) page numbers (1-3), ISSN 0033-2089

Rutkowski, L. (2009). Metody i techniki sztucznej inteligencji, Wydawnictwa Naukowo Techniczne PWN SA, ISBN 978-83-01-15731-9, Warsaw

Fabijański, P. & Łagoda, R. (2007). Digital Model of Series Resonant Converter with Piezoelectric Ceramic Transducers and Fuzzy Logic Control, In: *Adaptive and Natural Computing Algorithms - 8th International Conference, ICANNGA 2007, Warsaw, Poland 2007, Proceedings, Part I*, Beliczynski, B.; Dzielinski, A.; Iwanowski, M. & Ribeiro B., page numbers (642-648), Springer, ISBN-13 978-3-540-71589-4, ISBN-10 540-71589-4, Springer Berlin Heidelberg New York

Fabijański, P. & Łagoda, R. (2007). Genetic identification of parameters the piezoelectric ceramic transducers for cleaning system, In: *Recent Advances in Mechatronics*, Jablonski, R.; Turkowski, M. & Szewczyk, R., page numbers (16-21), Springer, ISBN-13 978-3-540-73955-5, Springer Berlin Heidelberg New York

Łagoda, R. & Fabijański, P. (2008). On Line PID Controller Using Genetic Algorithm and DSP PC Board, *Proceding of 13th International Power Electronics and Motion Control Conference EPE-PEMC2008*, CD-ROM, ISBN: 978-1-4244-1742-1 (CD-ROM) , Poland, September 2008, IEEE Catalog Number CFPO834A-CDR, Poznań

6. Appendix

Table 1 shows selected results of detailed studies of the piezoelectric ceramic transducer mechanical resonance frequency of 42.9 kHz. Measurements were performed impedance meter HP 4192 type IMPEDANCE Analyzer. Indicated in bold letter omitted in the genetic algorithm performance admittance points beyond the main loop of the graph. The contents of Table 1 refers to Fig. 3.1, and Fig. 3.2.

f kHz	R kΩ	X kΩ	tg(φ)	φ rad	cos φ	sin φ	G mS	B mS
42,00	0,0721	-0,2338	3,2426	1,27166	0,2947	0,9556	1,204320	4,650189
42,04	0,0674	-0,2132	3,1659	1,26485	0,3012	0,9536	1,346922	4,223453
42,08	0,0703	-0,1925	2,7371	1,22052	0,3432	0,9393	1,674703	4,456680
42,12	0,0841	-0,1745	2,0750	1,12171	0,4341	0,9008	2,241050	4,963926
42,16	0,1006	-0,1809	1,7988	1,06342	0,4859	0,8740	2,347906	4,296001
42,20	0,0921	-0,1763	1,9138	1,08930	0,4631	0,8863	2,328693	3,289479
42,24	0,0888	-0,1483	1,6705	1,03138	0,5136	0,8580	2,971594	3,192256
42,28	0,1148	-0,1358	1,1834	0,86919	0,6454	0,7638	3,630303	3,907909
42,32	0,1515	-0,1403	0,9255	0,74675	0,7339	0,6793	3,554097	3,487659
42,36	0,1501	-0,2015	1,3423	0,93051	0,5974	0,8019	2,378189	3,725892
42,40	0,1047	-0,2014	1,9231	1,09127	0,4614	0,8872	2,032128	3,739428
42,44	0,1432	-0,1502	1,0492	0,80942	0,6899	0,7239	3,323991	4,162891
42,48	0,1212	-0,1918	1,5831	1,00742	0,5340	0,8455	2,353507	4,431376
42,52	0,1227	-0,1868	1,5228	0,98973	0,5489	0,8359	2,455637	4,700036
42,56	0,1041	-0,1800	1,7284	1,04629	0,5008	0,8656	2,408478	4,873619
42,60	0,0992	-0,1666	1,6791	1,03365	0,5117	0,8592	2,639118	4,997220
42,64	0,0980	-0,1478	1,5079	0,98522	0,5527	0,8334	3,116930	4,959969
42,68	0,0973	-0,1351	1,3884	0,94660	0,5844	0,8114	3,510281	4,685777
42,72	0,0974	-0,1229	1,2610	0,90034	0,6213	0,7835	3,962806	4,035905
42,76	0,1005	-0,1081	1,0752	0,82162	0,6810	0,7322	4,613098	2,849480
42,80	0,1057	-0,0924	0,8738	0,71814	0,7530	0,6580	5,362597	1,944590
42,84	0,1139	-0,0752	0,6599	0,58331	0,8346	0,5508	6,115813	2,348010
42,88	0,1295	-0,0571	0,4408	0,41517	0,9150	0,4033	6,464511	2,602020
42,92	0,1542	-0,0514	0,3331	0,32152	0,9488	0,3160	5,838341	1,397040
42,96	0,1582	-0,0704	0,4451	0,41880	0,9136	0,4067	5,274907	-0,305750
43,00	0,1374	-0,0578	0,4207	0,39819	0,9218	0,3878	6,185462	-0,238180
43,04	0,1365	-0,0271	0,1982	0,19569	0,9809	0,1944	7,047725	-0,525000
43,08	0,1509	0,0070	-0,0463	-0,04622	0,9989	-0,0462	6,610636	-1,104520
43,12	0,1880	0,0084	-0,0449	-0,04483	0,9990	-0,0448	5,309238	-1,992950
43,15	0,1585	0,0133	-0,0838	-0,08362	0,9965	-0,0835	6,263456	-1,955190
43,17	0,1577	0,0283	-0,1798	-0,17789	0,9842	-0,1770	6,143423	-1,362000
43,20	0,1697	0,0661	-0,3894	-0,37135	0,9318	-0,3629	5,117705	-1,591100
43,24	0,2042	0,1018	-0,4985	-0,46245	0,8950	-0,4461	3,922151	-1,756280
43,28	0,2579	0,1058	-0,4104	-0,38940	0,9251	-0,3796	3,319092	-1,271460
43,32	0,2395	0,1108	-0,4627	-0,43339	0,9075	-0,4200	3,438489	-0,718400
43,36	0,2638	0,1777	-0,6736	-0,59280	0,8294	-0,5587	2,607243	4,650189
43,40	0,3530	0,2199	-0,6231	-0,55720	0,8487	-0,5288	2,040678	4,223453

Table 1. Results of laboratory tests by the ultrasonic transducer sandwich of the resonance frequency around 43 kHz

8

Lead-free piezoelectric transducers for microelectronic wirebonding applications

K.W. Kwok, T. Lee, S.H. Choy and H.L.W. Chan
Department of Applied Physics, The Hong Kong Polytechnic University
Kowloon, Hong Kong, China

1. Introduction

Lead-free piezoelectric ceramics have been extensively studied recently for replacing the widely used lead-based piezoelectric materials for environmental protection reasons. Among various candidates, potassium sodium niobate-based and bismuth titanate-based ceramics are the two most promising alternatives. In this chapter, the use of $(K_{0.475}Na_{0.475}Li_{0.05})(Nb_{0.92}Ta_{0.05}Sb_{0.03})O_3$ added with 0.4wt% CeO_2 and 0.4wt% MnO_2 (abbreviated KNLNTS) and $0.885(Bi_{0.5}Na_{0.5})TiO_3$-$0.05(Bi_{0.5}K_{0.5})TiO_3$-$0.015(Bi_{0.5}Li_{0.5})TiO_3$-$0.05BaTiO_3$ (abbreviated as BNKLBT) lead-free piezoelectric ceramics as the driving elements of ultrasonic wirebonding transducers is reported. The fabrication and characterization of the ceramics, in the form of ring, and the wirebonding transducers are presented. The ceramic rings are fastened and pre-stressed through a pair of titanium alloy plates in the transducers. The vibration characteristics of the ceramic rings and the transducers are analyzed using a finite-element method (FEM). On the basis of the simulation results, the dimensions of the rings and the titanium alloy plates are determined. The lead-free transducers, operating at a frequency of ~65 kHz, exhibit comparable voltage rise and fall times as the commercial lead zirconate titanate (PZT) transducers. Because of the better matching of the acoustic impedances between the ceramic and titanium alloy, an effective transfer of vibration energy is achieved in the transducers, leading to a large axial vibration (~1.7 μm at 0.1 W). Moreover, the lead-free transducers exhibit a small lateral vibration (0.05 μm at 0.1 W), which is essential for producing a small or narrow bond. The transducers have successfully bonded the aluminum wire on the standard die and gold-plated PCB. The bonds are of good quality, having a smaller deformation ratio (as compared to the commercial PZT transducer) and high bond strength (exceeding the industrial requirement). These clearly show that the lead-free piezoelectric ceramics are promising candidates for replacing the lead-based ceramics as a driving element in the future generation of wire bonders.

2. Lead-free Piezoelectric Ceramics

Lead-based piezoelectric ceramics, represented by $Pb(Zr,Ti)O_3$ (PZT) and PZT-based multi-component materials, have been widely used for piezoelectric actuators, sensors, transducers as well as microelectronic devices due to their excellent piezoelectric properties.

However, they contain a large amount of lead (more than 60 wt%). As lead is highly toxic and will vaporize during processing of the ceramics, there is a rising concern about manufacture as well as disposal of products containing PZT. To date, the lead-free piezoelectric ceramics that have been extensively studied belong to the bismuth-layered and perovsike structures. Bismuth-layered ceramics are featured with low permittivity and high Curie temperature, and hence they have been widely studied for high-temperature sensor applications. For perovskite-type structures, $K_{0.5}Na_{0.5}NbO_3$-based and $Bi_{0.5}Na_{0.5}TiO_3$-based piezoelectric ceramics have attracted considerable attention because of their good piezoelectric properties, and have been considered the most promising alternatives to lead-based piezoelectric ceramics.

$K_{0.5}Na_{0.5}NbO_3$ (abbreviated as KNN) is a solid solution of ferroelectric $KNbO_3$ and antiferroelectric $NaNbO_3$, with the composition K/Na = 50/50 close to the morphotropic phase boundary. It has a high Curie temperature (above 400°C), good ferroelectric properties, and large electromechanical coupling coefficients. A dense and well-sintered KNN ceramic (e.g., prepared by the hot-pressing technique) possesses a high density (ρ= 4.46 g/cm^3) and good piezoelectric properties (piezoelectric coefficient d_{33} = 160 pC/N and planar-mode electromechanical coupling coefficient k_p = 0.45) (Jaeger & Egerton, 1962). However, because of the high volatility of alkaline elements at high temperatures, it is very difficult to obtain dense and well-sintered KNN ceramics using a conventional sintering process, and the ceramics usually exhibit worse piezoelectric properties (d_{33} = 80 pC/N, k_p = 0.36) and low density (ρ= 4.25 g/cm^3). A number of studies have been carried out to improve the sinterability and properties of KNN ceramics; these include the formation of solid solutions of KNN with other ABO_3-type ferroelectrics or nonferroelectrics, e.g., $LiNbO_3$ (Guo et al., 2004), $Bi_{0.5}Na_{0.5}TiO_3$ (Zuo et al., 2007), $BaTiO_3$ (Ahn et al., 2005), $Ba(Ti_{0.95}Zr_{0.05})O_3$ (Lin et al., 2007a), $SrTiO_3$ (Wang et al., 2005), $LiTaO_3$ (Guo et al., 2005), and $LiSbO_3$ (Lin et al., 2007b), the substitutions of analogous ions (e.g., Li^+, Sb^{5+} and Ta^{5+}) for the A-site K^+ and Na^+ or the B-site Nb^{5+} ions (Saito et al., 2004; Lin et al., 2008a), and the use of sintering aids, e.g., CuO (Lin et al., 2008b) and $K_{5.4}Cu_{1.3}Ta_{10}O_{29}$ (Matsubara et al., 2005). Among various KNN-based ceramics, KNN ceramics co-modified with Li and Ta have been widely studied, and good piezoelectric properties have been reported for several compositions, e.g., $(K_{0.5}Na_{0.5})_{0.96}Li_{0.04}(Nb_{0.90}Ta_{0.10})O_3$ and $(K_{0.5}Na_{0.5})_{0.97}Li_{0.03}(Nb_{0.80}Ta_{0.20})O_3$ (Saito et al., 2004). Recently, our studies have shown that CeO_2 and MnO_2 are effective in improving the piezoelectric properties of the $(K_{0.475}Na_{0.475}Li_{0.05})(Nb_{0.92}Ta_{0.05}Sb_{0.03})O_3$ ceramic (Lee et al., 2008; Lee, 2008). For the ceramic added with 0.4wt% CeO_2 and 0.4wt% MnO_2, the piezoelectric and dielectric properties become optimum, having a d_{33} of 200 pC/N, a relative permittivity ε_r of 1150, a loss tangent tan δ of 1.9%, a k_p of 0.43 and a mechanical quality factor Q_m of 80.

$Bi_{0.5}Na_{0.5}TiO_3$ (abbreviated as BNT) is a perovskite-structured ferroelectric having Bi^{3+} and Na^+ complex on the A site of ABO_3-type compounds with a rhombohedral symmetry. It has a high Curie temperature (T_c = 320°C) and strong ferroelectricity, exhibiting a relatively large remanent polarization (P_r = 38 $\mu C/cm^2$) and a high coercive field (E_c = 73 kV/cm) at room temperature (Takennaka et al., 1991). Since its discovery by Smolenskii et al. (1961), many studies have focused on this lead-free ferroelectric ceramic. However, because of its high conductivity and coercive field, the poling of the ceramic is difficult and inefficient. As a result, BNT ceramics usually exhibit much weaker piezoelectric properties (d_{33} = 73-80 pC/N) as compared with $Pb(Zr_{0.52}Ti_{0.48})O_3$ ceramics (d_{33} = 223 pC/N) (Yoshii et al., 2006). In order to improve the poling process and enhance the piezoelectric properties, a number of studies have been carried out; these include the formation of solid solutions of BNT with other ABO_3-type ferroelectrics or nonferroelectrics, e.g. $BNT-BaTiO_3$ (Takennaka et al., 1991; Xu et al., 2008), $BNT-Bi_{0.5}K_{0.5}TiO_3$ (Yoshii et al., 2006), $BNT-KNbO_3$ (Fan et al., 2008), $BNT-BiAlO_3$ (Yu & Ye, 2008), $BNT-SrTiO_3$ (Hiruma et al., 2008), BNT-KNN (Kounga et al., 2008), $BNT-Bi_{0.5}K_{0.5}TiO_3-KNbO_3$ (Fan et al., 2007), $BNT-Bi_{0.5}K_{0.5}TiO_3-BiFeO_3$ (Zhou et al., 2009) and $BNT-Bi_{0.5}K_{0.5}TiO_3-BaTiO_3$ (Nagata et al., 2003; Wang et al., 2004a; Li et al., 2005; Makiuchi et al., 2005), the substitutions of analogous ions for the A-site $(Bi_{0.5}Na_{0.5})^+$ or B-site Ti^{4+} ions, e.g. $(Bi_{1/2}Na_{1/2})Ti_{1-x}(Zn_{1/3}Nb_{2/3})xO_3$ (Zhou & Liu, 2008), $(Bi_{0.5}Na_{0.5})_{0.94}Ba_{0.06}Zr_yTi_{1-y}O_3$ (Yao et al., 2007) and $(Bi_{0.5}Na_{0.5})_{1-1.5x}Bi_xTiO_3$ (Wang et al., 2004b), and the doping of metal oxides, e.g. Ta-doped $0.94BNT–0.06BaTiO_3$ (Zuo et al., 2008) and CeO_2-doped $Bi_{0.5}Na_{0.44}K_{0.06}TiO_3$ (Li et al., 2007). Among various BNT-based ceramics, the ternary systems $(Bi_{0.5}Na_{0.5})TiO_3-(Bi_{0.5}K0.5)TiO_3-BaTiO_3$ have relatively high piezoelectric and good dielectric properties for compositions near the morphotropic phase boundary. Our studies have shown that after the doping of bismuth lithium titanate $(Bi_{0.5}Li_{0.5})TiO_3$, both the electromechanical coupling factors and mechanical quality factors of the ceramics can be further improved and the dielectric loss can be reduced (Choy et al., 2007).

In the following sections, the preparation and characterization of $(K_{0.475}Na_{0.475}Li_{0.05})(Nb_{0.92}Ta_{0.05}Sb_{0.03})O_3$ added with 0.4wt% CeO_2 and 0.4wt% MnO_2 (abbreviated KNLNTS) and $0.885(Bi_{0.5}Na_{0.5})TiO_3-0.05(Bi_{0.5}K_{0.5})TiO_3-0.015(Bi_{0.5}Li_{0.5})TiO_3-0.05BaTiO_3$ (abbreviated as BNKLBT) lead-free piezoelectric ceramics, in the form of ring, will be presented. The ceramic rings will be used, as the driving elements, to fabricate ultrasonic wirebonding transducers.

2.1 Preparation of Ceramic Rings
2.1.1 KNLNTS
The KNLNTS ceramic rings were prepared by a conventional mixed oxide method using analytical-grad metal oxides or carbonate powders: Na_2CO_3 (99.95%), K_2CO_3 (99%), Li_2CO_3 (99%), Nb_2O_5 (99.99%), Ta_2O_5 (99.99%), Sb_2O_5 (99%), MnO_2 (99.99%) and CeO_2 (99.9%). The powders in the stoichiometric ratio of the composition $(K_{0.475}Na_{0.475}Li_{0.05})(Nb_{0.92}Ta_{0.05}Sb_{0.03})O_3$ were first dried and mixed thoroughly in ethanol using zirconia balls for 10 h. After the calcination at 900°C for 2 h, 0.4wt% CeO_2 and 0.4wt% MnO_2 were added. The mixture was then ball-milled again for 10 h and mixed thoroughly with a poly(vinylalcohol) (PVA) binder solution, and then uniaxially pressed into ring samples. The samples were finally sintered at 1140°C for 4 h in air. Silver electrodes were fired on the top and bottom surfaces of the samples. The ceramic samples were poled under a dc field of 4 kV/mm at 150°C in a silicone oil bath for 30 min.

2.1.2 BNKLBT
The conventional mixed oxide technique was used to prepare the BNKLBT ceramic rings. Reagent-grade Bi_2O_3 (99.9%), Na_2CO_3 (99.95%), K_2CO_3 (99%), Li_2CO_3 (99+%), $BaCO_3$ and TiO_2 (99.9%) were used as raw materials. The raw materials were weighed according to the formula $0.885(Bi_{0.5}Na_{0.5})TiO_3-0.05(Bi_{0.5}K_{0.5})TiO_3-0.015(Bi_{0.5}Li_{0.5})TiO_3-0.05BaTiO_3$. The powder was ball milled in ethanol using zirconia balls for 10 h. Calcination was conducted at 800°C for 2 h. After calcination, the mixture was dried and PVA was added as a binder for granulation. The granulated powders were pressed into rings. The compacted rings were sintered at

1170°C for 2 h in air. Silver electrodes were applied on both surfaces of the rings and fired at 650°C. The samples were poled in silicone oil at room temperature under 4 kV/mm for 10 min.

2.2 Material Properties

The KNLNTS and BNKLBT ceramics have been characterized in detail in our previous works (Lee et al., 2009; Chan et al., 2008). Table 1 lists some of the general material parameters of the ceramics, while the material parameters (with conventional symbols) used for the FEM simulations are summarized in Table 2. The material parameters for the PZT (APC 840) ceramic which is currently used in the commercial wirebonding transducers are also listed in Table 1 for comparison. The ceramic rings prepared in this work were characterized and similar results were obtained. The density ρ of the ceramics was measured using the Archimedes method. The relative permittivity ε_r and loss tangent tan δ were measured using an impedance analyzer (HP 4294A) at 1 kHz. The piezoelectric coefficient d_{33} was measured using a piezo-d33 meter (ZJ-30, China).

Material parameters	KNLNTS	BNKLBT	PZT (APC 840)
Density ρ (kg/m³)	4600	5780	7516
ε_r (at 1 kHz)	1150	766	1112
tanδ (%)	1.9	1.61	0.75
k_t	0.32	0.524	0.357
k_p	0.43	0.328	0.59
k_{31}	0.26	0.188	0.32
d_{33} (pC/N)	200	163	254
Q_m (radial)	80	142.1	774.5
Poisson's ratio σ	0.39	0.278	0.395
Young's modulus Y	87	110.5	94.22
Acoustic impedance Za (MRayl)	30	26.4	35.7

Table 1. Piezoelectric, dielectric and mechanical properties of the KNLNTS, BNKLBT and PZT ceramics (Lee et al., 2009; Chan et al., 2008).

	c_{11}^E (GPa)	c_{12}^E (GPa)	c_{13}^E (GPa)	c_{33}^E (GPa)	c_{44}^E (GPa)	c_{66}^E (GPa)
KNLNTS	195	124	64	173	32.7	35.3
BNKLBT	146	53	41.8	147	47.4	46.7
	e_{15} (C/m²)	e_{31} (C/m²)	e_{33} (C/m²)	ε_{11}^S	ε_{33}^S	ρ
KNLNTS	8.4	-16.3	11.4	614	746	4630
BNKLBT	5.9	-3.91	14.6	258	444	5780

Table 2. Material parameters of the KNLNTS and BNKLBT ceramics (Lee et al., 2009; Chan et al., 2008).

3. Ultrasonic Wirebonding Transducers

An ultrasonic wirebonding transducer is a device to provide ultrasonic energy to form intermetallic bond between a fine aluminium/gold wire and the bond pad of a chip in a wirebonding process. The schematic diagram of an Uthe 70PTL ultrasonic wirebonding transducer is shown in Figure 1. The transducer consists of three main parts: an ultrasonic driver, a horn and a wedge. The driver is a bolt-tightened Langevin transducer which is used to convert electrical energy into axial vibration along its length. It consists of four piezoelectric ceramic rings, which are connected electrically in parallel and mechanically in series and sandwiched between the front and back plates by a pre-stressed screw. The driver is operated at its fundamental axial mode (i.e. half-wave resonance) to provide maximum power efficiency. The horn has an exponential profile which couples and amplifies the axial vibration of the driver to provide a maximum vibration across the horn tip. The amplified vibration is then transmitted to the wedge tip, resulting in an oscillatory shear force parallel to the bonding wire and the bond pad surface. The wedge is fixed into a hole drilled at the horn tip by a small screw inserted axially so that it can be replaced for different IC packages and periodic quality assurance. The transducer is designed to operate at the 1.5-wavelength axial mode, i.e. the transducer (including the driver, the horn and the wedge) resonances at the second axial mode. For most of the commercial ultrasonic bonding systems, the operation frequency is around 65 kHz.

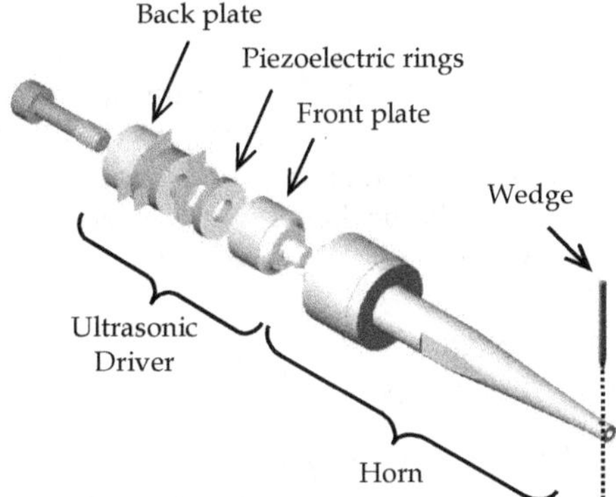

Fig. 1. Explosive diagram of the ultrasonic wirebonding transducer (Lee et al., 2009).

4. Computational Analysis of Vibration Characteristics

The overall performance of the transducer is affected not only by the dimensions of the components (which determine the resonance characteristics), but also by the material properties of the components, including the piezoelectric ceramic rings as the driving element and the front and back plates. Computer analysis of the vibration characteristics of the transducer hence becomes essential before the frication of a transducer with optimum performance. Finite element method (FEM) is a numerical simulation method commonly used for studying the performance of the transducer. In the following sections, FEM modeling, using a commercial code ANSYS 9.0, of the vibration characteristics of the KNLNTS ceramic rings and the corresponding KNLNTS transducers will be reported.

4.1 Ceramic Rings
4.1.1 Vibration Modes

The finite element model of the ceramic ring is shown in Figure 2a. The outer diameter of the KNLNTS ceramic ring was fixed at 13.6 mm (for matching with the commercial transducers), while the inner diameter and thickness were varied in the ranges of 2 to 9 mm and 1.5 to 3.5 mm, respectively. The ring was polarized along the thickness direction, and the electric field was applied on the top and bottom surfaces. The material parameters listed in Table 2 were used for the simulation. In general, five resonance modes are observed in the frequency range of 1 kHz to 2 MHz. The simulated vibration mode shapes at each resonance mode and the (out-plane) surface displacement along the radial direction are shown in Figures 2b-2f. As the KNLNTS ceramic has been assumed to be lossless in the simulations, only the relative displacement is plotted. To confirm the simulation results, the surface displacement of a ceramic ring was also measured using a Polytec laser vibrometer, giving the results plotted in Figure 2 for comparison. It can be seen that mode 1, mode 2 and mode 4 are associated with the three fundamental resonance modes, namely the radial mode, wall thickness mode and thickness mode, respectively. Mode 3 and mode 5 are associated with the complex resonance modes, resulting from the coupling between vibrations along thickness and radial directions. The simulation results have shown that if the wall thickness (outer radius - inner radius) and thickness of the ring are closer, the resonance responses become more complicated.

4.1.2 Dimension Optimization

In practice, the operation of a bolt-tightened Langevin transducer is based on the thickness mode vibration of the driving element. Accordingly, the effective electromechanical coupling coefficient k_{eff} for mode 4 is maximized. k_{eff} is a measure of the conversion efficiency between electrical and mechanical energy at a particular resonance mode, and is defined as (Berlincourt et al., 1964):

$$k_{eff} = \sqrt{1 - \left(f_r / f_a\right)^2} \qquad (3.1)$$

where f_r and f_a are the resonance and anti-resonance frequencies. Based on the simulated impedance spectra, f_r, f_a and hence k_{eff} were determined for ceramic rings with different inner diameters and thicknesses, giving the results shown in Figure 3. It can be seen that the calculated k_{eff} depends on the dimensions of the ring, showing a maximum value at an inner diameter of 5.2 mm and a thickness of 2.3 mm.

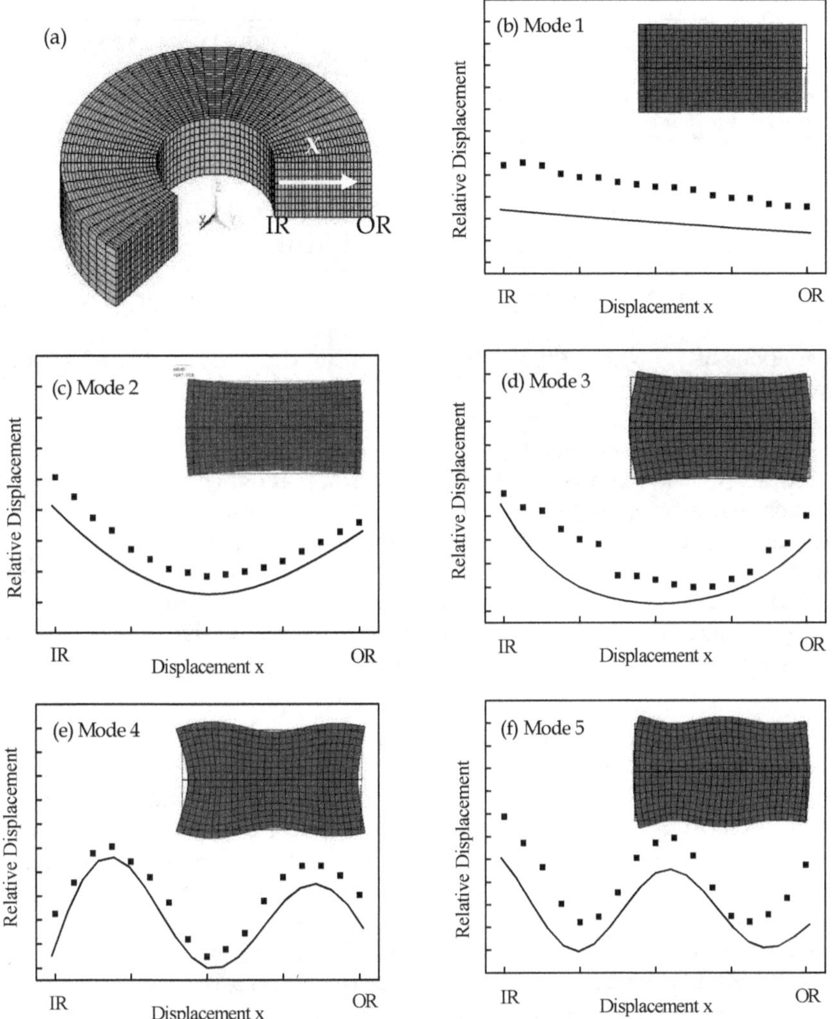

Fig. 2. Vibration mode shapes and the out-plane surface displacement profiles
(■, experimental values; —, FEM values) for a ceramic ring at various resonance modes
(Lee, 2008).

4.2 Wirebonding Transducers

As the transducer (except the wedge) had an axial-symmetry structure about its central (z)
axis, they were modeled as a 360° sector of elements with symmetry boundary conditions
applied the z axis. Similar to the commercial transducers, the front and back plates were
designed to have the same length. The shape and dimensions of the horn and wedge
currently used in the commercial transducer were used. All the components, including the
piezoelectric ceramic ring, were assumed to be lossless in the simulations, and the metal

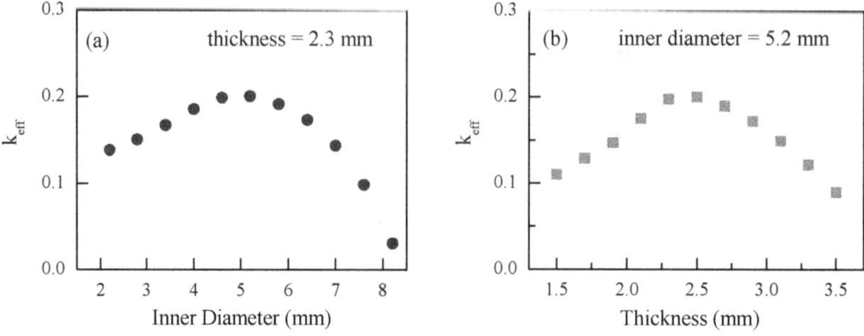

Fig. 3. Variations of k_{eff} with (a) inner diameter and (b) thickness for the KNLNTS ceramic ring (Lee, 2008).

components were further assumed to be isotropic. For the ease of meshing, the thread of the pre-stressed screw was ignored, and the screw was simplified to be a cylindrical rod. Simplification was also made to the front plate where the thread and threaded bore were removed. The ring-shaped copper electrodes were assumed to be very thin, and thus were not included in the simulations. All the components in the assembly were assumed to have perfect mechanical coupling to each other. The h-type approach was applied to the simulations, and good convergence of the results was reached by successively increasing the number of solid elements and the number of nodes.

4.2.1 Optimization of the Front and Back Plates

Stainless steel is widely used for the front and back plates in the commercial PZT transducers. It is not expensive and has an acoustic impedance (46 MRayl) relatively close to that of PZT (Table 1). For transducer applications, a good match of the acoustic impedance between the ceramic driving elements and metal components is important, as it can promote the coupling of ultrasonic energy. As shown in Table 1, the KNLNTS ceramic has a lower acoustic impedance (30 MRayl) as compared to PZT. The coupling of ultrasonic energy may hence become inferior. Therefore, titanium alloy was used for the front and back plates in fabricating the KNLNTS transducer (abbreviated as KNLNTS-Ti transducer). The acoustic impedance of titanium alloy is about 27 MRayl, which is much closer to that of the KNLNTS ceramic. It is expected that the energy coupling and hence the performance of the KNLNTS transducer can be improved. For comparison purposes, KNLNTS transducers using stainless steel front and back plates (abbreviated as KNLNTS-SS transducer) were also fabricated.

Using the material properties listed in Table 2 and the optimum dimensions of the ceramic ring (outer diameter = 13.6 mm, inner diameter = 5.2 mm and thickness = 2.3 mm), the resonance frequencies of the KNLNTS-Ti transducer were simulated and the corresponding resonance modes were examined. The material parameters of the metal parts, including stainless steel and titanium alloy, used for the calculation can be found elsewhere (Lee, 2008). Based on the simulation results, the optimum dimensions of the front and back plates for fabricating a wirebonding transducer with an operation frequency near 65 kHz were determined. The optimum length for the front and back plates for the KNLNTS-Ti transducer is 10.5 m, while that for the KNLNTS-SS transducer is 8 mm. It is noted that the length of the stainless steel plates currently used in the commercial wirebonding PZT transducers is also 8 mm.

4.2.2 Vibration modes

The simulations have showed that there are a large number of resonance modes of the KNLNTS-Ti transducer. For example, 90 modes are found in the frequency range up to 150 kHz. However, some of them, in particular those at high frequencies, are very weak, giving a very small effective electromechanical coupling coefficient (k_{eff} < 0.05). These modes cannot be excited effectively by the electrical driving signal in practice, so they are not considered. On the other hand, some of them are found to be excluded by the electrical boundary conditions and hence cannot be excited in practice.

Table 3 summarizes the effective modes for the KNLNTS-Ti transducer in the vicinity of the operation frequency (65 kHz). The modes are named according to their sequence from the low frequency end. Based on the simulated mode shapes, the modes can be categorized into three main groups; namely the axial mode, the torsional mode, and the complex flexural mode. Mode 34 is the axial mode and has a resonance frequency of 65.798 kHz, suggesting that it is the operation mode. It is the 1.5 wavelength axial mode and has a maximum k_{eff} of 0.317. The deformed shape and the axial profile of the transducer are plotted in Figure 4. The axial motion of the horn tip transmits to the wedge and causes the wedge to vibrate in a flexural manner. As shown in Figure 4b, the axial nodal point is located at the mounting barrel and the axial motion, after amplified by the horn profile, exhibits a maximum axial vibration at the horn tip. Physically, a pure axial excitation (of the ceramic ring) produces an axial front-to-back motion (along the length direction of the transducer) in the ultrasonic horn, which in turn, causes a large displacement at the tip of the wedge. As the motion is essentially in line with the wire to be bonded, the vibration mode is suitable for the operation.

The mode shapes for modes 31 to 38 are shown in Figure 5. Apart from the axial mode, the transducer can vibrate in other complex modes. For example, mode 33 is a flexural mode, making the wedge to vibrate in the left and right direction. At mode 37, the transducer vibrates in the up-and-down direction. As all the modes shown in Table 3 have a non-zero

Mode number	f_r (Hz)	f_a (Hz)	k_{eff}	Mode shape
31	53512	54889	0.220	1λ Axial mode
32	57798	58327	0.130	Complex mode
33	62502	62515	0.020	Flexural mode
34	65798	69395	0.317	1.5 λ Axial mode
35	67146	67358	0.079	Flexural mode
36	67407	67552	0.065	Torsional mode
37	69739	70900	0.180	Flexural mode
38	70019	70019	0.000	Barrel mode

Table 3. Modal results of the KNLNTS-Ti transducer (Lee, 2008).

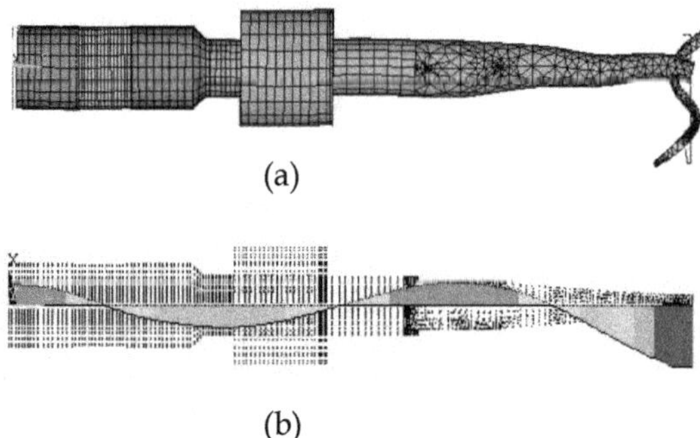

(a)

(b)

Fig. 4. (a) Deformed shape and (b) the axial vibration profile of the transducer at the 1.5-λ axial mode (Lee, 2008; Lee et al., 2009).

k_{eff}, they can be excited in practice. However, the lateral flexural motions from side to side and the torsional motion are not desirable for the applications, as they would tend to deteriorate the bond quality as a result of diminishing the purity of the axial excitation through the loss of the input energy to other vibrations.

5. Lead-Free Ultrasonic Wirebonding Transducers

Similar to KNLNTS, BNKLBT ceramic rings were also used to fabricate the wirebonding transducers with either the titanium alloy front and back plates (abbreviated as BNKLBT-Ti transducer) or the stainless steel front and back plates (abbreviated as BNKLBT-SS transducer). BNKLBT also has a low acoustic impedance (26 MRayl), which is closer to that of titanium alloy (27 MRayl) than PZT (46 MRayl). The BNKLBT ceramic rings had an outer diameter of 12.7 mm, an inner diameter of 5.1 mm and a thickness of 2.3 mm. Following similar procedures, the optimum length for the front and back plates for the BNKLBT-Ti transducer was determined as 10 m, while that for the BNKLBT-SS transducer was 8 mm. For comparison purposes, PZT transducer, with stainless steel front and back plates, (abbreviated as PZT-SS) was fabricated. The PZT ceramic rings (APC 840), supplied by American Piezo Ceramics Inc., are the same as those currently used in the commercial wirebonding transducers. A photograph of the ultrasonic drivers for the KNLNTS-Ti and KNLNTS-SS transducers is shown in Figure 6a, while a photograph of the KNLNTS-Ti transducer is shown in Figure 6b.

5.1 Electrical Characteristics

The electrical impedance and phase spectra (measured using an HP 4294A impedance analyzer) of the KNLNTS-Ti, KNLNTS-SS, BNKLBT-Ti, BNKLBT-SS and PZT-SS transducers are shown in Figure 7. It can be seen that the strongest resonance mode for all the transducers occurs in the frequency range of 62-68 kHz, i.e. in vicinity of the designed operating frequency 65 kHz. At this resonance frequency, the transducer vibrates in the 1.5-λ axial mode, providing a large front-to-back vibration at the tip of the wedge. As compared to the PZT-SS transducer, the lead-free transducers exhibit less spurious vibrations; but the operating mode is weaker and the impedance at resonance frequency (Z_f) is higher. The observed Z_f for the KNLNTS-Ti, KNLNTS-SS, BNKLBT-Ti and BNKLBT-SS transducers are 120 Ω, 200 Ω and 130 Ω and 130 Ω, respectively, while that for the PZT-SS transducer is 20 Ω.

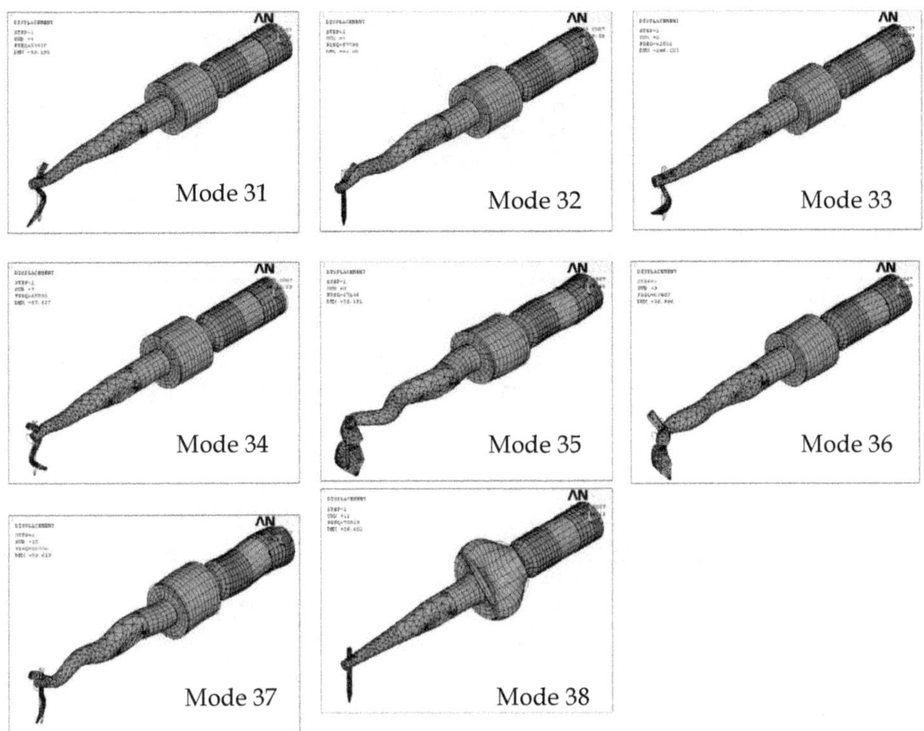

Fig. 5. Mode shapes for the resonance modes around 65 kHz (Lee, 2008).

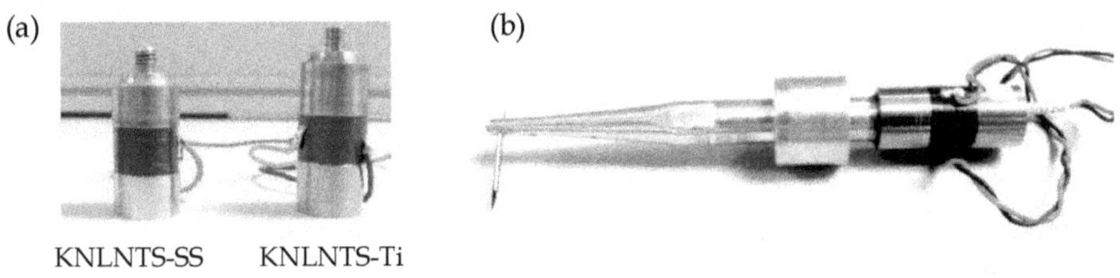

Fig. 6. Photographs of (a) the drivers for the KNLNTS transducers, and (b) the KNLNTS-Ti wirebonding transducer prototype (Lee et al., 2009).

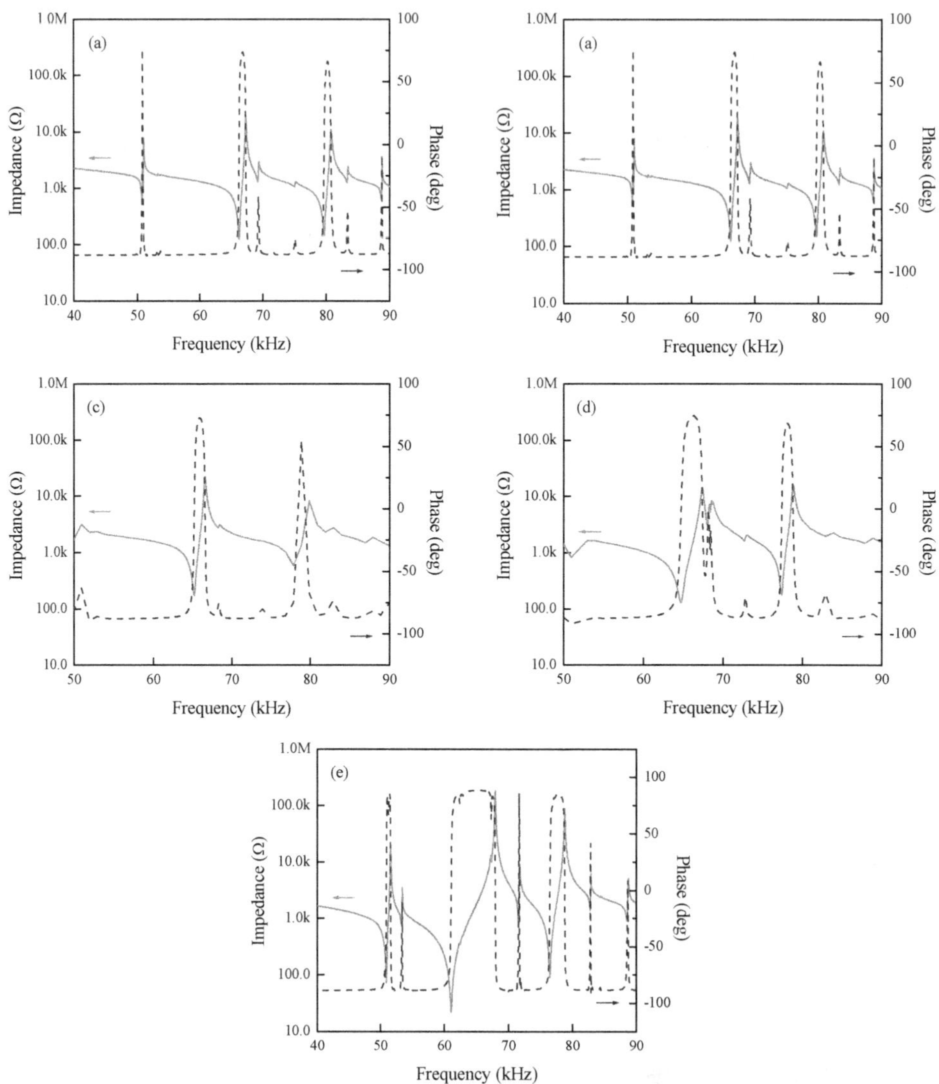

Fig. 7. Frequency plot of the electrical impedance and phase angle for the transducer prototypes: (a) KNLNTS-Ti, (b) KNLNTS-SS, (c) BNKLBT-Ti, (d) BNKLBT-SS; and (e) PZT-SS (Lee et al., 2009; Chan et al., 2008).

5.2 Vibrational Characteristics

As the vibration of the wedge tip is directly related to the wirebonding operation, the vibration characteristic of the wedge was important and hence was investigated. The transducer was affixed to a sample holder with similar clamping conditions to a wirebonding machine. An ultrasonic generator operating in a digital phase-locked-loop

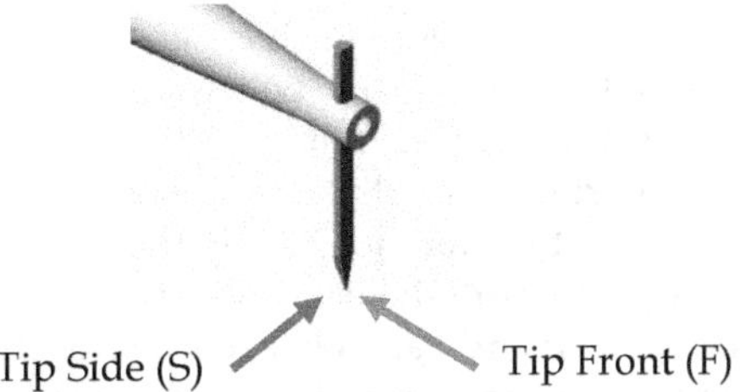

Fig. 8. Schematic diagram showing the positions of the vibration measurement. The arrows indicate the incidence of the laser beam and hence the vibration direction (Lee et al., 2009).

mode was used to drive the transducers at the corresponding resonance frequency and a constant power of 0.1 W. The vibrations at the front of the wedge tip (or tip front) and the side of the wedge tip (or tip side), were measured, using a Polytec laser Doppler vibrometer (OFV-303 and OFV-3001 controller processor). Figure 8 shows schematically the positions of the measurements.

Figure 9 shows, as an example, the temporal profile of the axial displacement at the tip front of the KNLNTS-Ti transducer. During the bonding process, envelop of sine waves at the transducer resonance frequency is sent to the transducer. As it takes time for the transducer

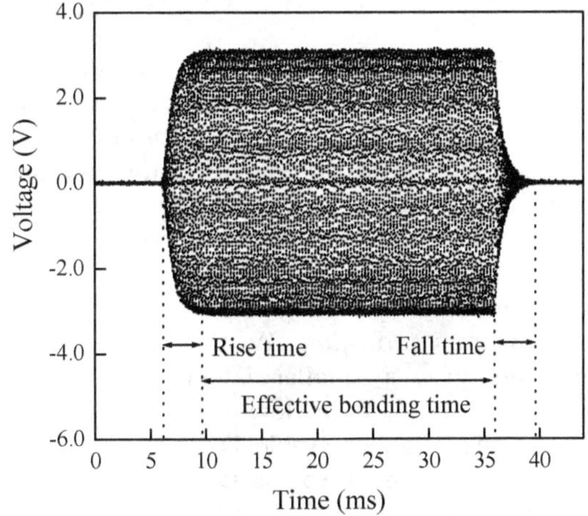

Fig. 9. Axial vibration response at the tip front of the KNLNTS-Ti transducer (Lee, 2008).

to response to the excitation for increasing or decreasing the vibration amplitude to the desired level, there exists response times called the rise time or fall time as shown in the temporal profile (Fig. 9). It has been found that both the rise and fall times for the lead-free transducers are about a few milliseconds, which are comparable to those of the PZT-SS transducer.

The observed vibration amplitudes at different positions of the transducers are compared in

Table 4. It can be seen that the vibration amplitudes at the tip front of the KNLNTS-Ti and BNKLBT-Ti transducers are larger (~65%) than those of the KNLNTS-SS and BNKLBT-SS transducers. This should be attributed to the better match of the acoustic impedances between the lead-free ceramics and the titanium alloy plates. By the converse piezoelectric effect, the piezoelectric ceramic ring converts electrical energy to mechanical energy. The mechanical energy is the source of the vibration, firstly transferred to the metal plates in the driver and then transferred to the horn. The acoustic impedance between the piezoelectric ceramics and the metal plates (front and back plates) is a critical factor that determined the efficiency of energy transfer in the ultrasonic wirebonding transducer. A better match of the acoustics impedances results in an increase in efficiency and hence an improvement in axial vibration performance.

	Tip Front (μm)	Tip Side (μm)	F/S ratio
KNLNTS-Ti	1.728	0.046	37.6
KNLNTS-SS	1.093	0.035	31.2
BNKLBT-Ti	1.669	0.082	20.2
BNKLBT-SS	0.956	0.085	11.3
PZT-SS	1.693	0.191	8.9

Table 4. Vibration amplitudes of the transducers driven at the corresponding resonance frequency and a constant power of 0.1W (Lee et al., 2009; Chan et al., 2008).

With the use of the titanium alloy front and back plates, both the lead-free transducers exhibit a large vibration at the tip front, ~ 1.7 μm, which is very close to that of the PZT-SS transducer (Table 4). However, the PZT-SS transducer produces a larger lateral vibration, giving a larger vibration amplitude at the tip side (0.191 μm) and thus a smaller ratio of the axial displacement to the lateral displacement, especially the F/S ratio. It should be noted that the actual vibrations of the wedge tip during wire-bonding may be different from the measurements on an un-loaded wedge tip (as shown in Table 4). Nevertheless, it is apparent that a large lateral vibration will produce a large bond width which is not desirable for real applications, in particular for miniaturization. As shown in Table 4, both the KNLNTS-Ti and BNKLBT-Ti transducers give a smaller lateral vibration (0.046 and 0.082 μm, respectively). This should be attributed to the low electromechanical coupling coefficient in the lateral direction (k_{31}) of the ceramics (Table 1). The transfer of energy to the lateral vibration at the driving element is small, so the lateral vibration of the tip is effectively reduced. Similar results have been reported for the transducers using 1-3 ceramic/polymer composites as the driving element (Chong et al., 2004). k_{31} of a 1-3 composite is generally reduced, so the performance of the composite transducer is enhanced.

The output power of the ultrasonic wirebonding transducer can be varied in different bonding conditions and the output power is determined by the electrical input power. It is important to have a transducer with high stability for industrial applications. The vibrations at the tip front and tip side for the transducers at different powers are shown in Figures 10. In general, all the vibration displacements increase with increasing power. However, it is seen that the F/S ratio remains almost unchanged at different powers.

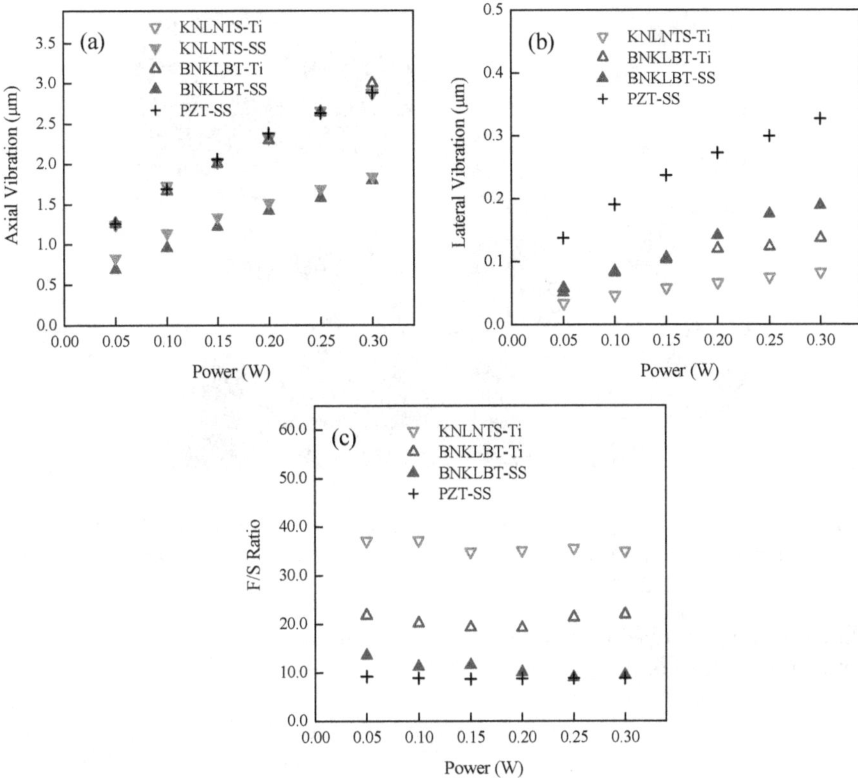

Fig. 10. Variations of (a) the vibration at the tip front, (b) the vibration at the tip side and (c) the F/S ration with input power. (Lee, 2008; Chan et al., 2008).

5.3 Wire-bonding Performance

The KNLNTS-Ti transducer was mounted on an automatic wire bonder (ASM AB530) for evaluating the wire-bonding performance (Fig. 11). A print circuit board (PCB) consisting of a standard die (ASM 96AAA) and a number of gold-plated bond pads was used for the evaluation. In each test, an aluminum wire with a diameter of 25 μm was first bonded on the die (1st bond) and then on the bond pad of the PCB (2nd bond) under a bond force of 0.30 N. The transducer was driven at 320 mW for the 1st bond and 390 mW for the 2nd bond, both for a duration of 20 ms. The quality of the bonded wire was examined by an microscope with a 200-times magnification for a visual inspection and then by a pull tester for a standard pull test. The visual inspection involves visual checking of the appearance and deformation ratio of each bond point, while the pull test is a destructive test aiming to measure the pull force of each bonded wire by pulling it at its loop top vertically until rupture occurs. The term good bond is defined as a wire bond possessing both higher pull force and smaller deformation ratio with smaller standard deviations. The most desirable case is breaking at the wire span followed by breaking at the bond neck. For a poor bond, it may be completely peeled off from the bonding surface in the pull test because of the weak

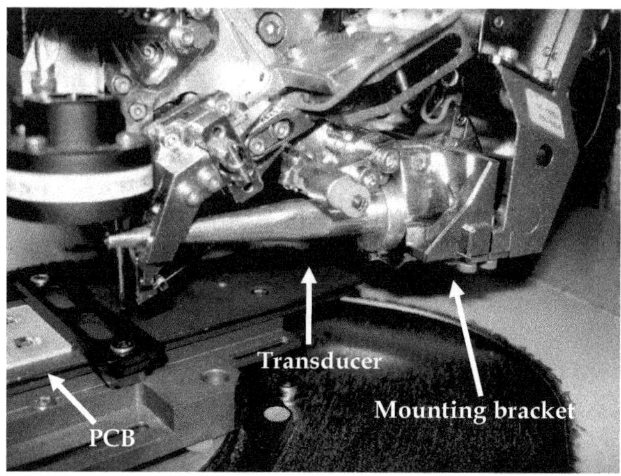

Fig. 11. Photograph showing a KNLNTS-Ti transducer mounted in the bracket of the bondhead assembly of an ASM model AB530 automatic wire bonder for bonding gold-plated printed circuit boards (Lee, 2008).

intermetallic adhesion between the wire and the bonding surface. For the nonstick bond, the wire cannot stick on the bonding surface on the completion of the whole bonding period.

The SEM micrographs of the bonds are shown in Figures 12a-c. The wire deformation is usually characterized by a ratio of the bond width to the diameter of the wire D (25 μm). The deformation ratio produced by the KNLNTS-Ti transducer is about 1.60D, which is relatively small as compared to that produced by the (commercial) PZT-SS transducers (1.94D). This is in good agreement with the vibration amplitude measurements (Table 4) and should be attributed to the small lateral vibration of the KNLNTS-Ti transducer. There was no bond failure during bonding, and most of the failures for the pull test were neck break (Fig. 12d) with a few exception of weld off. This indicates that the adhesion in the interfaces of the bond wire and the bonding surface is good. The averaged pull strength (the maximum force applied to the wire in the pull test) is about 0.5 N, which clearly exceeds the industrial requirement (0.4 N). All these results clearly show that the KNLNTS-Ti transducer is ready for use in the wirebonding applications.

6. Conclusion

Lead-free KNLNTS and BNKLBT piezoelectric ceramic rings have been successfully prepared and used as the driving elements for fabricating ultrasonic wirebonding transducers. In order to improve the energy transfer between different parts of the transducer, titanium alloy has been used to fabricate the front and back plates. The dimensions of the ceramic rings and the titanium alloy plates have been optimized to give an operation frequency of ~65 kHz. Because of the better matching of the acoustic impedances, an effective transfer of vibration energy is achieved in the KNLNTS-Ti and BNKLBT-Ti transducers, leading to a large axial vibration (~1.7 μm at 0.1 W). Probably attributed to the low planar electromechanical coupling coefficient k_{31} (0.26 and 0.188,

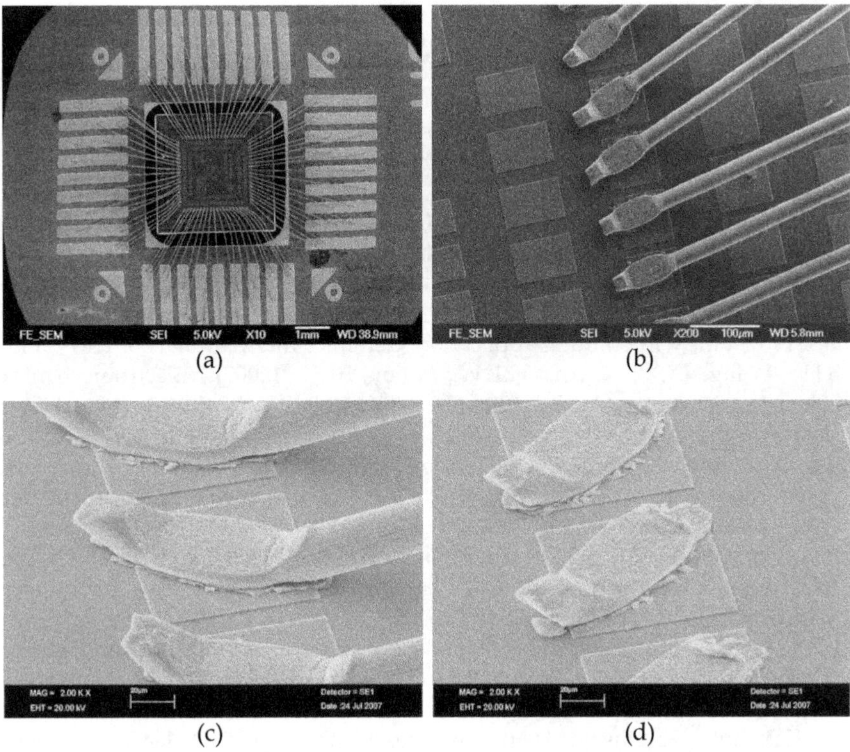

Fig. 12. SEM micrographs of (a) a chip bonded with wires, (b) the top view of the 1st bond, (c) the side view of the 1st bond, and (d) the neck break of 1st bond after the pull test (Lee et al., 2009).

respectively), the lead-free transducers exhibits a small lateral vibration (0.046 and 0.082 μm, respectively, at 0.1 W), which is smaller than that of the PZT transducer (0.191 μm) and is essential for producing a small or narrow bond. Moreover, the KNLNTS-Ti transducer has successfully bonded the aluminum wire on the standard die and gold-plated PCB. The bonds are of good quality, having a small deformation ratio (1.60D vs. 1.94D for the PZT transducer) and high bond strength (> 0.4 N). These clearly show that the KNLNTS and BNKLBT lead-free piezoelectric ceramics are promising candidates for replacing the lead-based ceramics as a driving element in the future generation of wire bonders.

Acknowledgments

The authors would like to acknowledge supports from the Research Committee (A/C code 1-ZV6Y) and the Centre for Smart Materials of The Hong Kong Polytechnic University.

7. References

Ahn, C.W.; Song, H.C.; Nahm, S.; Park, S.H.; Uchino, K.; Priya, S.; Lee, H.G.; Kang, N.K.(2005). Effect of MnO_2 on the piezoelectric properties of $(1-x)(Na_{0.5}K_{0.5})NbO_3$-$xBaTiO_3$ ceramics. Japanese Journal of Applied Physics Part 2 44: L1361-L1364.

Berlincourt, D.A.; Curran, D.R.; Jaffe, H. (1964). Piezoelectric and piezomagnetic materials and their function in transducer. Physical Acoustics. W. P. Mason. New York, Academic Press. 1, 169.

Chan, H.L.W.; Choy, S.H.; Chong, C.P.; Li, H.L.; Liu, P.C.K. (2008). Bismuth sodium titanate based lead-free ultrasonic transducer for microelectronics wirebonding applications. Ceramics International 34: 773-777.

Chong, C.P.; Li, H.L.; Chan, H.L.W.; Liu, P.C.K. (2004). Study of 1–3 composite transducer for ultrasonic wirebonding application. Ceramics International 30: 1141–1146.

Choy, S.H.; Wang, X.X.; Chan, H.L.W.; Choy, C.L. (2007). Electromechanical and ferroelectric properties of $(Bi_{1/2}Na_{1/2})TiO_3$-$(Bi_{1/2}K_{1/2})TiO_3$-$(Bi_{1/2}Li_{1/2})TiO_3$-$BaTiO_3$ lead-free piezoelectric ceramics for accelerometer. Applied Physics A 89: 775-781.

Fan, G.; Lu, W.; Wang, X.; Liang, F. (2007). Morphotropic phase boundary and piezoelectric properties of $Bi_{0.5}Na_{0.5}TiO_3$–$Bi_{0.5}K_{0.5}TiO_3$–$KNbO_3$ lead-free piezoelectric ceramics. Applied Physics Letters 91: 202908.

Fan, G.; Lu, W.; Wang, X.; Liang, F.; Xiao, J. (2008). Phase transition behaviour and electromechanical properties of $(Na_{1/2}Bi_{1/2})TiO_3$-$KNbO_3$ lead-free piezoelectric ceramics. Journal of Physics D 41: 035403.

Guo, Y.P.; Kakimoto, K.; Ohsato, H. (2004). Phase transitional behavior and piezoelectric properties of $(Na_{0.5}K_{0.5})NbO_3$-$LiNbO_3$ ceramics. Applied Physics Letters 85: 4121-4123.

Guo, Y.P.; Kakimoto, K; Ohsato, H. (2005). $(Na_{0.5}K_{0.5})NbO_3$-$LiTaO_3$ lead-free piezoelectric ceramics. Materials Letters 59: 241-244.

Hiruma, Y.; Imai, Y.; Watanabe, Y.; Nagata, H.; Takenaka, T. (2008). Large electrostrain near the phase transition temperature of $(Bi_{0.5}Na_{0.5})TiO_3$-$SrTiO_3$ ferroelectric ceramics. Applied Physics Letters 92: 262904.

Jaeger, R.E. and Egerton, L. (1962). Hot pressing of potassium-sodium niobates. Journal of the American Ceramic Society 45: 209-213.

Kounga, A.B.; Zhang, S.T.; Jo, W.; Granzow, T.; Rodel, J. (2008). Morphotropic phase boundary in $(1-x)Bi_{0.5}Na_{0.5}TiO_3$–$xK_{0.5}Na_{0.5}NbO_3$ lead-free piezoceramics. Applied Physics Letters 92: 222902.

Lee, T. (2008). Development of lead-free ultrasonic wirebonding transducers for microelectronic packaging applications. MPhil Thesis: The Hong Kong Polytechnic University.

Lee, T.; Kwok, K.W.; Chan, H.L.W. (2008). Preparation and piezoelectric properties of CeO_2-added $(Na_{0.475}K_{0.475}Li_{0.05})(Nb_{0.92}Ta_{0.05}Sb_{0.03})O_3$ lead-free ceramics. Journal of Physics D 41: 155402.

Lee, T.; Kwok, K.W.; Li, H.L.; Chan, H.L.W. (2009). Lead-free alkaline niobate-based transducer for ultrasonic wirebonding applications. Sensors Actuator A 150: 267-271.

Li, Y.; Chen, W.; Xu, Q.; Zhou, J.; Wang, Y.; Sun, H. (2007). Piezoelectric and dielectric properties of CeO_2-doped $Bi_{0.5}Na_{0.44}K_{0.06}TiO_3$ lead-free ceramics. Ceramics International 33: 95–99.

Li, Y.M.; Chen, W.; Xu, Q.; Zhou, J.; Gu, X.; Fang, S. (2005). Electromechanical and dielectric properties of $Na_{0.5}Bi_{0.5}TiO_3$–$K_{0.5}Bi_{0.5}TiO_3$–$BaTiO_3$ lead-free ceramics. Materials Chemistry and Physics 94: 328–332.

Lin, D.M.; Kwok, K.W.; Lam, K.H.; Chan, H.L.W. (2007a). Dielectric and piezoelectric properties of $(K_{0.5}Na_{0.5})NbO_3$-$Ba(Ti_{0.95}Zr_{0.05})O_3$ lead-free ceramics. Applied Physics Letters 91: 143513.

Lin, D.M.; Kwok, K.W.; Chan, H.L.W. (2007b). Structure and electrical properties of $K_{0.5}Na_{0.5}NbO_3$–$LiSbO_3$ lead-free piezoelectric ceramics. Journal of Applied Physics 101: 074111.

Lin, D.M.; Kwok, K.W.; Chan, H.L.W. (2008a). Phase transition and electrical properties of $(K_{0.5}Na_{0.5})(Nb_{1-x}Ta_x)O_3$ lead-free piezoelectric ceramics. Applied Physics A 91: 167-171.

Lin, D.M.; Kwok, K.W.; Chan, H.L.W. (2008b). Piezoelectric and ferroelectric properties of Cu-doped $K_{0.5}Na_{0.5}NbO_3$ lead-free ceramics. Journal of Physics D 41: 045401.

Makiuchi, Y.; Aoyagi, R.; Hiruma, Y.; Nagata, H.; Takenaka, T. (2005). $(Bi_{1/2}Na_{1/2})TiO_3$-$(Bi_{1/2}K_{1/2})TiO_3$-$BaTiO_3$-based lead-free piezoelectric ceramics. Japanese Journal of Applied Physics Part 1 44 (6B): 4350-4353.

Matsubara, M.; Kikuta, K.; Hirano, S. (2005). Piezoelectric properties of $(K_{0.5}Na_{0.5})(Nb_{1-x}Ta_x)O_3$-$K_{5.4}Cu_{1.3}Ta_{10}O_{29}$. Journal of Applied Physics 97: 114105.

Nagata, H.; Yoshida, M.; Makiuchi, Y.; Takenaka, T. (2003). Large piezoelectric constant and high Curie temperature of lead-free piezoelectric ceramic ternary system based on bismuth sodium titanate-bismuth potassium titanate-barium titanate near the morphotropic phase boundary. Japanese Journal of Applied Physics Part 1 42 (12): 7401-7403.

Saito, Y.; Takao, H.; Tani, T.; Nonoyama, T.; Takatori, K.; Homma, T.; Nagaya, T.; Nakamura, M. (2004). Lead-free piezoceramics. Nature 432: 84-87.

Smolenskii, G.A.; Isupov, V.A.; Agranovskaya, A.I.; Krainik, N.N. (1961). New ferroelectrics of complex composition. Soviet Physics-Solid State (English Translation) 2 (11): 2651-2654.

Takennaka, T.; Maruyama, K.; Sakata, K. (1991). $(Ba_{1/2}Na_{1/2})TiO_3$-$BaTiO_3$ system for lead-free piezoelectric ceramics. Japanese Journal of Applied Physics Part 2 30 (9B): 2236-2239.

Wang, X.X.; Choy, S.H.; Tang, X.G.; Chan, H.L.W. (2005). Dielectric behavior and microstructure of $(Bi_{1/2}Na_{1/2})TiO_3$–$(Bi_{1/2}K_{1/2})TiO_3$–$BaTiO_3$ lead-free piezoelectric ceramics. Journal of Applied Physics 97: 104101.

Wang, X.X.; Tang, X.G.; Chan, H.L.W. (2004a). Electromechanical and ferroelectric properties of $Bi_{0.5}Na_{0.5}TiO_3$-$Bi_{0.5}K_{0.5}TiO_3$-$BaTiO_3$ lead-free piezoelectric ceramics. Applied Physics Letters 85: 91–93.

Wang, X.X.; Kwok, K.W.; Tang, X.G.; Chan H.L.W.; Choy, C.L. (2004b). Electromechanical properties and dielectric behavior of $(Bi_{1/2}Na_{1/2})_{1-1.5x}Bi_xTiO_3$ lead-free piezoelectric ceramics. Solid State Communications 129: 319-323.

Xu, C.G.; Lin, D.M.; Kwok, K.W. (2008). Structure, electrical properties and depolarization temperature of $(Bi_{0.5}Na_{0.5})TiO_3$-$BaTiO_3$ lead-free piezoelectric ceramics. Solid State Sciences 10: 934-940.

Yao, Y.Q.; Tseng, T.Y.; Chou, C.C.; Chen, H.H.D. (2007). Phase transition and piezoelectric property of $(Bi_{0.5}Na_{0.5})_{0.94}Ba_{0.06}Zr_yTi_{1-y}O_3$ (y = 0 – 0.04) ceramics. Journal of Applied Physics 102: 094102.

Yoshii, K.; Hiruma, Y.; Nagata, H.; Takenaka, T. (2006). Electrical properties and depolarization temperature of $Bi_{1/2}Na_{1/2}TiO_3$-$B_{1/2}K_{1/2}TiO_3$ lead-free piezoelectric ceramics. Japanese Journal of Applied Physics Part 2 45: 4493–4496.

Yu H.; Ye, Z.G. (2008). Dielectric, Ferroelectric and piezoelectric properties of the lead-free $(1-x)(Na_{0.5}Bi_{0.5})TiO_3$-$xBiAlO_3$ solid solution. Applied Physics Letters 93: 112902.

Zhou C.; Liu, X. (2008). Effect of B-Site substitution of complex ions on dielectric and piezoelectric properties in $(Bi_{1/2}Na_{1/2})TiO_3$ piezoelectric ceramics. Materials Chemistry and Physics 108: 413–416.

Zhou, C.; Liu, X.; Li, W.; Yuan, C. (2009). Structure and piezoelectric properties of $Bi_{0.5}Na_{0.5}TiO_3$–$Bi_{0.5}K_{0.5}TiO_3$–$BiFeO_3$ lead-free piezoelectric ceramics. Materials Chemistry and Physics 114: 832–836.

Zuo, R.; Fang, X.; Ye, C. (2007). Phase structures and electrical properties of new lead-free $(Na_{0.5}K_{0.5})NbO_3$-$(Bi_{0.5}Na_{0.5})TiO_3$ ceramics. Applied Physics Letters 90: 092904.

Zuo, R.; Ye, C.; Fang, X.; Li, J. (2008). Tantalum doped $0.94Bi_{0.5}Na_{0.5}TiO_3$–$0.06BaTiO_3$ piezoelectric ceramics. Journal of the European Ceramic Society 28: 871–877.

9

Piezoelectric material-based energy harvesting devices: advances of SSH optimization techniques (1999-2009)

Elie Lefeuvre[1], Mickaël Lallart[2], Claude Richard[2] and Daniel Guyomar[2]
[1]*Univ. Paris-Sud*
[2]*INSA de Lyon*
France

1. Introduction

Energy harvesting, also known as power harvesting or energy scavenging, consists in using ambient energy to power small electronic or electrical devices. That includes thermoelectrics, piezoelectrics and electrodynamics, among other options, which begin now to be used in a growing variety of applications.

This chapter focuses on piezoelectrics, whose use for the purpose of generating electricity is actually not a recent idea (McLean Nicolson A., 1931). Piezoelectric materials have high power densities, but setting up large amount of these materials for high power applications is technically difficult. Consequently, they cannot compete with electromagnetic devices for most of usual applications of electrical energy production. Nonetheless, piezoelectric materials have recently encountered renewed interest for miniaturized energy harvesting devices from ambient vibrations. Drastic increase of international research efforts has been observed since last ten years in this field (Anton et al., 2007), and commercial products are becoming available (Energy Harvesting Forum (Online)). Such energy harvesting devices aim at replacing batteries in very low power electronic devices such as wireless sensors networks and control systems. Their use for recharging batteries of mobile electronic devices is also under development (Krikke, 2005). Maintenance free, self-powered devices for their lifetime has become an industrial objective for the next few years. However, there are further mountains to climb from self-powered devices become commonplace.

A deepened analysis of the papers published during last decade in the field of piezoelectric energy harvesting reveals that significant part of the works focused on mechanical optimization without taking into account constraints of the electrical side of the problem. Contrariwise, another part of these works has studied optimization from the electrical point of view without taking into account the effects reciprocally induced on the mechanical side. As a result, the solutions proposed were often non-optimal and sometimes very far from concerns of practical applications. Effective applications of piezoelectric energy harvesting concept actually require overall analysis of the problem (Mitcheson et al, 2007; Lefeuvre et al, 2009). To highlight all the aspects of this multi-physics optimization problem, this chapter exposes first the different energy conversions involved from ambient vibrations to usable electricity. The effects of couplings which link each stage of the energy conversion process are clearly identified.

Based on generic single degree of freedom model of the electromechanical part, effective methods for analyzing power output and energy conversion cycles of piezoelectric energy harvesting devices have been presented (Guyomar et al, 2005; Shu et al, 2006). Such analysis show in particular the importance of properly processing the current and the voltage delivered by the piezoelectric material for enabling effective energy conversion process. This side of the power optimization problem has been broadly investigated by the authors of this chapter. They have in particular proposed novel techniques which greatly improve energy transfers compared to conventional techniques (Lefeuvre et al, 2004). The so-called SSH techniques (SSH stands for "Synchronized Switching for Harvesting") were derived from piezoelectric semi-passive techniques formerly developed by Richard et al. (Richard et al., 1999) for vibration damping.

Parallel-SSHI and Series-SSHI were the first techniques proposed by the authors for the purpose of energy harvesting. Then, other techniques such as SECE and DSSH were developed (Lefeuvre et al, 2006a; Lallart et al, 2008c). Practical interest of these techniques has been demonstrated through several prototypes, with important efforts for lowering consumption of electronic circuitry needed for implementation (Lefeuvre et al, 2006b; Lallart et al, 2008b). Perspectives of Microsystems applications has motivated developments specifically dedicated to very-low piezoelectric voltages (Makiara et al, 2006; Lallart et al, 2008a; Garbuio et al, 2009).

This chapter presents the evolution of the synchronized switching approach, from the original techniques to the last advances in this domain. The different techniques are compared in term of theoretical performances, and practical implementation issues are discussed. Finally, future trends, challenges and development perspectives for piezoelectric energy harvesting are outlined in the conclusion of this chapter.

2. Energy conversion steps

Take into account of each step of the energy conversion chain appear essential for effective optimization of energy harvesting devices. For this purpose, accurate analysis and modelling of mechanical and electrical interactions related to energy conversions which occur at each stage of the system are required.

General diagram of mechanical-to-electrical energy conversion through piezoelectric effect is depicted in Fig. 1. Input mechanical energy may have various origins, such as shocks or vibrations, with various frequency spectrums. This mechanical energy is transmitted to the piezoelectric material through an important element of the device, so called "mechanical structure", which may act as a band-pass filter (in steady state operation), but also as intermediate mechanical energy storage tank (in impulse mode of operation). Variations of the strain/stress of the piezoelectric material enable conversion from mechanical to electrical energy: an alternating electric charge is generated on the piezoelectric electrodes by converse effect.

It must be noted that resulting AC piezoelectric voltage and current are neither suitable for
energy storage devices nor for small electronic or electrical devices. These devices actually
require DC voltage. So, another element must be included in the energy conversion chain:
the so called "electrical interface", whose basic function is AC to DC conversion of electrical
energy. In addition, this electrical interface may ensure the function of output voltage
regulation. But the most profitable effect which may be induced by this last interface is
overall optimization and improvement of the electromechanical energy conversion.

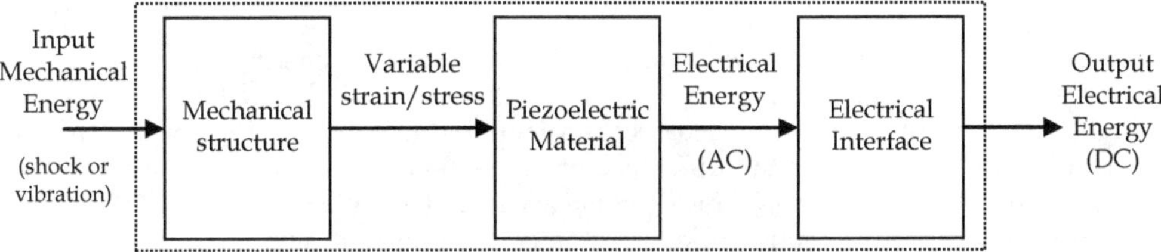

Fig. 1. General energy conversion diagram of piezoelectric energy harvesting devices.

Optimization of the energy conversion process based on the principle of impedance
matching has been proposed by Ottman et al. (Ottman et al., 2002). This principle is actually
a well known method for determining the impedance of the electric load which maximizes
the power of an electrical generator knowing its internal impedance. Indeed,
electromechanical structure of the energy harvesting devices can be considered as
equivalent to an electrical circuit composed of an electrical current generator connected to
an "internal" impedance network. Since actual electric load usually don't behave like the
desired "optimal load", emulation of the optimal load impedance can be achieved using
DC-DC switching mode electronic power converters included in the "Electrical Interface"
block. Several energy-efficient solutions designed for the purpose of emulation of the ideal
load impedance have been proposed (Ottman et al, 2003; Roundy et al, 2004; Lefeuvre et al,
2006b). The SSH optimization techniques, which are based on fundamentally different
approach, are also part of the "Electrical Interface" block.

3. Electromechanical modelling

Large diversity of mechanical structures associated to various piezoelectric materials has been
investigated for energy harvesting. Most common ones are cantilever and diaphragm, such as
depicted on Fig. 2. These structures are mechanically excited by ambient vibrations which
creates the base motion. Forces induced by inertial effect and stiffness of the structure thus
result in strain/stress variations of the piezoelectric material. Such vibrating electromechanical
devices can be modelled as single degree of freedom mass+spring+damper+piezo coupling
near a resonance frequency, as shown on Fig. 3 (Richards et al., 2004; Guyomar et al., 2005;
Badel et al., 2007). In this approach, an effective mass M is bounded on a spring of effective
stiffness K_E, on a damper of viscous coefficient C and on a piezoelectric element characterized
by effective coefficient α and capacitance C_0. These effective coefficients depend on materials
physical characteristics and on design of energy harvesters. They may be derived using modal
analysis (Wang et al., 1999; Badel et al., 2007).

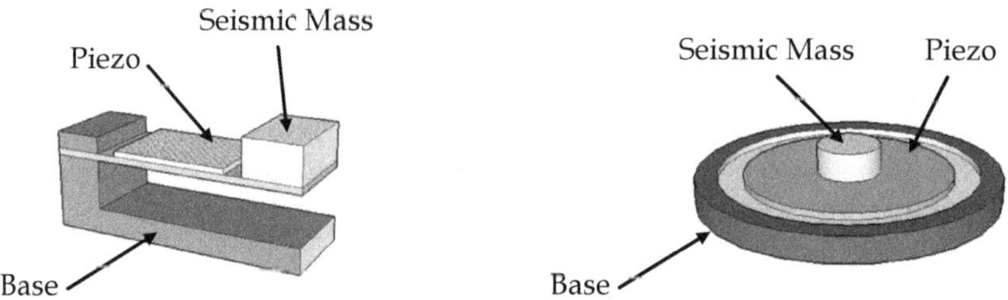

Fig. 2. Cantilever and diaphragm electromechanical structures for energy harvesting.

Dynamic equation of the electromechanical structure is given by (1), where the effective mechanical displacement u is the difference of the seismic mass displacement u_2 and the generator's base displacement u_1. This equation shows in particular the action of the inertial force on the system, which is the product of the effective mass M and the base mechanical acceleration $\ddot{u}_1$. The effective piezoelectric coefficient α defines the relation of the effective piezoelectric force F_P and the piezoelectric voltage V. This coefficient also defines the relation of the piezoelectric internal current I with the mechanical speed $\dot{u}$, as written in (2).

$$M \cdot \ddot{u} + C \cdot \dot{u} + K \cdot u + F_P = M \cdot \ddot{u}_1 \qquad (1)$$

$$\begin{cases} F_P = \alpha \cdot V \\ I = \alpha \cdot \dot{u} \end{cases} \qquad (2)$$

Owing to its simplicity, such an analytical model of the electromechanical part is very convenient for energy conversion analysis. This is actually the simplest way for taking into account all the electromechanical interactions.

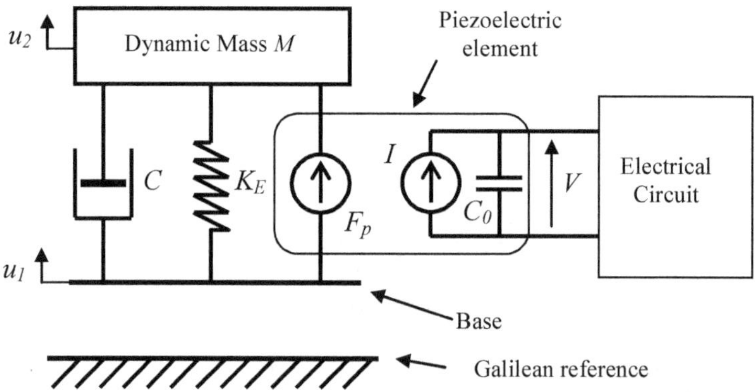

Fig. 3. Single degree of freedom electromechanical model.

4. Standard AC-DC interface

The electrical energy generated by the electromechanical part of the energy harvester, whose modeling was presented in the previous section, is transmitted here to the circuit represented on Fig. 4 (a). The typical application considered here is the recharging of an electrochemical battery. The battery needs stabilized DC voltage while the vibrating piezoelectric element generates an AC voltage (Fig. 4 (b)). So, the electrical interface connected between the piezoelectric element and the battery must ensure electrical compatibility. The so called "standard" interface is the simplest way to achieve this function: the piezoelectric voltage is first rectified by the diode bridge, and then the controller (DC-DC Converter) ensures the power optimization and voltage regulation.

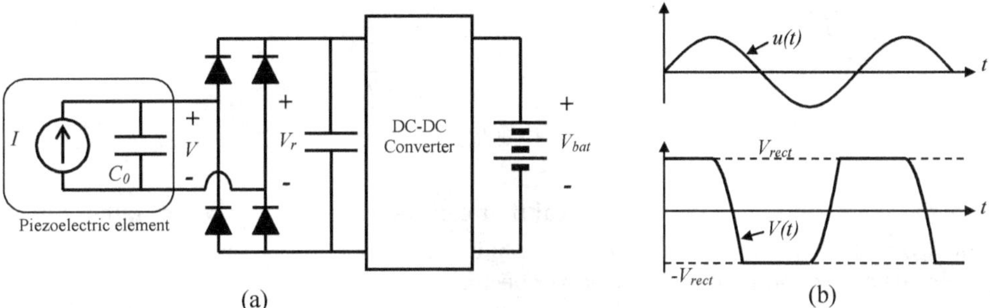

Fig. 4. (a) Standard electrical interface. (b) Typical waveforms of displacement and voltages.

4.1 Steady-state analysis
The model equations (1) and (2) are solved at the resonance frequency of the single-mode resonator. The base excitation of the energy harvester is considered here with sinusoidal displacement u_1 of magnitude U_1 and angular frequency ω (3). As shown on Fig. 4 (b), nonlinearities are induced by the electrical circuit on the piezoelectric voltage V. In spite of this, the displacement u of the electromechanical part is assumed to be sinusoidal. Indeed, effective electromechanical coupling is generally not important enough in practice to observe the impact of the voltage nonlinearities on the displacement.

$$u_1 = U_1 \sin(\omega \cdot t + \varphi) \tag{3}$$

For the sake of simplifying the analysis, the DC part of the electric load composed of the DC-DC converter and the electrochemical battery (Fig. 4 (a)) is modeled in the following developments by an equivalent resistor R which consumes same power as the actual load: $P = V_r^2 / R$. Losses due to the voltage drop across the diodes of the rectifier are neglected. Electrical equations of this circuit (Ottman et al., 2001; Lefeuvre et al., 2006b) and equations (1) (2) (3) lead to the following expression (4) of the power P supplied by the energy harvester at the resonance angular frequency ω_r (Guyomar et al. 2005, Lefeuvre et al. 2006a):

$$P = \frac{R\alpha^2}{\left(RC_0\omega_r + (\pi/2)\right)^2} \cdot \frac{M^2\omega_r^4 U_1^2}{\left(C + 2R\alpha^2/\left(RC_0\omega_r + (\pi/2)\right)^2\right)^2} \tag{4}$$

One or two different values can be found for the optimal load resistance which maximizes the power, depending on the electromechanical figure of merit of the electromechanical part. This figure of merit is defined as the product of the global coupling k squared and the electromechanical quality factor Q_M:

$$k^2 Q_M = \frac{\alpha^2}{\omega_r C C_0} \tag{5}$$

For weak values of the electromechanical figure of merit, $k^2 Q_M \leq \pi$, only one optimal value R_{opt} of the load equivalent resistance maximizes the power. Expressions of R_{opt} and the corresponding maximal power are the following:

$$\begin{cases} R_{opt} = \dfrac{\pi}{2 C_0 \omega_r} \\ P_{max} = \dfrac{\pi}{2} \cdot \dfrac{k^2 Q_M}{\left(\pi + k^2 Q_M\right)^2} \dfrac{M^2 \omega_r^4 U_1^2}{C} \end{cases} \tag{6}$$

For higher values of the electromechanical figure of merit, $k^2 Q_M \geq \pi$, two different optimal value R_{opt} of the load equivalent resistance maximizes the power. In this case, expressions of R_{opt} and the corresponding maximal power become:

$$\begin{cases} R_{opt} = \dfrac{2\alpha - \pi C_0 C \pm 2\alpha \sqrt{\alpha^2 - \pi C_0 C \omega_r}}{2C \left(C_0 \omega_r\right)^2} \\ P_{lim} = \dfrac{M^2 \omega_r^4 U_1^2}{8C} \end{cases} \tag{7}$$

Thus, according to these results, maximization of the power generated by the energy harvester can be summarized as choosing an optimal value for the load equivalent resistance R. The value of this resistor is determined by the electromechanical characteristics of the system and by the excitation frequency.

4.2 Resistor emulation using DC-DC converters

DC-DC switching mode power converters technology is well known for its high efficiency, which typically reach 80% to 99%. They enable accurate dynamic control of their power flow through their active power switches and dedicated regulation circuitry. Thus, such DC-DC converter technology appears ideal for efficient emulation of the optimal load resistor, while transferring most of the harvested energy to the actual load. Control strategy and related regulation circuitry must be designed in such a way that the input voltage and the input current of the converter vary proportionally. Such function is, by principle, achievable with any switching mode converter topology, but with some input and output voltage limitations.

In the context of our application, the input voltage range is directly related to the considered level of vibration, while output voltage range is determined by the characteristics of the electrochemical battery. Ottman et al. have for instance presented solutions for resistor emulation using buck DC-DC converter (Ottman et al., 2002; Ottman et al., 2003). Such converter architecture has intrinsically output voltage lower than input voltage. For this reason, energy harvesting cannot be optimized at low level of vibration. In other words, this converter may be used if the open circuit piezoelectric voltage is most of the time higher than the battery voltage. Conversely, the boost converter architecture has output voltage higher than input voltage. So, it may be used for energy harvesting of very low level vibrations. The buck-boost DC-DC converter architecture represented on Fig 5 (a) offers more degrees of freedom, with output voltage higher or lower than input voltage.

Theoretical study of this circuit and experimental evaluation for optimization of piezoelectric energy harvester has been presented by Lefeuvre et al. (Lefeuvre et al., 2006b). In this study, the control unit of this circuit turns on and off the MOSFET transistor at constant switching frequency. Variation of the duty cycle enables to control the power flow. Compared to other topologies, the main advantage of this buck-boost converter is the possibility of emulation of resistance with extremely simple control circuitry, and hence the power consumption of this circuitry is drastically reduced. Owing to this property, experimental results confirmed the possibility of optimally recharging a 4.8 V NiMh battery (2x2.4 V) with piezoelectric voltage magnitude in the range of 1.6 V to 5.5 V (Fig. 5 (b)). Overall efficiency of this interface, taking into account the rectifier losses, the buck-boost converter losses and the control circuit consumption was found to be between 71% and 79% for output power starting from 200 µW up to 1.5 mW. The emulated load resistance could be fixed at the value which was found to be optimal (12 kΩ) at resonance frequency of the piezoelectric cantilever bimorph (66.4 Hz) with acceleration values of the mechanical excitation in the range of 0.5 to 2 g.

Generally speaking, the main challenge in the design of optimization interfaces and related control circuitry is to reach reasonably high efficiency (Mitcheson et al., 2007). In other words, the electrical power losses are critical and it is generally difficult to reduce them below acceptable level compared to the power generated by the piezoelectric material. The control circuitry often appears to be the most consuming part. For this reason, DC-DC converters which enable load resistance emulation with extremely simple control principles are of great interest.

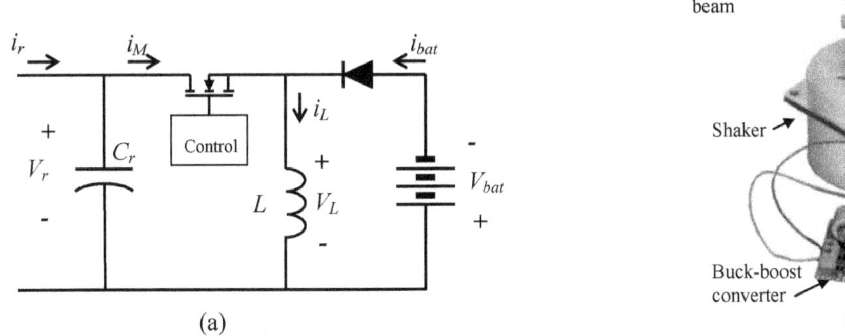

Fig. 5. (a) Buck-boost converter architecture. (b) Experimental setup.

5. The SSH techniques

5.1 Origin of the approach

The SSH techniques have their origins in nonlinear semi-passive techniques formerly developed for structural vibration damping using piezoelectric transducers. The first one, so called SSDS technique, is schematically depicted in Fig. 6 (a). In this simple circuit, the piezoelectric electrodes are connected to a switch. The control principle is to close for a very short time period this switch in each time a maximum of electrostatic energy is reached. As the switching time is very brief, the piezoelectric element is almost always in open-circuit conditions. So, the maxima occur when the displacement reach an extremum (Richard et al., 1999). When the switch is closed, there is therefore a fast dissipation of the electrostatic energy by joule effect in the resistance of the circuit. The typical voltage, displacement and speed waveforms are depicted on Fig. 6 (a) in the case of sinusoidal vibration. Such a dissipation of energy induces damping of the mechanical structure, whose attenuation A_{SSDS} expressed in Eq. (8) is derived from the electromechanical model (1), (2), (3) and (5). According to (8), SSDS attenuation depends on the electromechanical figure of merit defined in (5).

$$A_{SSDS} = \frac{1}{1 + \dfrac{4}{\pi} k^2 Q_M} \tag{8}$$

The second semi-passive damping technique presented here, so called "SSDI" technique, only differs from the previous one by an inductor added in series with the switch, as depicted in Fig. 6 (b). The principle of control is quite similar to SSDS, but instead of removing the piezoelectric charge and resetting the voltage to zero, the effect of the switch is a quick inversion of the piezoelectric voltage obtained by electrical oscillation, as shown on Fig. 6 (b). The time duration of the inversion is typically determined by the half-pseudo-period (9) of the resonant electric circuit composed of the piezoelectric capacitor C_0 and the inductor L. The voltage inversion is not perfect due to internal losses of the circuit, and is characterized by the inversion ratio γ defined in (10). In this second case, the mechanical vibration attenuation A_{SSDI} at the resonance frequency is given in (11) (Richard et al., 2000).

$$t_i = \pi \sqrt{LC_0} \tag{9}$$

$$\gamma = \left| \frac{V_{after}}{V_{before}} \right| = e^{-(\pi/2Q_i)} \tag{10}$$

$$A_{SSDI} = \frac{1}{1 + \dfrac{4}{\pi} \dfrac{1+\gamma}{1-\gamma} k^2 Q_M} \tag{11}$$

Physical interpretation of these nonlinear treatments applied to the piezoelectric voltage is that they create piezoelectric forces that are opposed to the speed and the excitation force at resonance frequency. A deepened analysis shows that the piezoelectric force is similar to a dry friction (badel et al. 2007). But the difference with pure dry friction is that the piezoelectric force magnitude is proportional to the displacement magnitude. The magnitude of this force depends on the electromechanical figure of merit, but also on the inversion ratio in the case of the SSDI, whose inductor artificially magnifies the piezoelectric voltage amplitude.

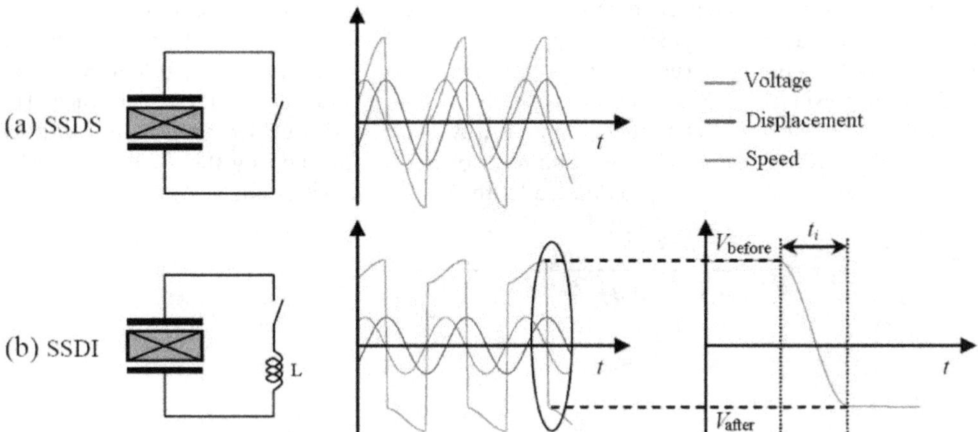

Fig. 6. Block schematic and associated waveforms of (a) SSDS and (b) SSDI techniques.

5.2 Parallel SSHI technique

The parallel SSHI technique was the first SSH method derived from SSD (Lefeuvre et al., 2004; Guyomar et al., 2005). In comparison to the standard AC-DC interface described in section 4, the switching system of the SSDI circuit (Fig. 6 (b)) is simply added in parallel with the piezoelectric element to obtain the parallel SSHI circuit represented in Fig. 7 (a).

Typical voltage waveforms are shown in Fig. 7 (b) in the case of sinusoidal mechanical excitation of the energy harvester. At time t_0, the displacement u reaches a minimum and the bridge cease to conduct. This coincides with the beginning of the piezoelectric voltage inversion through the switch S and the inductor L. Then, the voltage V varies proportionally to the displacement u until the DC voltage V_R is reached. So, the rectifier bridge conducts and transfers electrical energy to the battery. Half of the energy conversion cycle ends at time t_2, and the second half-cycle happens symmetrically, until the end of the period.

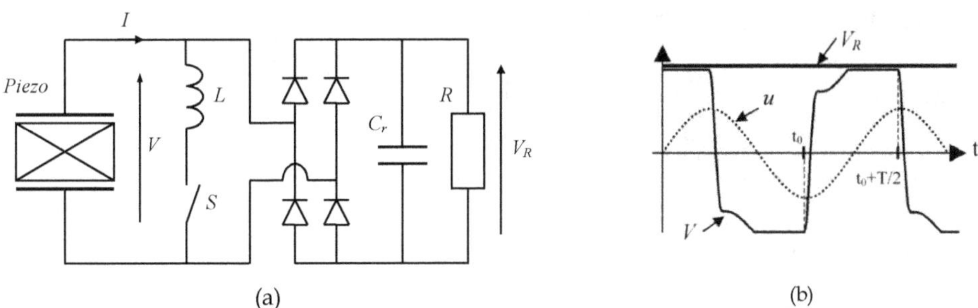

Fig. 7. (a) Circuit schematic and (b) typical waveforms of the parallel SSHI.

As in the case of the standard AC-DC interface, the DC load is modelled by a resistor R for the sake of simplifying calculation of the generated power. However, operation of the energy harvester is unmodified if the load resistor is replaced by a DC-DC converter followed by a battery, as depicted in Fig. 4 (a) in the case of the standard circuit.

Electrical equations of the parallel SSHI circuit and electromechanical equations (1), (2) with sinusoidal base excitation displacement (3) lead to the expression (12) of the power P supplied by the energy harvester at the resonance angular frequency ω_r (Guyomar et al. 2005). This power reaches a maximum for the optimal value of the load resistance. The optimal load resistance cannot be analytically expressed but its value can only be numerically derived from (12). The maximum of power supplied by the energy harvester with its optimal load is well approximated by the analytical expression (13).

$$P = \frac{4R\alpha^2}{\left(RC_0\omega_r(1-\gamma)+\pi\right)^2} \frac{M^2\omega_r^4 U_1^2}{\left(C + \frac{4R\alpha^2}{\pi}\frac{RC_0\omega_r(1-\gamma^2)+2\pi}{\left(RC_0\omega_r(1-\gamma)+\pi\right)^2}\right)^2} \tag{12}$$

$$P_{max} \approx \frac{k^2 Q_M}{\pi(1-\gamma)+8k^2 Q_M}\frac{M^2\omega_r^4 U_1^2}{C} \tag{13}$$

5.3 Series SSHI technique

The series SSHI is similar to the parallel SSHI technique presented previously, but the switching system is connected in series to the piezoelectric element instead of being connected in parallel. The circuit used by this technique is represented on Fig. 8 (a). It was found by another approach and described under another name by Taylor et al. (Taylor et al., 2001). The switching system is operated in the same way as for the parallel SSHI technique. Thus, the piezoelectric element is always in open-circuit configuration except during the voltage inversion sequences. So, the piezoelectric voltage variation is an image of the mechanical displacement u except when the switching device is turned on. Typical voltage and displacement waveforms in case of sinusoidal mechanical excitation are shown in Fig. 8 (b).

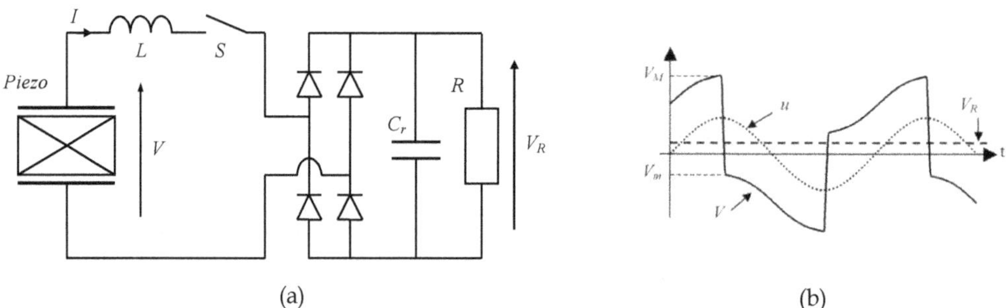

(a)   (b)

Fig. 8. (a) Circuit schematic and (b) typical waveforms of the series SSHI.

Electrical equations of the series SSHI circuit and electromechanical equations (1), (2) with sinusoidal base excitation displacement (3) lead to the expression (14) of the power P

supplied by the energy harvester at the resonance angular frequency ω_r. Equation (15) is an analytical approximation of the maximum power supplied to the optimal load.

$$P = \frac{4R\alpha^2(1+\gamma)^2}{\left(2RC_0\omega_r(1+\gamma)+\pi(1-\gamma)\right)^2} \frac{M^2\omega_r^4 U_1^2}{\left(C+\dfrac{4\alpha^2}{C_0\omega_r}\dfrac{1+\gamma}{2RC_0\omega_r(1+\gamma)+\pi(1-\gamma)}\right)^2} \tag{14}$$

$$P_{max} \approx \frac{k^2 Q_M}{2\pi\dfrac{1-\gamma}{1+\gamma}+8k^2 Q_M}\frac{M^2\omega_r^4 U_1^2}{C} \tag{15}$$

5.4 SECE technique

The SECE technique basically consists in extracting promptly and entirely the electric energy converted by the piezoelectric element on each extremum of the mechanical displacement u. From the standpoint of the piezoelectric element, the effect is identical to the SSDS technique presented in section 5.1. But instead of being dissipated by Joule effect, this electrical energy is transferred to the load (Lefeuvre et al., 2006a). The circuit used for this technique is composed of a bridge rectifier, a flyback-type switching mode power converter and an electric load (Fig. 9 (a)). The power converter is controlled by the on and off states of the transistor T. When the piezoelectric voltage V reaches an extremum, the transistor is turned on and the electrostatic energy of the piezoelectric element begins to be transferred to the coupled inductor L. When voltage V passes at zero, all the electrostatic energy has been removed from the piezoelectric element and the transistor is turned off. The coupled inductor value is chosen so that duration of the energy extraction sequence is very short compared to the vibration period. Thus, the piezo is on open-circuit configuration most of the time and its voltage is a piecewise function of the mechanical displacement u. Typical voltage and displacement waveforms are depicted in Fig. 9 (b) in case of sinusoidal mechanical excitation.

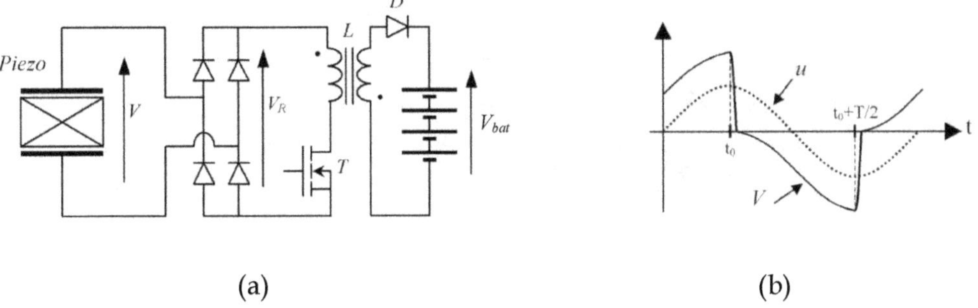

(a)   (b)

Fig. 9. (a) Schematic and (b) typical waveforms of SECE.

Energy transfer sequences of the SECE circuit and electromechanical equations (1), (2) with sinusoidal base excitation displacement (3) and the figure of merit definition (5) lead to the expression (16) of the power P supplied by the energy harvester at the resonance angular frequency ω_r.

$$P - \frac{2}{\pi} \frac{k^2 Q_M}{\left(1 + \frac{4}{\pi} k^2 Q_M\right)^2} \frac{M^2 \omega_r^4 U_1^2}{C} \tag{16}$$

Compared to the techniques previously presented, the power harvested using the SECE technique is not influenced by the characteristics of the electric load (voltage, resistance …). In other words, in this particular case there is no need for additional resistor emulation circuit for power optimization.

5.5 Comparison of harvested powers

The four techniques can be compared by observing in each case the evolution of the power harvested power as a function of the electromechanical figure of merit $k^2 Q_M$. Fig. 10 shows the evolution of the normalized harvested power in comparison to the maximum power which can be harvested using the standard circuit. These magnitudes are plotted for the standard, Parallel SSHI, Series SSHI and SECE techniques. In each case, the load is chosen to maximize the power supplied by the energy harvester and γ has been fixed at 0.76 for the Series and Parallel SSHI.

These charts show that the harvested power has the same limit P_{lim} (defined by Eq. (7)). This limit is reached for $k^2 Q_M > \pi$ in the case of the standard technique, whereas for the Parallel and Series SSHI techniques the power tends asymptotically towards this value. And for the SECE technique the power reaches this maximum for $k^2 Q_M = \pi/4$.

In practice, if we consider that the quantity of used piezoelectric material is roughly proportional to the figure of merit $k^2 Q_M$, then these results mean that to harvest a certain percentage of the maximum recoverable power P_{lim}, SSH techniques permits reducing dramatically the quantity of piezoelectric material in comparison to the standard DC technique. In this sense, SECE appears theoretically like the most interesting technique. However, Series and Parallel SSHI may be better in practice, depending on the losses of the circuits used for implementing of the different techniques. But compared to the standard technique, reduction by a factor 3, or more, of the piezoelectric material amount is generally possible using SSH techniques.

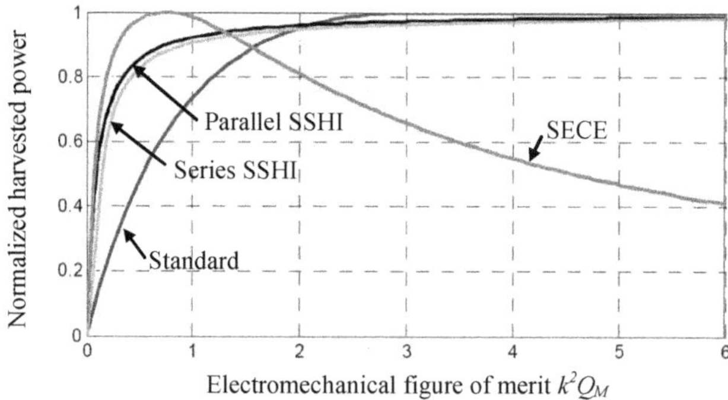

Fig. 10. Harvested powers as a function of the electromechanical figure of merit.

5.6 Deepened analysis of SSH techniques

Further to the original SSH techniques emergence, several works were proposed for deepening their analysis in various case of operation.

Badel et al. (2005) studied the problem of piezoelectric energy harvesting in pulsed operation. This kind of operation typically happens when the energy harvester is excited by mechanical shocks and impacts. Contrary to the case of continuous excitation, the vibration amplitude of the energy harvester decreases more or less rapidly after each excitation shock, depending on the system mechanical losses and also on additional damping effect induced by the energy harvesting process. So, the efficiency of energy harvesting in pulsed operation mode is strongly dependant of the speed of conversion of the mechanical energy into electrical energy. It has been theoretically shown and experimentally verified that the Parallel SSHI is the most effective technique, with typical improvement by factors 2.5 to 5 of the amount of energy harvested by comparison to the standard technique.

Lefeuvre et al. (2007) analyzed the problem harvesting energy from broadband vibration using multimodal resonant electromechanical structure. Comparison of the standard and SECE techniques showed that both techniques allows to effectively to harvesting energy in this case of operation. However, SECE exhibits two major advantages. First, contrary to the standard technique, SECE don't need adaptation of the load for optimal energy conversion. It must be pointed out that such load adaptation, which depends on the vibration frequency, is peculiarly complicated to achieve in multimodal and broadband cases of operations. Second, SECE is more efficient and can harvest up to 4 times more energy than the standard technique for electromechanical systems with relatively small figures of merit ($k^2Q_M < 2$).

Shu et al. (2006; 2007; 2009) compared the effect of frequency deviation from resonance on the performances of the Standard, Series SSHI and Parallel SSHI techniques. They demonstrated that both Series and Parallel SSHI interfaces significantly improve the energy harvester bandwidth. In addition, the Parallel and Series SSHI behavior were shown to be similar to that of a strongly coupled standard system operated respectively at the short circuit resonance and at the open circuit resonance.

6. Evolution of the SSH techniques

During these last years, several novel energy harvesting techniques derived from the original synchronized switching principle have expanded the "SSH family". Main advances concern improvement of efficiency of the electrical interfaces in the case of low piezoelectric voltages, together with improvement of the power harvested in the case of weak electromechanical figures of merit.

6.1 SSH interfaces for low piezoelectric voltages

As mentioned before, design of energy efficient electrical interfaces is a major issue for energy harvesting. This problem is technologically difficult to solve at very low power levels, typically below tens of microwatts, because of power losses of electronic components.

In the case of low voltages, the main problem comes from voltage drop across components, such as threshold voltage of the diodes used for voltage rectification, which significantly diminish efficiency of energy harvesting devices. This problem appears at low vibration level which results in a low piezoelectric voltage magnitude, typically below 1 or 2 volts. This problem becomes critical in the case of ultralow voltage such as that of MEMS-based (MicroElectroMechanical Systems) energy harvesting devices, whose magnitude reaches barely a few tens of milivolts. Indeed, such voltages are lower than the threshold voltage of the diodes commonly available.

An approach for reducing the voltage drops consists basically to limit the number of diodes used for voltage rectification. In this context, Makiara et al. (2006) have proposed a half-bridge circuit for the Parallel SSHI technique. The voltage rectifier requires only two diodes instead of four in the original Parallel SSHI circuit (Fig. 11 (a)). As a consequence, power loss inherent to the diodes threshold voltage is divided by two. The control principle of the switch is the following: when the piezoelectric voltage is maximal, corresponding to a maximum of mechanical displacement, the switch is toggled on "point 2". And, conversely, when the piezoelectric voltage is minimal, the switch is toggled on "point 1". However, despite very simple control principle, truly self-powered implementation of this half-bridge Parallel SSHI circuit is difficult in practice. Indeed, holding on the electronic switch during a half period of vibration consumes a lot of energy compared to the brief on state required by the original Parallel SSHI. This is why theoretical power gain brought by this circuit could have been verified using external power source for the control circuitry only.
Another half-bridge circuit was proposed by Lallart et al. (2008a) for implementation of the Series SSHI technique. The switches control principle is the following: when the piezoelectric voltage reaches a maximum the switch $S1$ is briefly turned on (during a time period equivalent to a half of the resonance period of the LC_0 circuit) and conversely when this voltage reaches a minimum the switch S2 is briefly turned off. Compared to the original Series SSHI, voltage drop due to the rectifier is limited to the diode threshold voltage V_D instead of $2V_D$. This makes a big difference in terms of efficiency when the piezoelectric voltage magnitude is only a few times bigger than V_D. Experimental validation of this circuit with truly self-powered electronic control circuitry showed for instance more than 60% increase of the power supplied by the energy harvesting device compared to the original Series SSHI with piezoelectric voltage magnitude below 2 V.

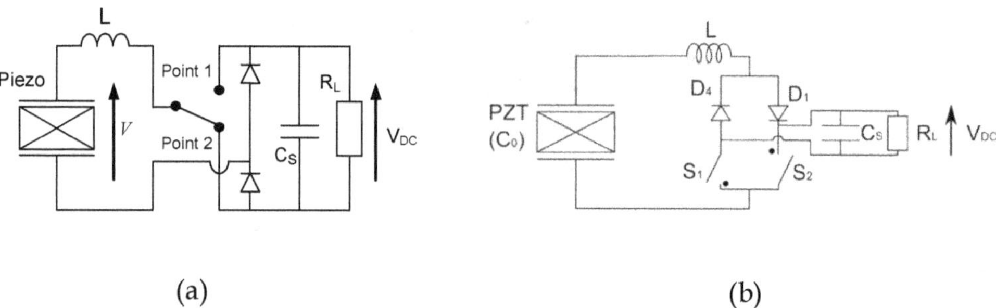

Fig. 11. Half-bridge SSH circuits: (a) Parallel SSHI (b) Series SSHI.

Ultimate reduction of the losses due to the threshold voltage of the diodes was brought by using the SSHI-MR circuit proposed by Garbuio et al (2009) ("MR" stands for Magnetic Rectifier). This SSHI-MR circuit is depicted in Fig. 12. Typical waveforms of this technique are very similar to those of Series SSHI. The main difference of the circuit comes from the magnetic transformer T which replaces the inductor L of the Series SSHI. This transformer has two primary windings and one secondary winding. The coupling factor m is chosen to be far greater than one. Each primary coil is connected in series with one switch ($S1$ or $S1'$). The secondary winding is connected to the smoothing capacitance C_S in parallel with the load R_L through a diode D that ensures a proper charge flow. When the voltage or the displacement is maximal (respectively, minimal), the switch $S1$ (respectively, $S1'$) is closed. As a consequence, electrical resonance occurs between the piezoelectric capacitance C_0 and the primary leakage inductance of transformer. The switch is opened after a half of the time period of resonance for the inversion of the piezoelectric voltage, exactly like in the Series SSHI technique. The inversion is characterized by the inversion factor γ in the same way as the series SSHI inversion coefficient.

The use of the transformer T in association with the single diode rectifier brings significant gain in term of efficiency. Indeed, the losses due to the diode threshold voltage V_D are here proportional to V_D/m compared with $2V_D$ for the original Series SSHI circuit. Thus, energy harvesting with extremely low voltage levels is possible. And, as the current is very low, the losses in the transformer are negligible. Moreover, the transformer enables to increase the output voltage, which is often desirable for the compatibility with electronic circuits.

Experimental validation of this circuit showed the possibility of harvesting energy from piezoelectric voltages as low as 30 mV, making possible operation of the energy harvesting devices in environments having extremely low levels of vibration.

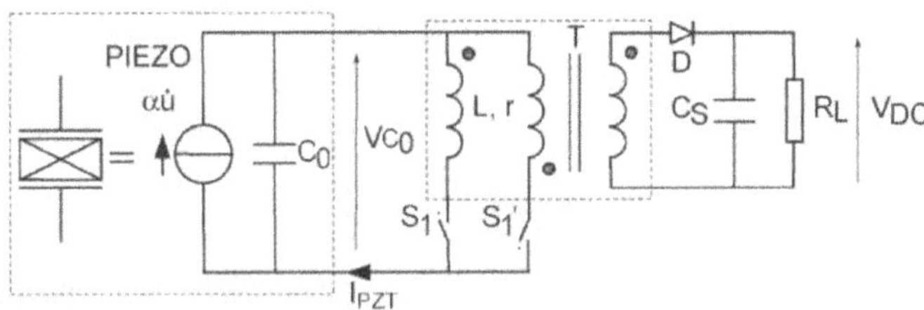

Fig. 12. SSHI-MR circuit.

6.2 Power improvement in the domain of weak electromechanical figures of merit

The DSSH circuit (DSSH stands for "Double Synchronized Switch Harvesting") is an association of the Series SSHI circuit and the buck-boost version of the SECE circuit (Lallart et al., 2008c), as depicted in Fig. 13 (a). An intermediate energy storage capacitor C_{int} is included, which brings an additional degree of freedom in the control of the energy conversion process. This intermediate energy tank is used here for controlling the trade-off between energy harvesting and the intrinsic damping effect of the energy harvesting process. In addition, use of the SECE circuit makes the harvested power optimal whatever the load characteristics (i.e. no influence of the load equivalent resistance).

The control principle of this circuit can be decomposed into four different steps. First step: the switches $S1$ and $S2$ are open and no energy transfers from the piezoelectric element to the load. Second step: when the piezoelectric voltage V_0 is maximal (or minimal), $S1$ is closed and a part of the piezoelectric energy is transferred to the intermediate capacitor C_{int} while voltage V_0 is reversed, like in the case of the original Series SSHI technique. Third step: $S1$ is opened, and $S2$ is closed until all the energy of C_{int} is transferred into the inductor $L2$. Fourth step: $S2$ is opened and all the energy of $L2$ is transferred to the load (modeled here by a storage capacitor C_S and a resistor R_L). Thus, the switching sequence is synchronized with the piezoelectric voltage (or displacement) extrema, so that the energy conversion cycle is very near to that of the Series SSHI.

Detailed analysis of the DSSH technique shows that the energy conversion is optimal when the value of C_{int} is very large compared to that of the internal capacitance of the piezoelectric element C_0 in the case of very weak values of the figure of merit. For higher values of the figures of merit, C_{int} value must be chosen as defined in the following equations:

$$\begin{cases} C_{int} \gg C_0 & for \quad k^2 Q_M \leq \dfrac{4}{\pi}\dfrac{1-\gamma_0}{1+\gamma_0} \\[3mm] C_{int} = C_0 \dfrac{2\pi}{\pi(\gamma_0-1)+4k^2 Q_M(1+\gamma_0)} & for \quad k^2 Q_M > \dfrac{4}{\pi}\dfrac{1-\gamma_0}{1+\gamma_0} \end{cases} \tag{17}$$

with

$$\gamma_0 = e^{-\xi_{sw}/\sqrt{1-\xi_{sw}^2}} \qquad and \qquad \xi_{sw} = \frac{1}{2}r_1\sqrt{\frac{C_0 C_{int}}{L_1(C_0+C_{int})}} \tag{18}$$

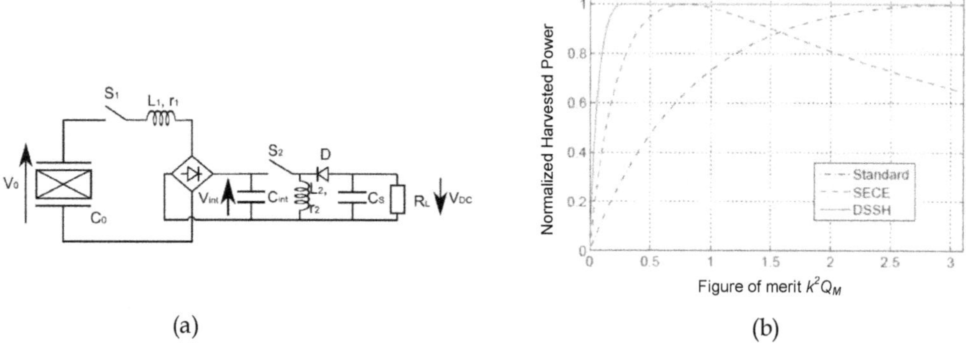

(a)   (b)

Fig. 13. (a) DSSH circuit, (b) power as a function of the electromechanical figure of merit.

In these conditions, the power P harvested using the DSSH technique is optimal and is can be expressed by the following equations:

$$\begin{cases} P = \dfrac{2\pi \cdot k^2 Q_M \left(1 - \gamma_0^2\right)}{\left(\pi\left(1 - \gamma_0\right) + 4k^2 Q_M \left(1 + \gamma_0\right)\right)^2} \dfrac{M^2 \omega_r^4 U_1^2}{C} & \text{for} \quad k^2 Q_M \leq \dfrac{4}{\pi}\dfrac{1 - \gamma_0}{1 + \gamma_0} \\[3mm] P = \dfrac{M^2 \omega_r^4 U_1^2}{8C} & \text{for} \quad k^2 Q_M > \dfrac{4}{\pi}\dfrac{1 - \gamma_0}{1 + \gamma_0} \end{cases} \qquad (19)$$

Variation of the normalized power harvested by the DSSH, SECE and Standard techniques as a function of the figure of merit $k^2 Q_M$ is compared in Fig. 13 (b). These charts clearly show the interest of the DSSH technique, whose power is much more important than that of the two other techniques for very weak values of the electromechanical figure of merit. From another point of view, using the DSSH technique reduces the constraints of design of the electromechanical part in terms of quality factor and electromechanical coupling. In particular, the amount of piezoelectric material can be drastically reduced compared to energy harvesting devices using the standard technique.
The DSSH technique presents significant advantages for energy harvesting, but its practical implementation is a little bit more delicate than that of the SECE or the SSHI techniques. Truly self-powered control circuit of the DSSH has been demonstrated, and despites the cumulated losses of the two conversion stages, performances of this technique remain much better in practice than that of the standard and SECE techniques in the domain of weak figures of merit.

7. Conclusion

Piezoelectric materials are of major interest for high efficiency, low-power energy harvesting from ambient vibration. Owing to decrease of power consumption of electronic devices, increasing variety applications could benefit of such "infinite" lifespan micro-sources of energy.

Based on a single degree of freedom {spring + mass + damper} mechanical system associated to the piezoelectric electromechanical coupling, this chapter analysed first the electromechanical power converted in association with the so-called standard electrical interface, commonly found in piezoelectric energy harvesting devices. Then, the original SSH techniques, derived from semi-passive piezoelectric damping techniques, were presented. Compared to the Standard electrical energy processing, SSHI (series and parallel) and SECE techniques enable significant improvement of harvested power. With the seismic mechanical structure considered here, the power is in particular strongly increased at weak values of the electromechanical figure of merit $k^2 Q_M$. As a result, compared to the standard approach, the amount of piezoelectric material can be reduced by factor 4 or more, thus enabling the design of compact, high efficiency energy harvesting devices. Reduction of the power dependence to the excitation frequency is another advantage brought by these techniques. Practical interest of the proposed approach has been shown in an energy harvesting devices designed for supplying the nodes of a wireless network used for structural health monitoring (Guyomar et al., 2007). Interest of these techniques for pyroelectric energy harvesting has also been demonstrated (Sebald et al., 2008).

This paper presented then the techniques recently derived from the original SSH techniques. Half-bridge parallel and series SSHI interfaces have increased the efficiency of the related techniques in the case of low piezoelectric voltages, making possible the energy harvesting from low-level vibrations. The SSHI-MR circuit has then even more reduced the problems related to voltage drops, so that energy can be harvested in the case of piezoelectric voltages as low as a few tens of millivolts. The DSSH interface, which was derived from both the series SSHI and SECE circuits, showed the possibility of dramatically increasing the power harvested using electromechanical structures with extremely weak figures of merit. The general principle of processing the piezoelectric voltage synchronously with the mechanical excitation offers large degrees of freedom for power improvement as well as reduction of the sensitivity to variations of the load and to drift of the excitation frequency, so it is highly probable that novel interfaces based on the original SSH principle will be proposed in the future.

Current challenges in the field of piezoelectric energy harvesting concern the robustness to frequency mistuning and the improvement of the power harvested from broadband vibrations. Indeed, efficient piezoelectric energy harvesters operate in resonant mode. Consequently they exhibit narrow bandwidth and only a small part of available energy can be harvested. Among solutions investigated to address this problem, the most promising way is based on bistable electromechanical structures (Eturk et al, 2009; Stanton et al, 2009). Compared to usual resonators, these non linear resonators exhibit enlarged bandwidth while keeping advantage of high power conversion capability. Ongoing works aim at evaluating the interest of associating the SSH techniques with the non linear piezoelectric resonators.

8. References

Anton, S. R. & Sodano, H. A. (2007). A review of power harvesting usingpiezoelectric materials (2003–2006). *Smart Materials and Structures*. 16, R1–R21.

Badel, A.; Lefeuvre, E.; Richard, C. and Guyomar, D. (2005). Efficiency enhancement of a piezoelectric energy harvesting device in pulsed operation by synchronous charge inversion. *Journal of Intelligent Materials Systems and Structures*, Vol. 16, N°10, 889-901.

Badel, A.; Lagache, M; Guyomar, D.; Lefeuvre, E. & Richard, C. (2007). Finite element and simple lumped modeling for flexural nonlinear semi-passive damping. *Journal of Intelligent Material Systems and Structures*, Vol. 18, Issue 7, 727-742.

Energy Harvesting Forum. Energy Harvesting Electronic solutions for Wireless Sensor networks and Control Systems. (Online). http://www.energyharvesting.net/

Erturk, A.; Hoffmann, J. & Inman, D. J. (2009). A piezomagnetoelastic structure for broadband vibration energy harvesting. *Applied Physics Letters*, 94.

Garbuio, L.; Lallart, M.; Guyomar, D.; Richard, C. & Audigier D. (2009). Mechanical Energy Harvester With Ultralow Threshold Rectification Based on SSHI Nonlinear Technique. *IEEE Transactions on Industrial Electronics*, Vol. 56, No. 4, 1048-1056.

Guyomar, D.; Badel, A.; Lefeuvre, E. & Richard, C. (2005). Towards Energy Harvesting using Active Materials and Conversion Improvement by Nonlinear Processing. *IEEE Transactions on UFFC*, vol. 52, no. 4, 584-595.

Guyomar, D.; Jayet, Y.; Petit, L.; Lefeuvre, E.; Monnier, T.; Richard; C. & Lallart, M. (2007). Synchronized switch harvesting applied to selfpowered smart systems: Piezoactive microgenerators for autonomous wireless transmitters. *Sensors and Actuators A: Physical*, Vol. 138, N°1, 151-160.

Horowitz, S.; Kasyap, A.; Liu, F.; Johnson, D.; Nishida, T.; Ngo, K. Sheplak, M. & Cattafesta, L. (2002). Technology Development for Self-Powered Sensors. *AIAA 1st Control Flow Conference Proceedings*, pp. 2-10, St. Louis, June 2002, AIAA, Reston.

Krikke D. (2005). Sunrise for Energy Harvesting Products. *IEEE Pervasive Computing*, Issue Date: January 2005, 4-8.

Lallart, M. & Guyomar, D. (2008a). An optimized self-powered switching circuit for non-linear energy harvesting with low voltage output. *Smart Materials and Structures*, 17, N°3, 1-8.

Lallart, M.; Lefeuvre, E.; Richard, C. & Guyomar, D. (2008b). Self-Powered Circuit for Broadband, Multimodal Piezoelectric Vibration Control. *Sensors and Actuators A: Physical*, vol. 143, N°2, 377-382.

Lallart, M.; Garbuio, L.; Petit, L.; Richard C. & Guyomar D. (2008c). Double Synchronized Switch Harvesting (DSSH): A New Energy Harvesting Scheme for Efficient Energy Extraction. *IEEE Transactions on UFFC*, Vol. 55, No. 10, 2119-2130.

Lallart, M.; Garbuio, L.; Petit, L.; Richard C. & Guyomar D. (2010). Low-cost piezoelectric voltage inverter for outstanding performance in piezoelectric energy harvesting. *IEEE Transactions on UFFC*, vol. 57, no. 2, 281-291.

Lefeuvre, E.; Badel, A.; Richard, C. & Guyomar, D. (2004). High Performance Piezoelectric Vibration Energy Reclamation. *SPIE Proceedings*, Vol. 5390, 379-387.

Lefeuvre, E.; Badel, A.; Richard, C. & Guyomar, D. (2006a). A comparison between several vibration-powered piezoelectric generators for standalone systems. *Sensors and Actuators A: Physical*, vol. 126, issue 2, 405-416.

Lefeuvre, E.; Audigier, D.; Richard, C. & and Guyomar, D. (2006b). Buck-Boost Converter for Sensorless Power Optimization of Piezoelectric Energy Harvester. *IEEE Transactions on Power Electronics*, Vol. 22, N°5, 2018-2025.

Lefeuvre, E.; Badel, A.; Richard, C. & Guyomar, D. (2007). Energy harvesting using piezoelectric materials: case of random vibrations. *Journal of Electroceramics*, Vol. 19, 349-355.

Lefeuvre, E.; Sebald, G.; Guyomar, D.; Lallart, M. & Richard, C. (2009). Materials, structures and power interfaces for efficient piezoelectric energy harvesting. *Journal of Electroceramics*, vol. 22, 171-179.

Makihara, K.; Onoda, J. & Miyakawa, T. (2006). Low energy dissipation electric circuit for energy harvesting. *Smart Materials and Structures*, 15, 1493–1498.

McLean Nicolson A. (1931). Piezoelectric Crystal Converter. US Patent N°1975517.

Mitcheson, P. D.; Green T. C. & Yeatman E. M. (2007). Power processing circuits for electromagnetic, electrostatic and piezoelectric inertial energy scavengers. *Microsystem Technologies* 13:1629–1635.

Ottman, G. K.; Bhatt, A. C.; Hofmann, H. & Lesieutre, G. A. (2002). Adaptive piezoelectric energy harvesting circuit for wireless remote power supply. *IEEE Transactions on Power Electronics*, vol. 17, 669–676.

Ottman, G. K.; Bhatt, A. C.; Hofmann, H. & Lesieutre, G. A. (2003). Optimized Piezoelectric Energy Harvesting Circuit Using Step-Down Converter in Discontinuous Conduction Mode. *IEEE Transactions on Power Electronics*, vol. 18, 696–703.

Richard, C.; Guyomar, D.; Audigier, D. & Ching, G. (1999). Semi-passive damping using continuous switching of a piezoelectric device. *SPIE Proceedings*, Vol. 3672, 104-111.

Richard, C.; Guyomar, D.; Audigier, D. & Bassaler, H. (2000). Enhanced semi-passive damping using continuous switching of a piezoelectric device on an inductor. *SPIE Proceedings*, Vol. 3989, 288-299.

Richards, C. D.; Anderson, M. J.; Bahr D. F. & Richards, R. F. (2004). Efficiency of energy conversion for devices containing a piezoelectric component. *Journal of Micromechanics and Microengineering*, Vol. 14, 717-721.

Roundy, S. & Wright, P. K. (2004). A piezoelectric vibration based generator for wireless electronics. *Smart Materials and Structures*, 13, 1131–1142.

Sebald, G.; Lefeuvre, E. & Guyomar, D. (2008). Pyroelectric energy conversion: optimization principles. *IEEE Transaction on Ultrasonics, Ferroelectrics and Frequency Control*, Vol. 55, n°3, 538-551.

Shu, Y. C. & Lien, I. C. (2006). Analysis of power output for piezoelectric energy harvesting systems. *Smart Materials and Structures*, 15, 1499–1512.

Shu, Y. C. & Lien, I. C. (2007). An improved analysis of the SSHI interface in piezoelectric energy harvesting. *Smart Materials and Structures*, 16, 2253-2264.

Shu, Y. C.; Lien, I. C.; Wu, W. J. and Shiu, S. M. (2009). Comparisons between parallel- and series-SSHI interfaces adopted by piezoelectric energy harvesting systems, *Proceedings of SPIE 16th International Symposium on Smart Structures and Materials*, vol. 7288, San Diego, California, april 2009.

Sodano, H. A.; Inman, D. J. & Park, G. (2004). A Review of Power Harvesting from Vibration using Piezoelectric Materials. *The Shock and Vibration Digest*, Vol. 36, No. 3, 197–205.

Stanton, S. C.; McGehee, C. C. & Mann, B. P. (2009). Reversible hysteresis for broadband magnetopiezoelastic energy harvesting. *Applied Physics Letters*, 95.

Taylor, G. W.; Burns, J. R.; Kammann, S. M.; Powers, W. B. & Welsh, T. R. (2001). The energy harvesting eel: a small subsurface ocean/river power generator. IEEE Journal of Ocean Engineering, Vol. 26, 539-547

Wang, Q. M. & Cross, L. E. (1999). Constitutive equations of symmetrical triple layer piezoelectric benders. *IEEE Transactions on Ultrasonics, Ferroelectrics and Frequency Control*, Vol. 46, 1343-1351.

10

Electrostrictive polymers as high-performance electroactive polymers for energy harvesting

Pierre-Jean Cottinet, Daniel Guyomar, Benoit Guiffard,
Laurent Lebrun and Chatchai Putson
INSA de Lyon (LGEF)
France

1. Introduction

In this time of technological advancements, conventional materials such as metals and alloys are being replaced by polymers in such fields as automobiles, aerospace, household goods, and electronics. Due to the tremendous advances in polymeric materials technology, various processing techniques have been developed that enable the production of polymers with tailor-made properties (mechanical, electrical, etc). Polymers enable new designs to be developed that are cost effective with small size and weights (Gurunathan et al., 1999).

Polymers have attractive properties compared to inorganic materials. They are lightweight, inexpensive, fracture tolerant, pliable, and easily processed and manufactured. They can be configured into complex shapes and their properties can be tailored according to demand (Gurunathan et al., 1999). With the rapid advances in materials used in science and technology, various materials with intelligence embedded at the molecular level are being developed at a fast pace. These smart materials can sense variations in the environment, process the information, and respond accordingly. Shape-memory alloys, piezoelectric materials, etc. fall in this category of intelligent materials (Zrínyi, 2000). Polymers that respond to external stimuli by changing shape or size have been known and studied for several decades. They respond to stimuli such as an electrical field, pH, a magnetic field, and light (Bar-Cohen, 2004).These intelligent polymers can collectively be called active polymers.

One of the significant applications of these active polymers is found in biomimetics – the practice of taking ideas and concepts from nature and implementing them in engineering and design. Various machines that imitate birds, fish, insects and even plants have been developed. With the increased emphasis on "green" technological solutions to contemporary problems, scientists started exploring the ultimate resource – nature – for solutions that have become highly optimized during the millions of years of evolution.

There are many types of active polymers with different controllable properties, due to a variety of stimuli. They can produce permanent or reversible responses; they can be passive or active by embedment in polymers, making smart structures. The resilience and toughness of the host polymer can be useful in the development of smart structures that have shape control and self-sensing capabilities (Bar-Cohen, 2004).

Polymers that change shape or size in response to electrical stimulus are called electroactive polymers (EAP) and are classified depending on the mechanism responsible for actuation as electronic EAPs (which are driven by electric field or coulomb forces) or ionic EAPs (which change shape by mobility or diffusion of ions and their conjugated substances). A list of leading electroactive polymers is shown in Table 1.

Ionic EAP	*Electronic EAP*
Ionic polymer gels (IPG)	Dielectric EAP
Ionic polymer metal composite (IPMC)	Electrostrictive graft elastomers
Conducting polymers (CP)	Electrostrictive paper
Carbon nanotubes (CNT)	Electro-viscoelastic elastomers
	Ferroelectric polymers
	Liquid crystal elastomers (LCE)

Table 1. List of leading EAP materials

The electronic EAPs such as electrostrictive, electrostatic, piezoelectric, and ferroelectric generally require high activation fields (>150V/μm) which are close to the breakdown level of the material. The property of these materials to hold the induced displacement, when a DC voltage is applied, makes them potential materials in robotic applications, and these materials can be operated in air without major constraints. The electronic EAPs also have high energy density as well as a rapid response time in the range of milliseconds. In general, these materials have a glass transition temperature inadequate for low temperature actuation applications.

In contrast, ionic EAP materials such as gels, ionic polymer-metal composites, conducting polymers, and carbon nanotubes require low driving voltages, nearly equal to 1–5V. One of the constraints of these materials is that they must be operated in a wet state or in solid electrolytes. Ionic EAPs predominantly produce bending actuation that induces relatively lower actuation forces than electronic EAPs. Often, operation in aqueous systems is plagued by the hydrolysis of water. Moreover, ionic EAPs have slow response characteristics compared to electronic EAPs. The amount of deformation of these materials is usually much more than electronic EAP materials, and the deformation mechanism bears more resemblance to a biological muscle deformation. The induced strain of both the electronic and ionic EAPs can be designed geometrically to bend, stretch, or contract (Bar-Cohen, 2004).

The principles of operation of EAP in actuator mode consist of applied electric field thought the thickness that contacting this one, and stretching in the area. As with many actuator technologies, electronic EAP are reversible and can be operated in generator mode. In this mode of operation, mechanical work is done against the electric field, and electrical energy is produced. Thus, the electronic EAP is acting as an electromechanical generator transducer in this mode of operation.

Technologically, the generator mode of electronic EAP is potentially as important as the actuator mode. Actuators are indeed pervasive in modern technology, yet the critical need for new energy systems, such as generators, may be more important than the sheer number

of possible applications. Moreover the current trend in electronic devices is their integration in most of common systems in order to extend the number of functions and to improve their reliability. The recent progresses in ultralow-power electronics allow powering complex systems using either batteries or environmental energy harvesting. Although large efforts in battery research have been made, such powering solution raises the problem of limited lifespan and complex recycling process. The current energy requirement thus leads to the research of other energy sources for mobile electronics. Strong research effort and industrial development deal with energy harvesting using piezoelectric materials, one of the most promising solutions for direct power supply and energy storage for low power wearable devices. However, those materials tend to be stiff and limited in mechanical strain abilities; so for many applications in which low frequency and large stroke mechanical excitations are available (such as human movement). Organic materials, however, are softer and more flexible; therefore, the input mechanical energy is considerably higher under the same mechanical force. Piezoelectric polymers such as PVDF, unfortunately, have a much lower piezoelectric coefficient compared to the piezoelectric ceramic materials. A study has shown that the energy harvesting is lower than with piezoelectric ceramic bimorphs (Liu, et al., 2004). Electrostatic-based systems, such as dielectric elastomers, typically require a very high electric field intensity (20-120 V/µm) to achieve a significant energy harvesting (Pelrine, et al., 2001). A recent research shown that the polyurethane and P(VDF-TrFE-CFE) which are an electrostrictive polymers were capable of generating strains above 10% under a moderate electric field (20V/µm)) (Guiffard, et al. 2006; Guiffard, et al., 2009), thus leading to them being considered as potential actuators. Furthermore, these materials are lightweight, very flexible, have low manufacturing costs and are easily moulded into any desired shapes. Moreover the mechanical energy density is comparable to that piezoelectric single crystals (Ren, et al., 2007).
There exist different methods for harvesting mechanical to electrical energy using elestrostrictive polymer, the first point of this chapter provide a brief overview of the most used methods. The second paragraph presents a method for enhancing the electromechanical responses of elestrostrictive polymer using conductive particules, with a presentation of the process of fabrication. The third points discusses pratical consideration such as circuit topologies, but also a comparison between the other technology for harvesting energy. In fact, electrostrictive polymer is much more competitive for some applications compared with others, and it is important to identify general advantage or attractive features of electrostrictive polymer for energy harvesting.

2. General principles of the electrostrictive polymer generator mode

Electrostriction is generally defined as a quadratic coupling between strain (S_{ij}) and polarization (P_m):

$$\begin{cases} E_m = \varepsilon_{mn}^{'T}.P_n + 2.Q_{klmn}.T_{kl}.P_n \\ S_{ij} = s_{ijkl}^{P}.T_{kl} + Q_{ijmn}.P_m.P_n \end{cases} \tag{1}$$

where s^P_{ijkl} is the elastic compliance, Q_{ijkl} is the polarization-related electrostriction coefficient, ε'^T_{ik} is the inverse of the linear dielectric permittivity, and T_{kl} is the stress. Assuming a linear relationship between the polarization and the electric field, the strain S_{ij}

and electric flux density D_i are expressed as independent variables of the electric field intensity E_k, E_l and stress T_{kl} by the constitutive relations according to equation (2)(Ren, et al., 2007):

$$\begin{cases} S_{ij} = M_{ijkl}.E_k.E_l + s_{ijkl}^{E}.T_{kl} \\ D_i = \varepsilon_{ik}^{T}.E_k + 2.M_{ijkl}.E_l.T_{kl} \end{cases} \tag{2}$$

here s_{ijkl}^{E} is the elastic compliance, M_{ijkl} is the electric-field-related electrostriction coefficient, and ε_{ik}^{T} is the linear dielectric permittivity.

There exist two methods for harvesting energy when using an electrostrictive material, the first consists in realizing of cycles, whereas the second involves in working in the pseudo piezoelectric behaviour.

2.1 Energy harvesting using cycles

This method, proposed by Y. Liu et al (Liu, et al., 2005), was inspired by the process for harvesting energy in the case of dielectric elastomers. The mechanical-to-electrical energy harvesting in electrostrictive materials is, as an example, illustrated in the mechanical stress/strain and electric field/flux density plots shown in Fig. 1. Initially, the material presented in Fig. 1 had no stress applied to it. When a stress was applied, the state traveled along path A. Finally, the applied stress was reduced. Due to the change in the electrical boundary conditions, the contraction path did not follow path A but path B. Both in the mechanical and electrical planes, the material state traversed a closed loop. In the mechanical plane, the rotation designated that the net energy flow went from the mechanical to the electrical. The areas enclosed in the loop of the mechanical and electrical planes were equal and corresponded to the converted energy density in units of J/m³.

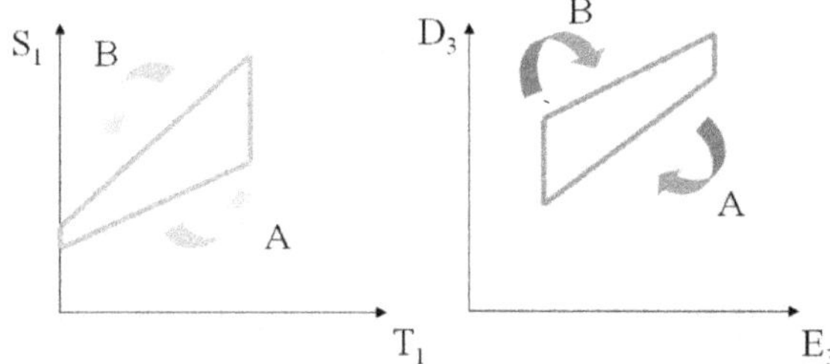

Fig. 1. An energy harvesting cycle

Ideally, the energy harvesting cycle consists of a largest possible loop, bounded only by the limitations of the material. Y. Liu et al. have analyzed electrical boundary conditions that can be applied in order to obtain an optimized energy harvesting. They demonstrated that elestrostrictive materials have significant electric energy densities that can be harvested. Of the electrical boundary conditions investigated in their work, the best energy harvesting density occurred when the electric field in the material was increased from zero to its maximum value at a maximum stress, and then returned to zero at a minimum stress (Fig.

2). They used the concept of a coupling factor, as defined in the IEEE standard of piezoelectrics (IEEE 1988), which is a useful figure of merit for energy harvesting, and is given in equation (3):

$$k = \sqrt{\frac{W_1}{W_1 + W_2}} \qquad (3)$$

Here $W_1 + W_2$ is the input mechanical energy density, and W_1 is the output electrical energy density.

A constant electric field E_0 exists from state 1 to state 2 as the stress is increased to T_{max}. From state 2 to state 3, the electric field is increased from E_0 to E_1, then kept constant until the stress is reduced from T_{max} to 0 from state 3 to state 4. At zero stress, the electric field is reduced to E_0, returning to state 1. In the dielectric-field plot, the paths 1-4 and 2-3 are not parallel, which is due to the stress dependence of the dielectric constant. The converted energy W_1 can be expressed by equation (4):

$$W_1 = T_{max}\left(E_1^2 - E_0^2\right) \qquad (4)$$

The input energy density $W_2 = \left(1/2\right)sT_{max}^2$, and thus the coupling factor, are given by equation (5):

$$k = \sqrt{\frac{M\left(E_1^2 - E_0^2\right)}{\dfrac{1}{2}sT_{max} + M\left(E_1^2 - E_0^2\right)}} \qquad (5)$$

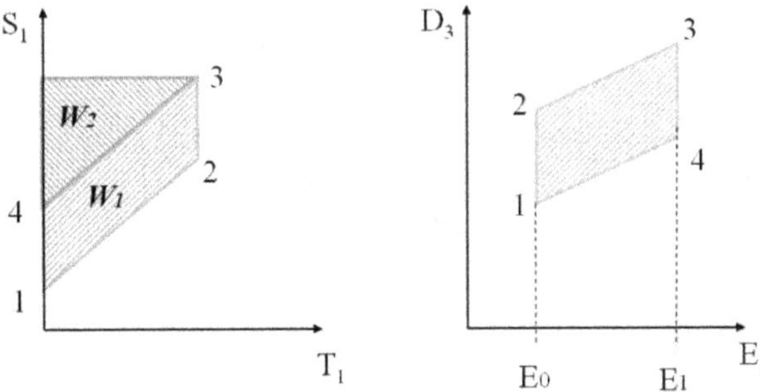

Fig. 2. An energy harvesting cycle under constant electrical field condition as the material is stressed and unstressed.

K. Ren et al (Ren, et al., 2007) investigated a means of using this method for harvesting energy. The experiment was carried out under quasistatic conditions at 1 Hz. The electric parameters were E_0=46 MV/m and E_1=67 MV/m, and with this technique, they were able to harvest 22.4 mJ/cm³ for a transverse strain of 2%. This can be compared to results reported in the literature, for piezopolymers and piezoceramics with a conventional energy harvesting scheme, in which the energy harvesting was below 5 mJ/cm³ (Poulin, et al., 2004).

2.2 Energy harvesting in pseudo piezoelectric behaviour

Another way of harvesting energy using electrostrictive polymers consists of working in the pseudo piezoelectric behavior. For this, the electrostrictive polymer was subjected to a DC-biased electric field. As the polymer was not piezoelectric, it was necessary to induce polarization with a DC bias in order to obtain a pseudo piezoelectric behavior (Guyomar, et al., 2009; Lebrun, et al., 2009; Cottinet, et al. 2010).

An isotropic electrostrictive polymer film contracts along the thickness direction (the electric field direction) and expands along the film direction when an electric field is applied across the thickness, assuming that only a nonzero stress is applied along the length of the film (Fig. 3). The constitutive relation (2) can then be simplified as:

$$\begin{cases} S_1 = M_{31}.E_3^2 + s_{11}^E.T_1 \\ D_3 = \varepsilon_{33}^T.E_3 + 2.M_{31}.E_3.T_1 \end{cases} \tag{6}$$

The current induced by the transverse vibration can be measured as $I = \int_A \dfrac{\partial D_3}{\partial t} dA$, where

A corresponds to the area of the electrostrictive polymer. The current produced by the polymer can thus be related to the strain and electric field by:

$$I = \int_A \left[\frac{\partial E_3}{\partial t}\left(\varepsilon_{33}^T + \frac{2.M_{31}.S_1 - 6.M_{31}^2.E_3^2}{s_{11}^E} \right) + \frac{2.M_{31}.\dfrac{\partial S_1}{\partial t}.E_3}{s_{11}^E} \right].dA \tag{7}$$

where $\partial E_3/\partial t$ and $\partial S_1/\partial t$ are the time derivates of the electrical field and strain. Since a DC can be given by:

$$I_0 = 2.M_{31}.Y.E_{dc}.\int_A \frac{\partial S_1}{\partial t}.dA \tag{8}$$

with $1/s_{11}^E = Y$, and where Y is the Young modulus.

Assuming a constant strain, the relation between the displacement and the strain S_1 in the polymer can be expressed according to equation (8):

$$S_1 = \frac{\Delta L}{L_0} = \frac{u}{L_0} \tag{9}$$

where $\Delta L = u$ is the amplitude of the transverse displacement, and L_0 is the initial value of the length.

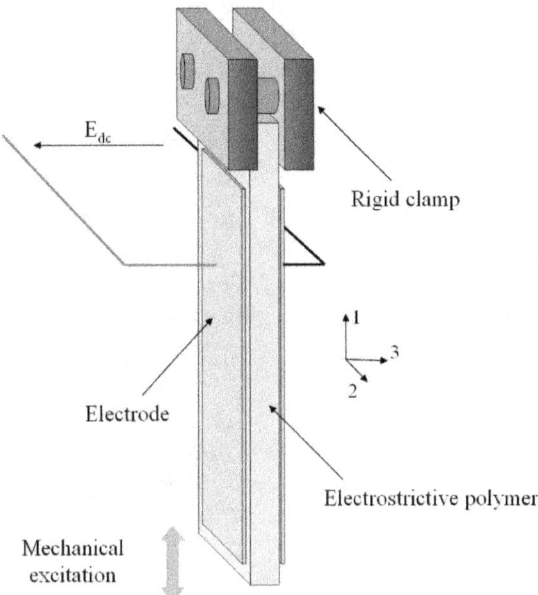

Fig. 3. Mechanical configuration of an electrostrictive polymer

The electric impedance of a polymer vibrating at a given frequency can be modeled by an equivalent electric circuit. Figure 4 displays the most commonly adopted form of an electrical scheme (Cottinet, et al. 2010), where C_p is the capacitance of the clamped polymer and $R_p(\omega)$ is a resistance representing the dielectric losses and conduction (static losses at zero frequency). Both these factors are functions of the frequency caused by the relaxation phenomenon. The third branch is the motional branch, modeled by a current source I_0 (7) that is capable of modeling the harvested current from vibrations.

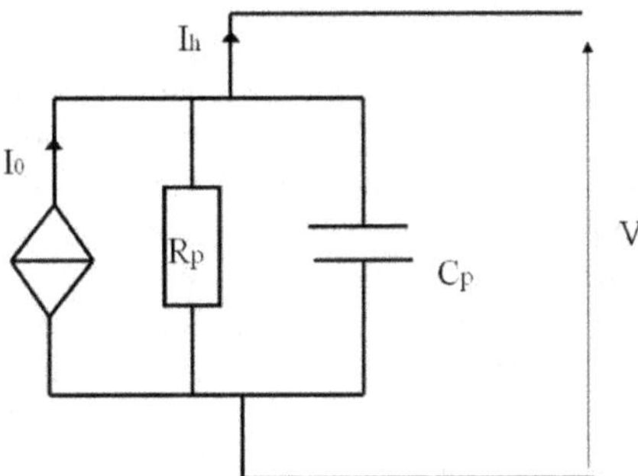

Fig. 4. The equivalent electric circuit of an electrostrictive polymer

A previous study (Cottinet, et al. 2010) has demonstrated that it was possible to neglect R_p. The dynamic model of the current can thus be simplified as:

$$I_h = \alpha\left(E_{dc}\right).\frac{\partial u}{\partial t} - C_p.\frac{\partial V}{\partial t} \qquad (10)$$

with $\alpha\left(E_{dc}\right) = \dfrac{2.M_{31}.Y.A.E_{dc}}{L_0}$ and where u is the displacement.

This expression is similar to the typical system of an equation in the case of the piezoelectricity (Badel, et al. 2005). Results obtained with this method are presented in section 4.

2.3 Conclusions

For both techniques presented above, the electrostrictive coefficient M appears to be an important parameter in order to increase the energy harvesting of the polymer. Therefore, the next section provides a description of the various methods available to increase this coefficient.

3. Enhancing the electrostrictive coefficient

The electromechanical transduction properties of any electrostrictive polymer are intrinsically regulated by the dielectric permittivity of the material. In fact, Eury et al. (Eury, et al., 1999] and Guillot et al. (Guillot, et al., 2003) have demonstrated that the electrostrictive coefficient (M) was proportional to $\varepsilon_0(\varepsilon_r-1)^2/(Y.\ \varepsilon_r)$. The aim of the present section is to provide an overview of the available methods for increasing the permittivity of an electrostrictive polymer. Modifying a polymer matrix in order to increase its dielectric permittivity means acting on the dipolar moments of the material and, therefore, on its polarization.

3.1 Methods for increasing the dielectric permittivity

Currently, a variety of methods are available in order to increase the dielectric permittivity of polymer materials. These may be classified into two main groups: those involving composites and those based on new synthetic polymers. The first approach concerns the dispersion of a filler into the polymer matrix. The second strategy, on the other hand, deals with the synthesis of new materials with tailored characteristics.

A composite is a heterogeneous substance consisting of two or more materials which does not lose the characteristics of each component. Moreover, this combination of materials brings about new desirable properties. Naturally occurring composites include tendon, bone, bamboo, rock, and many other biological and geological materials. Composite engineering has become a very common methodology in the field of materials for achieving a set of specific properties.

There exists a large choice of fillers that may increase the dielectric and conductive properties, among others, of a polymer. The use of high-permittivity inorganic fillers is a well-established approach to improve the dielectric constant of a polymer matrix (Mazur,

1995). In particular, powders of ferroelectric/piezoelectric ceramics, showing very high dielectric constants (ε_r=1000 for lead magnesium niobate- PMN), can, in principle, give rise to significant increments of the permittivity. Gallone et al. (Gallone, et al. 2007) have demonstrated that it was possible to obtain a fourfold increase in permittivity of a silicone elastomer at 10 Hz by charging the material with 30vol% lead magnesium niobate-diacrylate (PMM-PT).

Despite it being possible to considerably improve a material's dielectric properties by using ceramic fillers, such a method is not always suitable for enhancing the actuation or mechanical-to-electrical conversion properties (Szabo, et al., 2003). In fact, ferroelectric ceramic fillers are usually extremely stiff, which is likely to cause a loss of strain capabilities in the resultant composites as well as a loss of flexibility.

Insulators are not the sole candidate materials as suitable fillers. In fact, dielectric improvements can also be achieved with conductive fillers, as described in the following. The use of conductive fillers as a possible means to increase the dielectric permittivity is interesting since free charges not only contribute to conduction, but also possibly give rise to Maxwell-Wagner polarization. Such insulator-conductive composite systems are prone to show losses with a peroclative behavior, which may result in a dramatic increase of their conductivity at filler concentrations exceeding a certain threshold. The percolation threshold represents the filler concentration at which conducting paths are formed between particles in contact with each other in the matrix. The increase of the dielectric permittivity in a composite follows equation 11.

$$\varepsilon_{composite} = \varepsilon_{matrix} \left(\frac{V_{percolation} - V_{filler}}{V_{filler}} \right) \tag{11}$$

Here, $\varepsilon_{composite}$ and ε_{matrix} represent the dielectric permittivity of the composite and the matrix, respectively, $V_{percolation}$ is the filler percolation concentration, and $V_{fillers}$ is the filler concentration.

Unfortunately, the maximum increase in composite permittivity is achieved close to the percolation threshold (Dang, et al., 2002). In light of this fact, reducing the stiffening introduced by inorganic filler and simultaneously exploiting the dielectric enhancement when conductive fillers are introduced to a polymer matrix is very interesting. The work of Zucolotto et al. (Zucolotto, et al., 2004) demonstrated that the perocolation threshold can be lowered down to 5 wt% in a case of a styrene-ethylene-butylene terpolymer with carbon black particles, which is an evident advantage in terms of mechanical properties.

An ideal approach in order to obtain elastomers with specific improved dielectric properties is represented by a challenging synthesis of new molecular architectures. There exists various approaches for obtaining polymer-like blends of known polymers, or copolymerization,etc. Lehmann et al. developed a process for synthetically modifying the dielectric properties of liquid -crystalline elastomers; in this type of material, the polarization phenomena can be enhanced by the rearrangement of the lateral group chains and the creation of crystalline regions (Lehmann, et al., 2001).

Blending of different polymers can result in novel materials with potentially attractive properties. For example, Huang et al. (Huang, et al., 2004) proposed a blend of polyurethane and phthalocyanine, which also included PANI, and at 1 Hz, they obtained a permittivity of 800 for the composition 14-15-85 vol%PANI-Pc-PU.

The different methods available for enhancing the dielectric permittivity of polymers are listed in Table 2 which also gives the advantages and drawbacks of each technique. Random composites represent readily applicable approaches suitable for increasing the dielectric permittivity of elastomers. In the long run, the challenge consists in synthesizing a new highly polarizable polymer. All this research is necessary to achieve new generations of electrostrictive polymers, operating at lower electric fields.

	Type of fillers	*Advantages*	*Drawbacks*
Random composite	Dielectric	+ high dielectric permittivity	- large percentage of filler - increase in elastic modulus
Random composite	Conductive	+ high dielectric permittivity for low percentage of filler	- increase in conductivity - decrease of maximum voltage possible to apply.
Polymer blend	No fillers	+ very high dielectric permittivity + no problem of conductivity + no mechanical reinforcement	- process of realization complex

Table 2. Comparison between the different methods for enhancing the dielectric permittivity

3.2 Methods developed in the laboratory

Smart materials are primary elements in energy harvesting in that they represent the first stage of converting ambient vibrations into electrical energy. Consequently, the optimization of such a material is very important.

According to the amazing physical properties of nanofillers, e.g., carbon nanopowders or silicon carbide nanowires, random composites can be realized in the laboratory. Two types of commercially available polymers have been employed in this work: polyurethane and P(VDF-TrFE-CFE). Moreover, two composites were synthesized specifically for the study. These polyurethane composites were prepared in the laboratory, using a thermoplastic polyurethane - the 58887 TPU elastomer (Estane) - as the matrix. The neat polyurethane (PU) films as well as their filled counterparts were prepared by solution casting (Guiffard, et al. 2006; Guiffard, et al., 2009). The PU granules were dissolved in N,N-dimethylformamide (DMF) at 80°C for 45 minutes, and two types of inorganic fillers (i.e., a carbon black nanopowder (C) and silicon carbide (SiC) nanowires coated with a carbon layer) were individually ultrasonically dispersed in DMF. Subsequently, the SiC nanowire solution at 0.6 g/l was mixed with the dissolved PU under mechanical stirring at 80°C for 1 hour, after which this viscous mixture was poured onto glass plates and cured at 50°C for 24 hours to remove most of the solvent. Figure 5 summarizes the fabrication process of the composites.

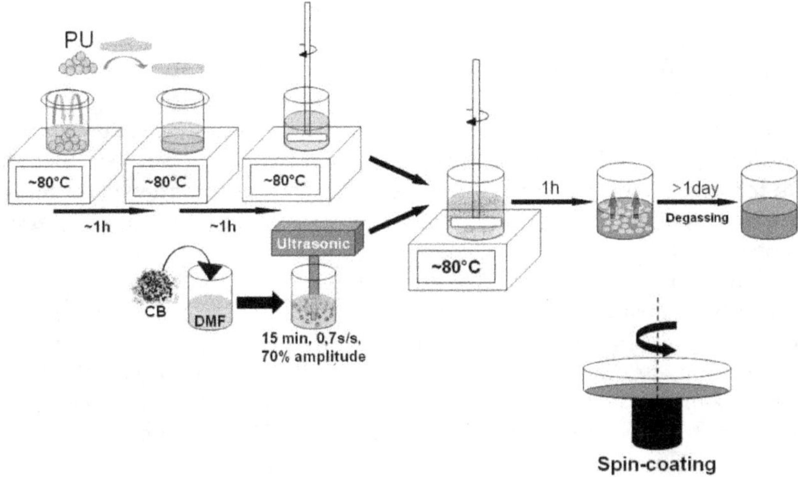

Fig. 5. Principle of realization of the polymer composite

3.3 Characteristics of the composites

The microstructure of the composites was investigated by scanning electron microscopy (SEM) on samples fractured in liquid nitrogen and sputtered with a thin gold layer prior to the SEM observations. The degree of filler dispersion in the polymer matrix and the binding between the nanocharges and the matrix determined the properties of the composite materials. SEM images of the fracture surfaces of the composites are presented in Fig. 6. Figures 6.a and 6.b respectively depict the dispersion of a carbon black nanopowder and SiC nanowires within the PU matrix. Both filler types were found to be well dispersed in the matrix which was in good agreement with the nonconduction state observed in the composites.

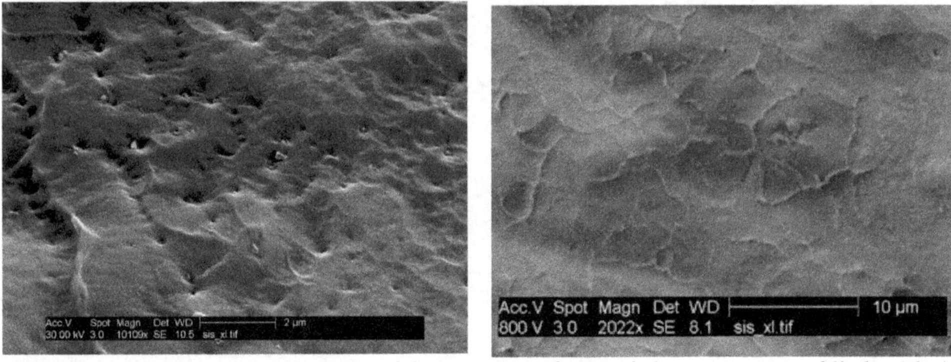

Fig. 6. (Left to right) Morphology of fractured surfaces of the composites filled with (a) a carbon black nanopowder fillers and (b) SiC nanowires.

In order to evaluate the contribution of space charge, the dielectric constant of the composites loaded with fillers was measured using an HP 4284A *LCR* meter over a broad range of frequencies (from 0.01 Hz to 1 MHz) at room temperature. Figure 7 shows the variation in dielectric constants for a pure PU material and filled composites versus frequency. A large decrease in the dielectric constant was observed at around 1 Hz for both composites when the frequency increased. Such a behavior is known to be due to the loss of one of the polarization contributions (interfacial polarization, orientation polarization, electronic polarization, atomic polarization, etc) of the dielectric constant value (Mitchell, 2004). Considering the value of the frequency, this decrease can be unambiguously attributed to the loss of the space-charge-induced interfacial polarization contribution. It can moreover been seen that the contribution of the space charge can be neglected for frequencies below 5 Hz.

As further shown in Fig. 7, the dielectric constant of the C-filled and SiC-filled nanocomposites was consistently higher than that of the pure PU composite. As expected, at higher frequencies, the gap between the values of the dielectric constant for pure PU as opposed to for the nanocomposites was not very high, which confirmed the fact that the filler content was low in comparison to the threshold value. Moreover, the space charge mechanism did not contribute to the dielectric constant. The incorporation of conductive charges probably also increased the space charge density in addition to intrinsically induced by the existence of soft and hard segments within the matrix. Similar observations have been reported by Dang et al., (Dang, et al., 2002), who assumed an additional contribution to the quantity of accumulated charge when fillers were used.

As previously reported, (Nurazreena, et al. 2006) the percolation threshold depends not only on the size, shape, and spatial distribution of the fillers within the polymeric matrix but also on the processing. It is clear that the percolation threshold must be different when employing SiC nanowires as opposed to a carbon black powder as fillers. This remark can explain the slight difference observed between the permittivity of the two composite types in addition to the fact that the two fillers had not been incorporated in equal content.

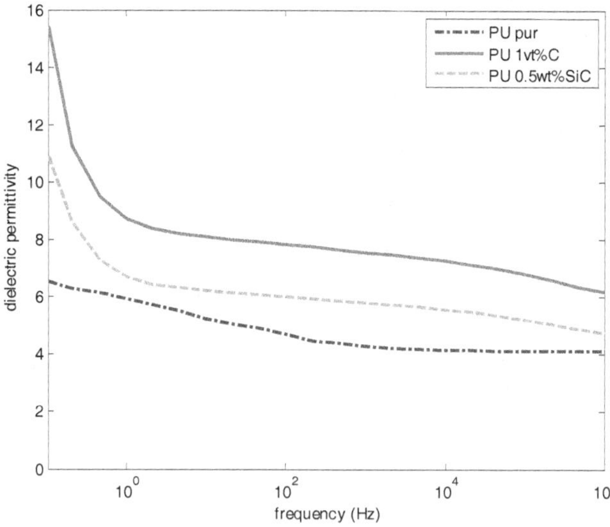

Fig. 7. The variation in the dielectric constant for a pure PU material and filled composite vs frequency.

4. Energy harvester characterization

This section describes the setup developed for characterizing the harvested power. It also includes a discussion of the obtained results and a comparison to the model mentioned in section 2.2.

4.1 Principle of measurement of the harvested power

Figure 8 provides a schematic representation of the setup developed for characterizing the power harvested by the polymer film. The mechanical system consisted of a shaker and a capacitive sensor. The shaker produced a vibration force in sinusoidal form, causing the sample to undergo a transverse vibration. The capacitive sensor (Fogale MC 940) recorded the transverse displacement of the sample from which the strain S_1 was calculated. The electrostrictive polymer was subjected to a DC biased electric field, produced by a function generator and amplified by a high-voltage power amplifier (Trek Model 10/10). As the polymer was not piezoelectric, it was necessary to induce a polarization with a DC bias in order to obtain a pseudo-piezoelectric behavior. The electroactive polymer was excited both electrically and mechanically, in order for its expansion and contraction to induce a current measured by the current amplifier (Keithley 617), thus giving rise to "an image" of the power harvesting by the polymer, due to electrical resistance (R_c). In this setup, the current was chosen as it is known to be less sensitive to the noise from the electrical network (50 Hz) and in order to avoid problems of impedance adaptation. All the data was monitored by an oscilloscope (Agilent DS0 6054A Mega zoom).

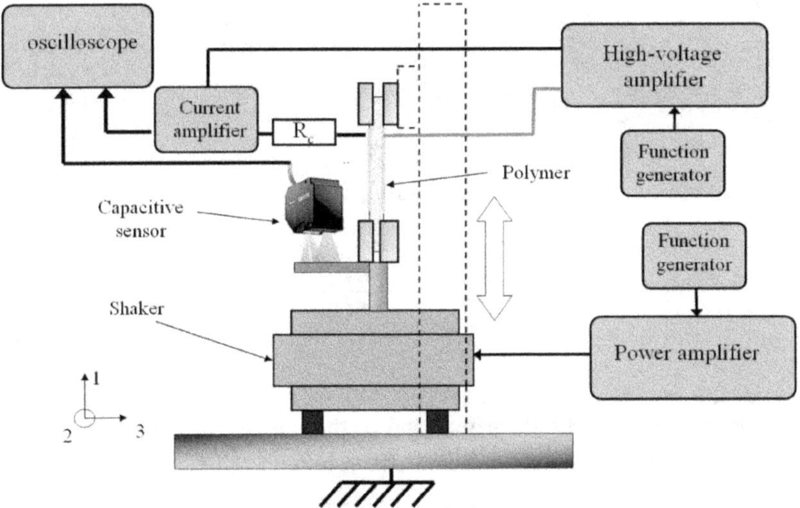

Fig. 8. A schematic of the experimental setup for the energy harvesting measurements.

A purely resistive load was directly connected to the electrostrictive element (Fig.8). In this case, the voltage on the load R_c was alternating. Considering equation (10) and the resistance of the load, the dynamic voltage on the electrostrictive element can be expressed in the frequency domain as a function of the displacement, the angular frequency ω.

$$\tilde{V} = \frac{\alpha\left(E_{dc}\right).R_c}{1+jR_c.C_p.\omega}\,j\omega.\tilde{u} \qquad (12)$$

Starting from equation (12), the harvested power can be written as a function of the displacement amplitude u_m:

$$P_{harvested\_AC} = \frac{\tilde{V}_m.\tilde{V}_m^*}{2.R_c} = \frac{\alpha\left(E_{dc}\right)^2.R_c}{1+\left(R_c.C_p.\omega\right)^2}\cdot\frac{\omega^2 u_m^2}{2} \qquad (13)$$

with V_m as the maximum voltage.

The harvested power reaches a maximum $P_{harvested\_AC\_max}$ for an optimal load R_{opt_AC}, and the optimal load resistance can be calculated according to :

$$\frac{\partial P_{harvested\_AC}}{\partial R_c} = \frac{1-\left(R_c.C_p.\omega\right)^2}{\left(1+\left(R_c.C_p.\omega\right)^2\right)^2}\cdot\frac{E_{dc}^2.\omega^2.u_m^2}{2} \qquad (14)$$

$\dfrac{\partial P_{harvested\_AC}}{\partial R_c} = 0$ when $R_C^2 = \dfrac{1}{C_p^2.\omega^2}$ and thus the power is at maximum for

$$R_c = \frac{1}{C_p.\omega} = R_{opt\_AC} \qquad (15)$$

Consequently, for the matched load the maximum power harvested is equal to

$$P_{harvested\_AC\_max} = \frac{\alpha\left(E_{dc}\right)^2.\omega.u_m^2}{4.C_p} \qquad (16)$$

4.2 Validation of the model

Preliminary measurements for the determination of the theoretical results of the harvested power have also been performed to evaluate the values of $\alpha\left(E_{dc}\right) = 2.M_{31}.Y.A.E_{dc}/L_0$.

Current measurements under short circuit conditions have been carried out for the former, as the short-circuit current magnitude I_0 is given by equation (7). Assuming a uniform strain in the material, the short-circuit current can be expressed as:

$$I_0 = 2.M_{31}.A.Y.E_{dc}.\omega.S_1 \qquad (17)$$

The short-circuit current was performed at 100 Hz for a constant strain of 0.5% and for different electric fields (Fig. 11). As expected, a linear relation between the current and the electric field was observed, and a previous study demonstrated the validation of this model (Lebrun, et al. 2009).

The choice of the frequency is not trivial, since most potential vibration sources have their fundamental vibration mode bellow 200 Hz, as demonstrated by Roundy et al. (Roundy, et al., 2003), and summarized in Table 3.

Vibration source	*Fundamental frequency vibration (Hz)*
Clothes dryer	121
Windows next to a busy road	100
People walking	1

Table 3. Vibration sources and their fundamental vibration mode

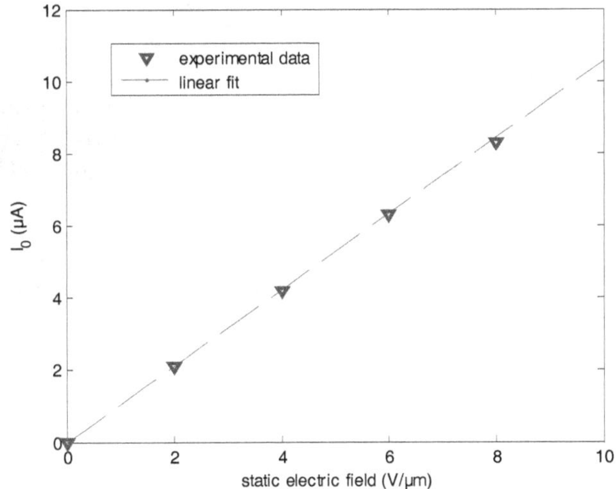

Fig. 9. The short-circuit current versus the static electric field for a constant strain of 0.5% at 100 Hz for P(VDF-TrFE-CFE).

In order to access the validity of the model of the harvested power presented in section 2.2 (eqs. (12) and (22)), various measurements were carried out. Figures 10 presents the power versus the electric field, for a constant electric load and strain (S_1=0.25%), as well as the power harvested as a function of the strain for a constant electric field (E_{dc}=5V/µm), with the same electric load (R_c=500kΩ). As expected from the model, a quadratic dependence between the power and the static electric field strain was observed. The experimental results thus validated the developed model for evaluating the harvested power.

To assess the validity of the model, various measurements were carried out. Figure 11 presents the power versus electric load for a given electric field (5 V/µm) and strain (0.2%) at 100 Hz. This data pointed out the existence of an optimal load resistance, as theoretically expected according to equation (15)

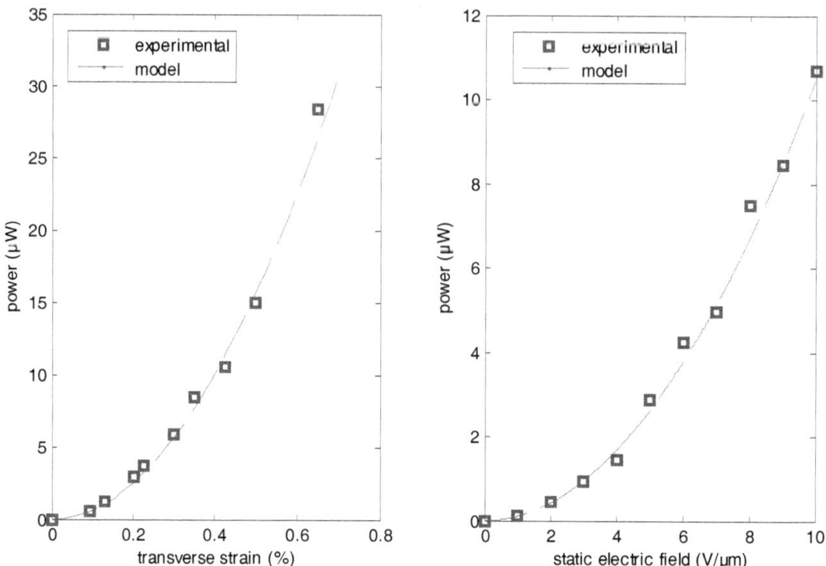

Fig. 10. (Left to right) The harvested power as a function of the static electric field, for a constant strain of S_1=0.2%, and the harvested power versus the transverse strain for a biased field of 5 V/μm, in both cases for a P(VDF-TrFE-CFE) at 100 Hz and Rc=500 kΩ.

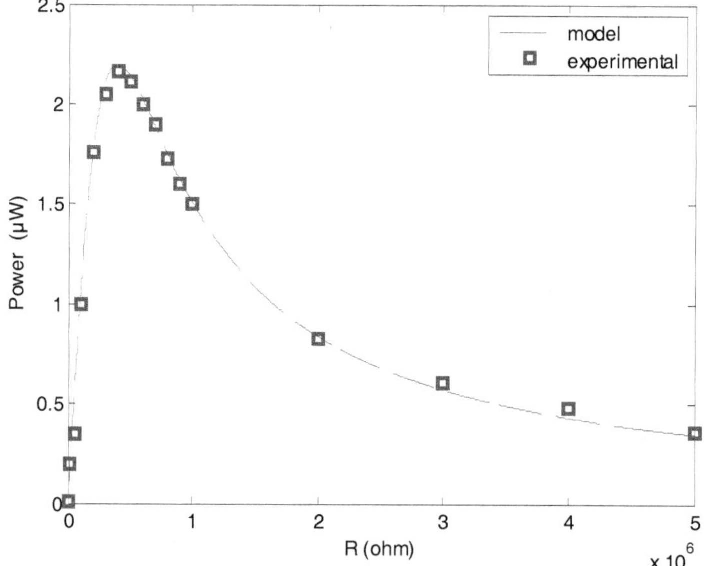

Fig. 11. The harvested power in AC for a constant electric field of 5 V/μm and a strain of 0.2% at 100 Hz.

4.3 Harvested power versus polymer

Table 4 gives the harvested power density for the polymer and composites at E_{dc}=5 V/μm, S_1=0.25% and 100 Hz, for their matched load. The power density for the polymer and composites was very low. Although this could be considered a disappointing result, one should keep in mind that in the electrostrictive case, the power was proportional to the square of the bias field (eq. 16), evaluated in Table 4 for a relatively low bias value. For example, doubling the value of the bias field to 10 V/μm (which is still quite low) would result in a power that was four times larger.

The output power of the composites with C nanofillers, SiC fillers, and pure PU under the conditions given below was estimated to be 1, 0.41, and 0.25 μW/cm³, respectively. The ratio between the estimated harvested power for the pure PU film and the nanofilled composites was equal to 3 in the case of C nanofillers and 1.4 for SiC fillers. This ratio was almost identical to the ratio of the square of the film permittivity, which was in good agreement with the fact that M_{31} is practically proportional to $\varepsilon_0(\varepsilon_r-1)^2/(Y. \varepsilon_r)$ and with the expression of power harvesting according to equation (16).

Obviously, the PU loaded with 1%C exhibited the highest dielectric permittivity and power density in the case of PU. However, with pure PU and for the same power density, an electric field of 10 V/μm was necessary, which was two times as high as for PU 1vl%C. The way of introducing the conductive particles into the polymer matrix was very interesting since it was possible to create a material capable of harvesting power under low electric fields. The highest power density of 180 μW/cm³ was obtained with the terpolymer matrix of P(VDF-TrFE-CFE). This seemed to be a very promising material, especially if nanoparticles were induced in the matrix.

Type	ε_r	Y (MPa)	Power harvested (μW/cm³)
Pure PU	4.8	40	0.25
PU 0.5%SiC	5.6	70	0.41
PU 1%C	8.4	40	1.0
P(VDF-TrFE-CFE)	42	500	180

Table 4. Material characteristics and harvested power density at 100 Hz for a static electric field of 5 V/μm and a transverse strain of 0.2%.

Nevertheless, the dielectric permittivity is not the only parameter to vary when attempting to optimize the harvested power; it depends on the application. For example, if a large strain of more than 100% is available, e.g., during human walking, a polymer able to support such a large strain before plastic transition is necessary. An example of such a material is PU, since it can be stretched up to 400%. This can be compared to the maximum for a P(VDF-TrFE-CFE), which is only 40%.

Two techniques exist for harvesting energy when utilizing electrostrictive polymers. The first was proposed by Ren et al. (Ren, et al., 2007) and consists of creating an energetic cycle, whereas the second involves applying a bias voltage for working in the pseudo piezoelectric behavior. Table 5 presents a comparison between the two methods. The energy harvesting based on the pseudo-piezolectric behavior was lower than the cycle-based energy harvesting. Although this could be considered a disappointing result, one should keep in mind that the strain in our case was ten times lower and the electric field was fourteen times lower. The model demonstrates that the energy harvesting was proportional to the square of these two conditions (strain and electric field), so for the same configuration, the energy density was 30 mJ/cm³ for a system working in the pseudo piezoelectric state. This density was approximately the same as in the case of Ren et al.

The great advantage of our method for mechanical energy harvesting consists in this technique requiring only a static field for its operation. The technique is consequently very simple to realize and implement, using for example batteries, or for an autonomous system, employing a piezoelectric material to produce the bias voltage.

Method	*Material*	S_1 *(%)*	E_3 *(V/μm)*	*Energy harvesting (J/cm³)*
pseudo-piezoelectric	P(VDF-TrFE-CFE)	0.2	5	$1.8 \cdot 10^{-6}$
cycle	P(VDF-TrFE) [Ren 2007]	2	67	$22.4 \cdot 10^{-3}$

Table 5. Comparison between two methods for harvesting energy.

4.4 Practical considerations

There exists a large number of detailed designs of electrostrictive polymer generators. An exhaustive description is beyond the scope of a single chapter, and furthermore, the technology is rapidly evolving. However, some general aspects can be considered based on the above analysis.

Figure 12 shows two general block diagrams of simplified electrostrictive generator circuits. Electrostrictive polymer, are often operated by a relatively high voltage (500V-2000V).

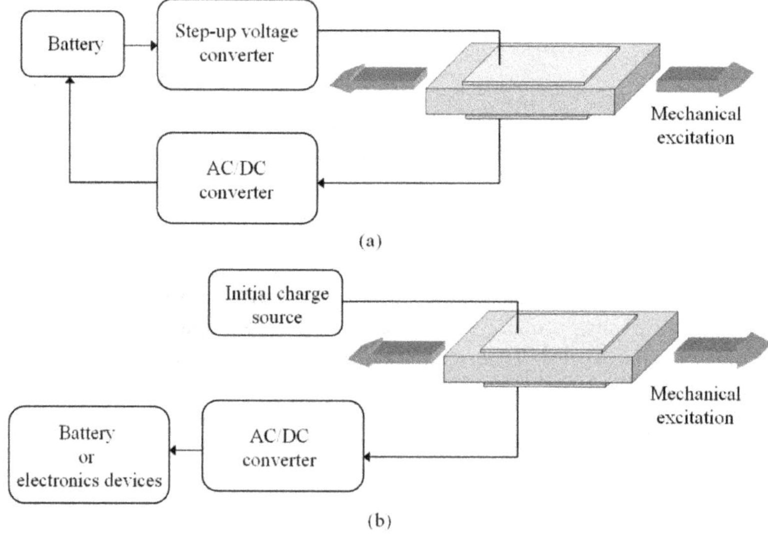

Fig. 12. Two conceptual circuit topologies for electrostrictive polymers generators

Portable applications are powered with lower voltages compatible with battery power. Moreover, in order to generate the bias voltage required for working in the pseudo-piezoelectric behavior, a step-up voltage converter has to be part of the generator circuit as in Fig. 12. Typically, they use high-speed switching with inductors or transformers, though other methods such as piezoelectric converters are also known. One such option is to use rechargeable low-voltage batteries for applying the bias electric field, as shown in Fig. 12 (a). However, a possible drawback of this design is that the static electric field must be raised in voltage relative to that of the battery. A step-up converter is needed in this case, which leads to increased costs. The important issue in this case is however not so much the added cost but the additional power loss from the conversion. Note that in configuration (b), only the generated energy has to be converted. The bias voltage was provided by means of other materials, like piezoelectric ceramics, which rendered it possible to create an autonomous system without batteries.

Electrostrictive polymer leakage is another practical consideration in power generation. Such leakage is a direct loss to the system. The importance of leakage phenomena depends very much on both the electrostrictive resistivity and the frequency operation. Leakage losses may influence the choice of polymer, but good dielectric materials can generally be identified to address this issue for most applications. For example, PU has such an outstanding resistivity that leakage losses may be insignificant (Cottinet, et al. 2010).

The biggest advantage of electrostrictive polymers might be the point that, in addition to a good dynamic range, they operate best at relatively long strokes and modest forces. This is a difficult part of the design space to address using conventional smart materials such as piezoelectrics ceramics. Moreover, it is a very common mechanical input available from a number of sources in the environment, such as human motion and wave power.

5. Applications of electrostrictive polymers for energy harvesting

Virtually any application where there is a need of electrical energy is a potential application for electrostrictive polymer generators. Figure 13, presents the orders of magnitude of the powers consumed by various CMOS electronic equipment that could be powered by miniature energy harvesting devices. This data shows the potential applications of electrostrictive polymers for energy harvesting. The harvested power density reported in this paper would be enough to power complex devices such as watches or RFID tags.

However, this material is much more competitive for certain applications as opposed to others, and it is important for a new technology to identify general and specific areas of application where it can be the most competitive. A general advantage and drawback of the various classes of EAPs for energy harvesting are summarized in Table 6. All the materials exhibit a high energy harvesting density, except for the ionic polymer. Nevertheless, the main advantage of this material is that it does not require any other polarization tension in order to operate. It is thus possible to realize autonomous systems without batteries.

The difference in energy harvesting between dielectric elastomers and electrostrictive polymers is not very large. The great advantage of dielectric elastomers is the possibility of having high strains of more than 400% as compared to electrostrictive polymers (100%). However, the former materials require higher electric fields, causing the size of the device to increase, in addition to the loss due to the conversion of the high voltage.

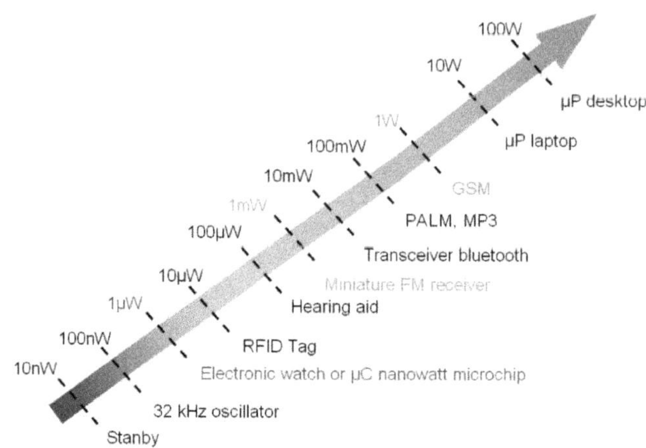

Fig. 13. The various powers consumed by CMOS electronic devices (Sebald, et al., 2008).

Type	Conditions	Energy harvesting	advantage	inconvenient
Dielectric elastomer - Silicone (Pelrine, et al. 2001) - Acrylique (Pelrine, et al. 2001)	?[3] , E^2=160MV/m ? , E^2=160MV/m	13mJ/g 400mJ/g	+ high energy density + cheaper + high strain	- high voltage for harvesting energy - significant loss - passive material (requires voltage to operate).
IPMC -Nafion® 117 (Brufau, et al. 2008)	S=0.2%	8nJ/cm³	+ active material	- expensive - works at low frequency
Electrostrictive polymer - P(VDF-TrFE) (Ren, et al., 2007) - P(VDF-TrFE-CFE)	S^1=2%, E^2=67MV/m S^1=0.2%, E^2=5MV/m	22.4mJ/cm³ 1.8µJ/cm³	+ high energy density for low electric field + low loss + high working frequency	- passive material

[1] value of the transverse strain for the energy harvesting density
[2] value of electric field for the energy harvesting density
[3] not indicated in the paper
? not indicate in the paper

Table 6. Non-exhaustive synoptic table of the different methods used for harvesting energy

Compared to piezoelectric generators, electrostrictive polymers may offer advantages such as lower cost, lighter weight, or smaller size. These materials are well suited for harvesting energy from human motion. Natural muscle, the driving force for human motion, typically works at low frequencies and is intrinsically linear; two characteristics where electrostrictive polymers offer advantages. They are also interesting for wave motions with relatively low frequencies of 0.1 Hz to 1 Hz, and high amplitudes (wave heights on the order of 1 m are common).

Many other interesting applications exist for electrostrictive polymer. Remote and/or wireless devices are growing in use; an ideal devices can harvest their own energy thereby eliminating the need for battery replacement. Electrostrictive polymers are well suited for these applications if mechanical energy is available, as might be the case in portable devices carried by people, animals, or automobiles. Moreover, it is possible to deposit such materials on large surface, which is very interesting for health monitoring in the wings of airplanes, for example.

In summary, the major advantages of this technology include the lighter weight, greater simplicity, and lower cost.

6. Conclusions

Electrostrictive polymer composites for energy harvesting have been developed and tested using different types of polymers. A good agreement between the modeled data and actual results was found. Various means of increasing the intrinsic characteristics of the polymers was also studied A comparison between the different techniques for harvesting energy has been realized, and demonstrated the simplicity and advantage of working in the pseudo-piezoelectric behavior.

Future work will concern the research of polymer matrices and novel fillers to increase the permittivity of the resultant composites and consequently their capabilities of harvesting electrical energy upon mechanical vibrations. As an example can be mentioned P(VDF-TrFE-CFE) filled with carbon particles. Another point of research within this topic will concern an optimization of electrical boundary conditions on our harvester by using a non-linear technique.

The present chapter demonstrates the potential for future application. In fact, one of the most important trends in the electronic equipment technology from its origins has been the reduction in size and the increase in functionality. Nowadays, small, handheld, though very powerful, devices are commercially available. They allow the user to play music, to wirelessly communicate or to compute practically everywhere. The size of such devices is becoming so small that the term wearable device is being used instead of portable device; they can be integrated in everyday objects such as watches, glasses, clothes, etc.

Electrostrictive polymers have demonstrated excellent performances. Numerous applications appear feasible, but challenges remain. Electrostrictive polymers seem to be more advantageous for applications requiring low or variable frequencies, low cost, and large areas. Currently, piezoelectric (PZT) materials are the most popular options for harvesting mechanical energy because of their compact configuration and compatibility with Micro-Electro-Mechanical Systems (MEMS). However, their inherent limitations include aging, depolarization, and brittleness. Finding ways to circumvent these limitations is the next exciting and puzzling challenge to achieve realistic EAP energy harvesters.

7. Acknowledgments

This work was supported by the DGA (Délégation Générale pour l'Armement).

8. References

Badel, A., Guyomar, D., Lefeuvre, E., and Richard, C. (2005). Efficiency enhancement of a piezoelectric energy harvesting device in pulsed operation by synchronous charge inversion. *Journal of Intelligent Material Systems and Structures*, Vol. 16, No.10, pp. (889-901), 1045-389X, 2005.

Bar-Cohen, Y. (2004) Electroactive Polymer (EAP) Actuators as Artificial Muscles (Reality, Potential, and Challenges). SPIE Press, Bellingham, ISBN: 9780819452979, Washington, USA.

Brufau-Penella, J., Puig-Vidal, M., Giannone, P., Graziani, S., and Strazzeri, S. (2008). Characterization of the harvesting capabilities of an ionic polymer metal composite device. *Smart materials and structures*, Vol. 17, No. 1, pp.(015009.1-015009.15), 0964-1726.

Cottinet, P.-J., Guyomar, D., Guiffard, B., Putson, C., and Lebrun, L.,. (2010). Modeling and Experimentation on an Electrostrictive Polymer Composite for Energy Harvesting *IEEE Transactions on ultrasonics, ferroelectrics, and frequency control*, Vol. 57, No. 4, pp., 0885-3010.

Dang, Z.-M., Fan, L.-Z., Shen, Y., Nan, C.-W. (2002). Study on dielectric behavior of a three-phase CF/(PVDF + BaTiO$_3$) composite. *Chemical physics letters*. Vol. 369, No 1-2, pp. (95-100), 0009-2614.

Eury, S., Yimniriun, R., Sundar, V., Moses, P.J. (1999). Converse electrostriction in polymers and composites *Materials Chemistry and Physics*. Vol 61, No.1, pp. (18-23), September 1999.

Gallone, G., Carpi, F., De Rossi, D., Levita, G. and Marchetti, A. (2007) Dielectric constant enhancement in a silicone elastomer filled with lead magnesium niobate–lead titanate. *Materials Science and Engineering: C*, Vol. 27, No. 1,(January 2007), pp. (110-116).

Guiffard, B., Seveyrat, L., Sebald, G., and Guyomar, D. (2006). Enhanced electric field-induced strain in non-percolative carbon nanopowder/polyurethane composites. *Journal of physics. D, Applied physics.* Vol. 39, No 14, pp. (3053-3057), 0022-3727.

Guiffard, B., Guyomar, D., Seveyrat, L., Chowanek, Y., Bechelany, M., Cornu, D., and Miele, P. (2009). Enhanced electroactive properties of polyurethane films loaded with carbon-coated SiC nanowires *Journal of physics. D, Applied physics.* Vol. 42, No 5, pp. (055503.1-055503.6), 0022-3727.

Guillot, F.M., Balizer, E. (2003). Electrostrictive effect in polyurethanes. *Journal of Applied Polymer Science.* Vol. 89, No.2, pp. (399-404), July 2003.

Gurunathan, K., Murugan, A.V., Marimuthu, R., Mulik, U.P., and Amalnerkar, D.P. (1999) Electrochemically synthesised conducting polymeric materials for applications towards technology in electronics, optoelectronics and energy storage devices *Materials Chemistry and Physics*, Vol. 61, No. 1, pp. (173–191), 0254-0584, 1999.

Guyomar, D., Lebrun, L., Putson, C., Cottinet, P.-J., Guiffard, B., and Muensit, S. (2009). Electrostrictive energy conversion in polyurethane nanocomposites *Journal of Applied Physics.* Vol. 106, No. 1, pp. (014910 – 014919), 0021-8979, 2009.

Huang, C., Zhang, Q.-M., deBotton, G. and Bhattacharya, K. (2004). All-organic dielectric-percolative three-component composite materials with high electromechanical response. *Applied Physics letters.* Vol. 84, No 22, pp. (1757632- 1757632).

IEEE Standard on piezoelectricity,(1988), ANSI/IEEE Standard 176-1987, 0-7381-2411-7.

Lebrun, L., Guyomar, D., Guiffard, B., Cottinet, P.-J. and Putson, C. (2009). The Characterisation of the harvesting capabilities of an electrostrictive polymer composite. *Sensors and Actuators A: Physical* Vol. 153, No. 2 ,pp. (251-257), August 2009.

Lehmann, W., Skupin, H., Tolksdorf, C., Gebharde, E., Zentel, R., Krüger, P., Lôsche, M, et Kremer, F. (2001). Giant lateral electrostriction in ferroelectric liquid-crystalline elastomers. *Nature*. Vol. 410, No 6827, pp. (447-450), 0028-0836.

Liu, Y., Ren, K., Hofmann, H. F. and Zhang, Q. M. (2004) Electrostrictive polymer for mechanical energy harvesting. *Proceedings SPIE, Int. Soc. Opt. Eng.*, Vol. 5385 ,pp. (17-28) , San Diego CA, March 2004, SPIE, Bellingham WA, USA.

Liu, Y., Ren. K., Hofmann, H., and Zhang, Q.M. (2005). Investigation of electrostrictive polymers for energy harvesting. *IEEE Transactions on ultrasonics, ferroelectrics, and frequency control*. Vol. 53, n°12, pp. (2411-2417), December 2005.

Mazur, K. (1995) Polymer-ferroelectric ceramic composites, In *Ferroelectric Polymers*, Nalwa, H. S. (Ed.), 539-610,CRC press, 9780824794682, New-York.

Mitchell, B.S. (2004) *An Introduction to Materials Engineering and Science for Chemical and Materials Engineers*, Wiley, 978-0-471-43623-2,New York.

Nurazreena, L. B., Ismail, H. (2006). Metal Filled High Density Polyethylene Composites – Electrical and Tensile Properties, *Journal of Thermoplastic Composite Materials*, Vol. 19, No. 4, pp. (413-425).

Pelrine, R., Kornbluh, R.D., Eckerle, J., Jeuck, P., Oh, S., Pei, Q., and Stanford S. (2001) Dielectric elastomers: Generator mode fundamentals and application *Proceedings SPIE, Int. Soc. Opt. Eng.*, Vol. 4329, pp. (148-156), San Diego CA, March 2001, SPIE, Bellingham WA,USA.

Poulin, G., Sarraute, E., Costa, F. (2004). Generation of electrical energy for portable devices. Comparative study of an electromagnetic and a piezoelectric system. *Sensors and Actuators A: physical*, Vol. 116, No. 13, pp. (461-471), 0924-4247 , 2004.

Ren, K., Liu, Y., Hofmann, H. and Zhang Q. M. (2007) An active energy harvesting scheme with an electroactive polymer *Applied Physcics Letters*. Vol. 91, No. 13, pp. (132910-132913), 2007.

Roundy, S., Xright, P.K., Rabaey, .J. (2003) A study of low level vibrations as a power source for wireless sensor nodes, *Ubiquitous Computing*, Vol. 26, No. 11, pp. (1131-1144), 2003.

Sebald, G., Lefeuvre, E., and Guyomar, D. (2008) Pyroelectric energy conversion: Optimization Principles. *IEEE Transactions on Ultrasonics, Ferroelectrics and frequency Control*, Vol. 55, No. 3,pp. (538-551) March 2008, 0885-3010.

Szabo, J. P., Hiltz, J. A., Cameron, C. G., Underhill, R. S., Massey, J., White B., Leidner J.(2003). Elastomeric composites with high dielectric constant for use in Maxwell stress actuators. *Proceedings of Eectroactive polymer actuators and devices (EAPAD)*, pp. (556-567), 0-8194-4856-7, San Diego CA, March 2003, SPIE, Bellingham WA, USA.

Zrínyi, M. (2000). Intelligent polymer gels controlled by magnetic fields. *Colloid & Polymer Science*, Vol. 278, pp.(98–103), 1435-1536, August 2000.

Zucolotto, V., Avlyanov, J. and Mattoso, H. C. (2004). Elastomeric conductive composite based on conducting polymer-modified carbon black. *Polymer composites*. Vol. 25, No 6, pp. (617-621), 0272-8397.

11

Ceramic-controlled piezoelectric: development, applications and potentiality in electrical and biomedical engineering

E. Suaste Gómez, C. O. González Morán and J. J. A. Flores Cuautle
Centro de Investigación y de Estudios Avanzados del Instituto Politécnico Nacional,
México

1. Introduction

In the solid-state materials there are the dielectrics called smart materials, which are those that have the ability to manifest qualitatively properties under the external influence, such as pyroelectric, piezoelectric and ferroelectric materials. Ferroelectric materials are those that exhibit spontaneously electrical polarization and the polarization has more than one possible equilibrium orientation, similar to ferromagnetic materials, ferroelectrics present "domains" or areas in which the polarization is oriented in certain crystallographic direction, by means of applying an intense electric field (E), it is possible to reorients the domains in the direction of the nearest electric field allowed by the material structure and inducing dipole moments inside matter. Independent of the structure of the constituent material a cylindrical symmetry in the polarized ceramics is induced (Jaffe & Cook, 1971).

The fact that an applied E field induces dipole moments inside material objects has two common physical mechanisms for the creation of these induced dipole moments inside matter. The first is the rotation of existing polarized region ("dipoles") inside the material into partial alignment with the applied field. These sub regions can be individual molecules, as in the polar molecules of water, or domains consisting of thousand of atoms, as ferroelectric ceramics. The second mechanism is the creation of dipoles by charge separation of initially nonpolar atoms or molecules (Wai-Kai, 2005); in this case, the electron clouds are pulled in one direction by the E-field, while the positive nuclei are pushed in the opposite direction as shown in Fig. 1.

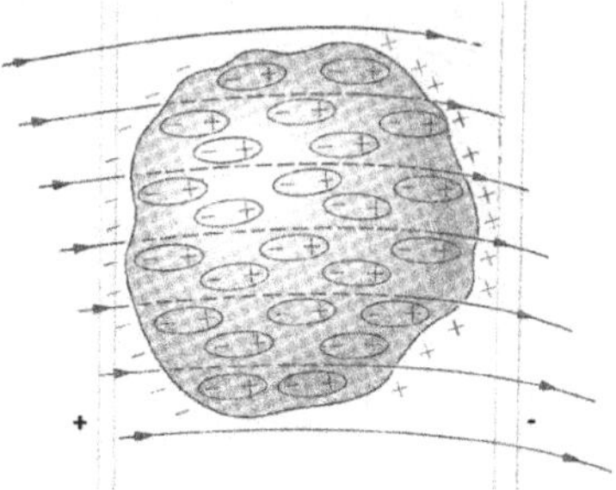

Fig. 1. Polarization of the material by an electric field to create dipoles through matter

Among such piezoelectric materials, Barium titanate (BaTiO$_3$) and Lead Lanthanum Zirconate Titanate (PLZT) are well known materials. BaTiO$_3$ is a ferroelectric classic perovskite structure and PLZT is a ferroelectric solid solution with wide-ranging material properties that depend on its composition (Moulson & Herbert, 2003; Plonska & Surowiak, 2006). Several studies have been carry out in order to study the influence of compositionally modified ceramic bodies, the addition of a variety of elements to obtain a doped ceramic is one of the most important research line, another line is the study of multilayer capacitors based on these materials, having these concepts in mind, a metallic wire implanted in the transversal way in the middle of a ferroelectric ceramic is use as control electrode, once upon the ceramic is poled this piezoelectric bulk is called ceramic-controlled piezoelectric (CCP), the metallic wire has more than one propose in the ceramic: using this wire implanted, a free ceramic face is obtained, this face is used in order measure optic and mechanic events, the wire in total immersion into the ceramic serve as control electrode which allows to develop several applications.

Recently, we have fabricated bulks of BaTiO$_3$ single plate and PLZT single plate, using a Pt-wire of 300 μm of diameter, this wire was implanted into ceramics to get a free face in order to measure optic and mechanic events. Our structure design includes a modified electrode configuration. The Pt-wire was chosen as implant because it possesses high resistance to chemical attack, it has excellent high-temperature characteristics (melting point 1768.3 ºC), and it has a stable electrical properties and thermal conductivity with small variations (Touloukian et al., 1970).

When CCP is polarized, the mobile charges in the Pt-wire accumulate on the surface until the field they produce completely cancels the external applied field within Pt-wire, thereby producing equilibrium. Then inside Pt-wire which is in electrical equilibrium, the electrical field is zero as shown in Fig. 2. By this statement, we really mean that the net charge is distributed over a section of the surface having a thickness of several atomic layers not in the geometrical sense (González & Suaste, 2009).

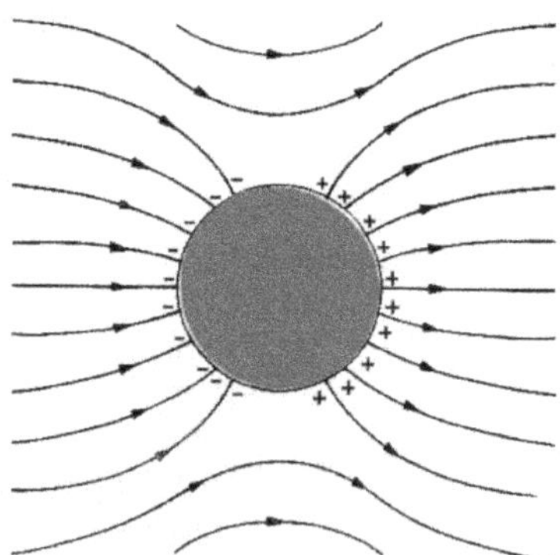

Fig. 2. Electric field waves a cross Pt-wire (top view).

The Pt-wire cylinder dimensions are 10 mm high and 0.3 mm diameter, this superficial area is 9.42 mm². The superficial area of one side face of the ceramic is 78.53 mm². When we compare this both superficial area, the ceramic is 8.33 times greater than Pt-wire superficial area.

Pt-wire implanted into PLZT ceramic bulk develops a singular performance in the CCP due to the domain´s coating on Pt-wire. This type of domain wall separates domains polarized perpendicularly to each other and when CCP is polarized the field should be such as to exert opposite torques on the polar axes of opposite domains, but not large enough as to permanently rotate the polarization. A similar effect is obtained by applying a suitable stress to the ceramic, as the piezoelectric effect has a different sign in antiparallel domains. When CCP is poled, over Pt-wire is originated a great concentration of domains due to dipoles were oriented over all external part of itself. These concentrations achieve free flux charges around Pt-wire when CCP is excited by stress or LASER illumination on the side face as shown in Fig. 3.

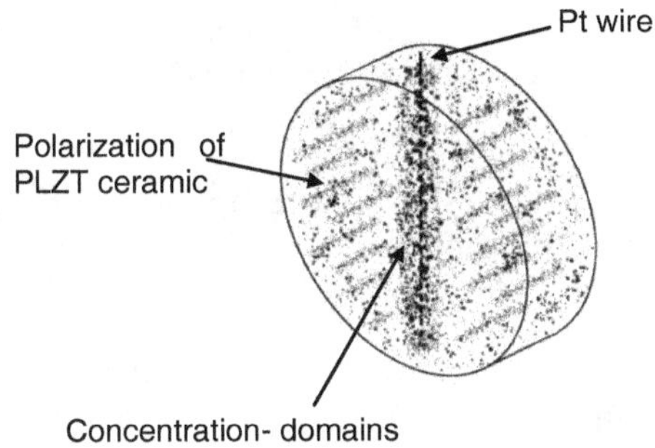

Fig. 3. Scheme from CCP and its distributed domains in around Pt-wire

Considering that ferroelectric and piezoelectric materials can be used as sensors and actuators. Piezoelectric pressure and acceleration sensors are now commercially available as well as a variety of piezo-vibrators. The development of ultrasonic motors for a variety of new applications has been exhibited and widespread in recent years. Besides, ferroelectric materials generally have excellent electro-optic properties due exhibit photovoltaic effects under near-ultraviolet light and thus seen as promising candidates for application in optical communications systems of the future using Pt implants.

When the PLZT is illuminated after poling, voltage and current can be generated due to the separation of photo induced electron and holes by its internal electric field. This is because the photovoltaic effect occurs within the material and is considered to be an optical property of the material itself and show application potentials for realizing energy transfer in micro electromechanical systems and optoelectronic devices (Sturman & Fridkin, 1992; Ichiki et al., 2004, 2005).

The sample fabrication, how the CCP behaves to the temperature on its dielectric constant (with and without implant), the ceramic structure, mechanical and optical performance, and some important applications are described in the following sections (Suaste et al., 2009).

2. Fabrication

The oxide-mixing route is the most used method for commercial purposes although it has several limitations, like difficulties in achieve microscopic compositional uniformity, is one of the cheaper methods to get ferroelectric ceramics, the method involve the general steps: mixing and grinding, calcination, grinding, shaping, densification, and finishing, which will be next briefly described.

Mixing: the raw powders are weighted in an appropriated portion with an allowance for the impurity and moisture content, the grain size is uniformed by grinding usually this part is wet by means of include some liquid to agglutinated the powder.

Calcination: the calcinations purpose is to begin the reaction by the firing the powders at temperatures around 800 – 1100° C depending of the material kind, this step not always is necessary, it depends on the material kind.

Shaping: in this stage the required dimensions and shape are forming by molding the powders and applying pressures that are around 75 - 300 MPa, is in this step that a Pt - wire is put into the middle of the ceramic in transversal way in order to obtain the third electrode.

Densification: also know as firing, it is made in a refractory recipient and at temperatures of 80 to 90 % of material melting temperature in which the constituent ions have the mobility for the solid state sintering process take place.

Finishing: the finishing involves the polishing, machining (if it is necessary), and metallizing.

In our case the ceramic design includes a modified electrode configuration described as follows: a Pt-wire as control electrode in total immersion into the ceramic, the function of this extra electrode is to provides mobile charges in the ceramic, Pt-wire implanted into Lanthanum modified PZT (PLZT) and Barium titanate ceramic bulk develops a particular performance in the CCP due to the domain's coating on wire and increases the analysis area as Fig. 4. shows.

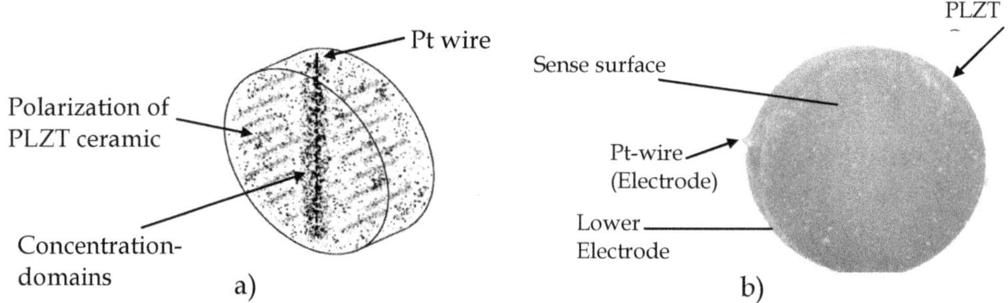

Fig. 4. a) Diagram of Pt-wire position in piezoceramic and domains representation, b) PLZT ceramic image

Energy Dispersive Spectroscopy (EDS) was made in order to verify the chemical interaction between Pt-wire and ceramic, this analysis was made in the Pt-wire ceramic boundary, Pt-wire and ceramic and reveals that no chemical reaction occurs as shown in Fig. 5.

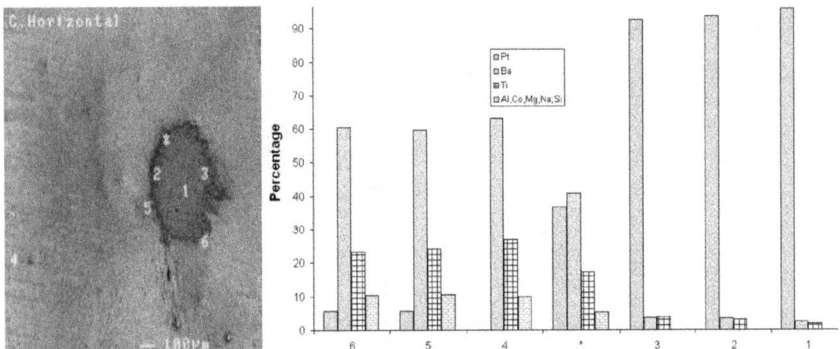

Fig. 5. Scanning Electron Microscopy (SEM) and EDS made in a barium titanate ceramic with Pt-wire implanted, numbers and (*) represent points in which an EDS was made.

Studies of the dielectric constant ε as function of temperature in BaTiO$_3$ and PLZT ceramic samples (with and without Pt-wire) were carried out. Curie temperature (T$_c$) was measured by the dielectric constant method.

Samples were heated at a rate of 10∘C/min, until 600∘C, the capacitance was measured at 1 kHz with a Beckman LM22A RLC bridge. The dielectric constant was determined by expression 1.

$$\varepsilon = \frac{Cl}{\varepsilon_0 A} \tag{1}$$

Where C is the capacitance in F, l is the thickness of the sample in m, A the sample area in m^2 and ε_0 the vacuum permittivity = 8.85 × 10^{-12} F/m.
Results are shown in Fig. 6 for BaTiO$_3$ and PLZT ceramics respectively and indicate that T$_c$ for BaTiO$_3$ remains equal and PLZT has a 15°C shift while dielectric constant drops from 3916 to 2846 and 6923 to 5889 with Pt-wire for BaTiO$_3$ and PLZT respectively.

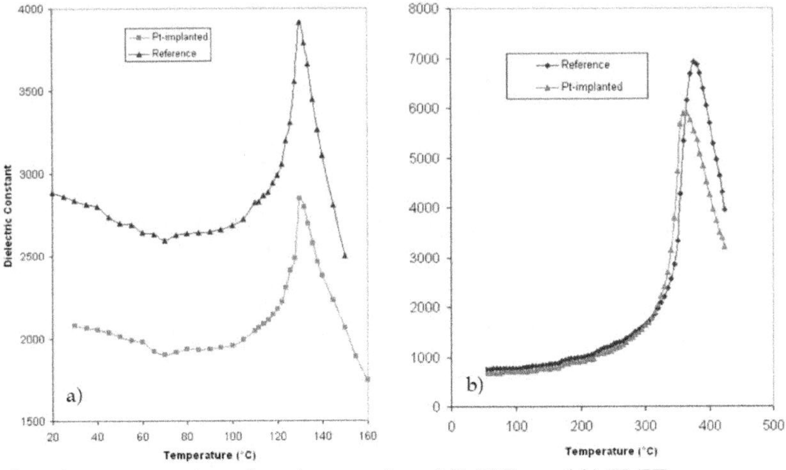

Fig. 6. Dielectric constant piezoelectric ceramics: a) BaTiO$_3$ and b) PLZT.

3. Experimental Setups

To validate and determine sensible characteristics of the CCP, two experimental setups (mechanical and optical) were used. Over a free side face of CCP were selected six random specific zones to get optical and mechanical signals as shown Fig. 7.

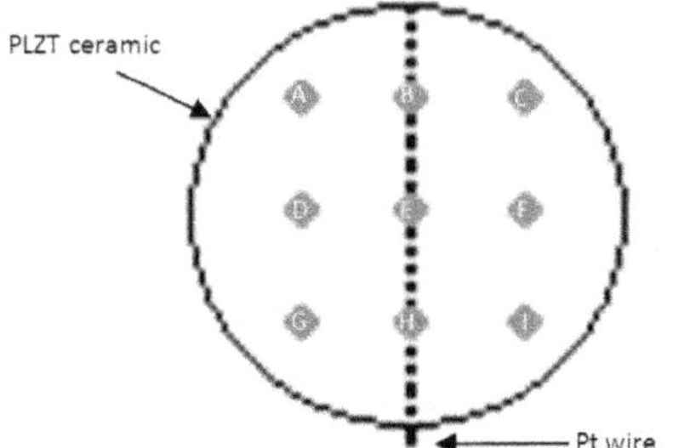

Fig. 7. Selected zones for realizing optical and mechanical analysis.

Optical setup

Figure 8 shows, the first experimental setup, to obtain the photovoltaic current magnitude, of those six zones, a fiber-coupled LASER system BWF1 mod, BWF-65015E/553369 modulated at 3 Hz of frequency with a wavelength of 650 nm and a maximum LASER illumination of 160 mW/cm^2 were used. We include a set of lenses to reduce the LASER beam spot to a 0.5 mm of diameter size. The CCP was placed on an X-Y translation microscope stage. The electrical contacts from CCP were connected to an SRS-Low Current Noise Preamplifier SR570 to get records. Those records were registered by an Oscilloscope Agilent DSO3062A measuring their peak to peak voltage to construct its graphics which the Z-axis is the magnitude of the photovoltaic current of each record.

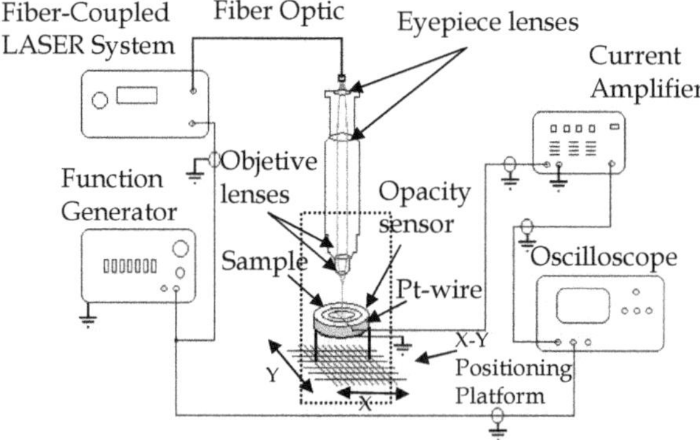

Fig. 8. Experimental setup to determine the ceramic sensibility of six random zones

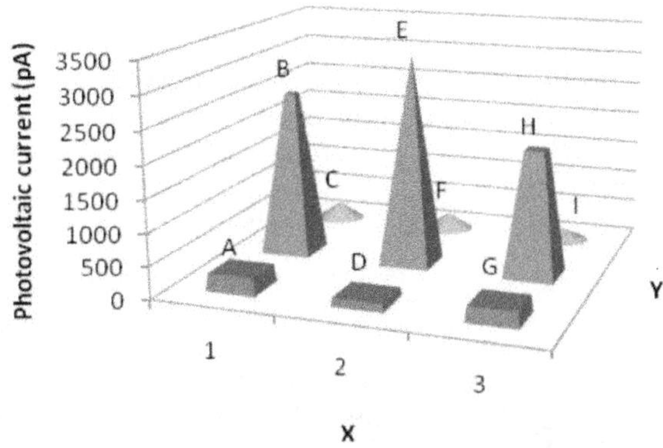

Fig. 9. Photovoltaic current generated by an illumination of 160 mW/cm² on six zones of CCP.

Mechanical setup

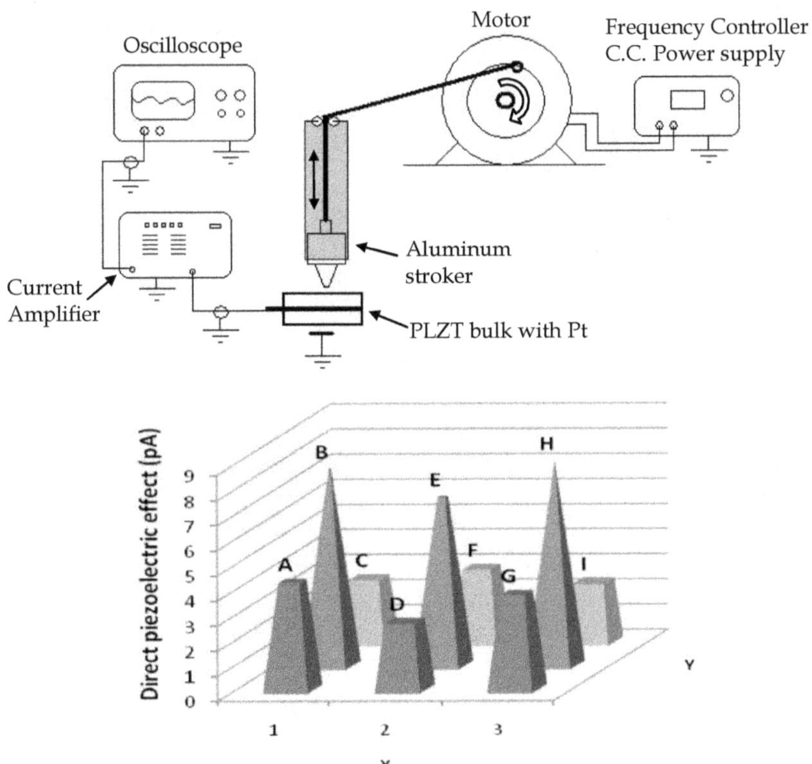

Fig. 10. Electrical current was generated by a force of 53 mN on six zones of CCP.

This setup consists of constant weight pulse (gram-force) at a frequency of 2.26 Hz over the CCP in the same zones such as optical setup. The components of this setup were on a 24V

CD control motor Micro Switch 33VM82-020-11. The CCP was struck in a pulsed form controlled by a mechanism developed in our laboratory the experimental setup is described in Fig. 10. The force applied was 53.9 mN and signals were record in a SRS-Low Current Noise Preamplifier. Those records were registered by an Oscilloscope Agilent measuring their peak to peak voltage to construct its graphics which the Z-axis is the magnitude of the photovoltaic current of each record.

4. Applications

Optical applications

Due to pyroelectric properties that CCP has, it is possible to develop low frequency sensors which allow detecting some light wavelengths from ultraviolet to infrared, as show in the next three experimental setups.

The first optical application consisted of obtaining an electrical signal of CCP between a side face and the Pt-wire electrode due to pulsed LASER beam that affected the opposed side face, for the stimulus it was used a He-Ne LASER and an optical chopper. The reference signal was obtained from controller chopper output and CCP signal was increased and recorded by a Low-Noise Current Preamplifier and Oscilloscope respectively.

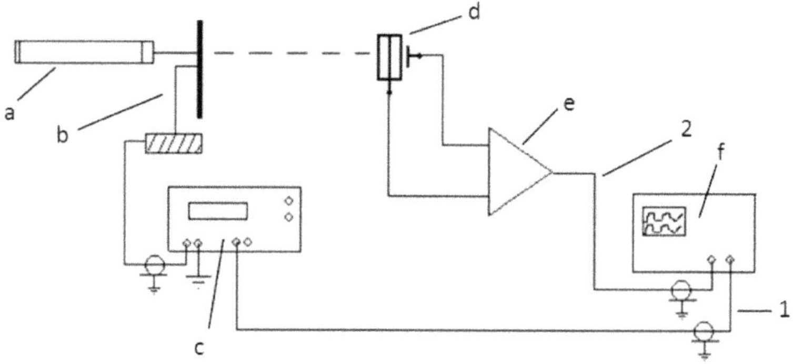

Fig. 11. CCP used as pulsed LASER beam detector. a) He-Ne LASER, b) Optical Chopper system, c) Chopper controller system d) CCP, e) Amplifier, f) Oscilloscope.

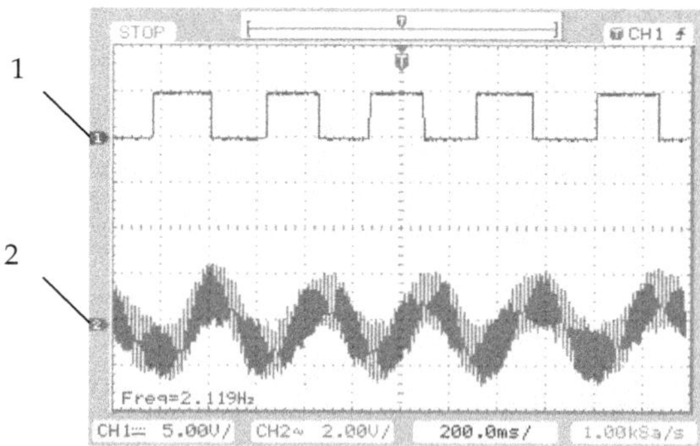

Fig. 12. Graphics from schematic setup, 1) Chopper pulse signal (reference) 2.11 Hz 2) Signal obtained from CCP 2.11 Hz

In the second optical application, CCP was used as visible light detector; a stroboscope was used to a 3.1 Hz, the light sparkles in one of the faces of CCP to a distance of 5 cm and on the sensor of a Standard Light Detector. Electrical signals were obtained, magnified and registered in an oscilloscope.

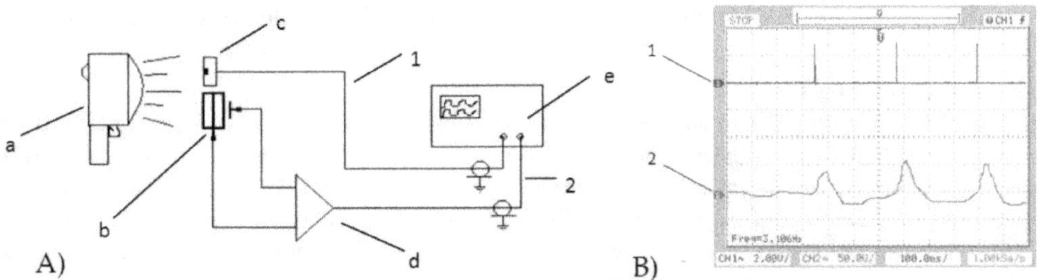

Fig. 13. CCP used as visible light detector. A) Schematic diagram, a) Stroboscope, b) CCP, c) Standard Light Detector, d) Amplifier, e) Oscilloscope, B) Graphics from schematic setup, 1) Reference signal from Standard Light Detector, 2) Amplified signal from CCP.

The third application is an acoustical-optical application where CCP is used as sound emitter; its electrical AC source was 60 Hz at 270 V Peak-Peak, this sound was detected by a metallized PVDF film which reflects He-Ne LASER beam to a Standard Light Detector DETN12 as shown in Fig. 14(A). The electrical responses from PVDF film and Standard Light Detector were observed in the oscilloscope, Fig. 14(B).

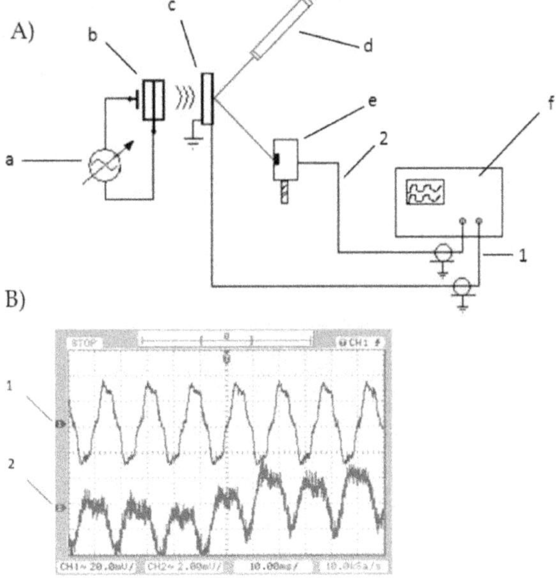

Fig. 14. CCP used as acoustical-optical sound emitter, detected and modulated optically with PVDF. A) Schematic diagram, a) 60 Hz at 270 V Peak-Peak source, b) CCP, c) Metallized PVDF film, d) He-Ne LASER, e) Standard Light Detector, f) Oscilloscope, B) Graphics from schematic setup, 1) Reference PVDF signal, 2) Signal obtained from Standard Light Detector.

In the optical applications, it is possible to be appreciated that He-Ne LASER beam in one of the side faces of CCP was register satisfactorily and that was compared with the chopper controller signal is illustrated in Fig. 11. Also, CCP is sensible to visible light on one of side faces of CCP that was compared with reference light detector and obtained a waveform as shown in Fig. 13. Additionally, on operation mode of acoustical-optical, CCP worked as a good sound emitter of AC signal having as a receiver a PVDF film, demonstrated in Fig. 14.

Mechanical Application

Mechanical application, CCP was used as impact sensor on a 24V CD control motor Micro Switch 33VM82-020-11 with dc analog tachometer included, CD motor was struck in a pulsed form controlled by a mechanism developed based on a relay system controlled by push bottom is illustrated in Fig. 15(A). The acquisition was realized from the tachometer output (reference signal) and CCP side face and Pt-wire electrode, both signals were registered in an oscilloscope and displayed in Fig. 15(B).

Mechanical, in Fig. 15(A) shows the CCP as well as a good vibrations sensor and the obtained graphics by CD motor tachometer and CCP were displayed in Fig. 15(B).

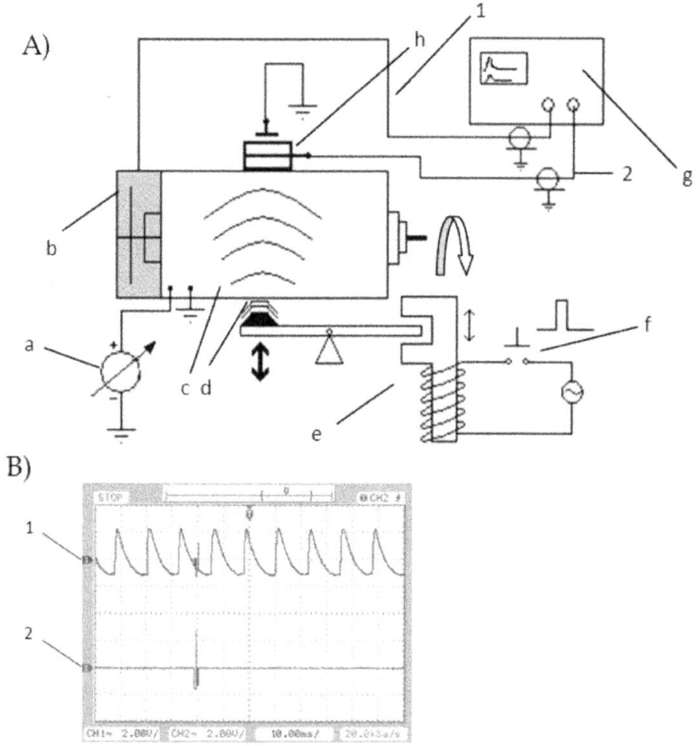

Fig. 15. CCP used as vibrations detector. A) Schematic diagram, a) DC Source, b) Tachometer, c) DC motor, d) struck zone, e) Impact pulsed device, f) Controlled push bottom, g) Oscilloscope, h) CCP, B) Graphics from schematic setup, 1) Signal from tachometer, 2) Signal from CCP.

Thermal application

Thermal application, CCP was used as pyroelectric sensor from 25 ºC to 80 ºC obtaining change of current. CCP was coupled onto external surface of a furnace with a hole, in Fig. 16(A), acquisition temperature is shown. To measure current signal, it was put as reference a digital thermometer and for obtaining of variation in electric charge (or current) in an amperemeter, illustrating the tendency of the variation of electrical charges of CCP as shown in Fig. 16(B). Thermal, also responds to the changes of temperature as pyroelectric sensor with an ascending tendency as is demonstrated in Fig. 16(B).

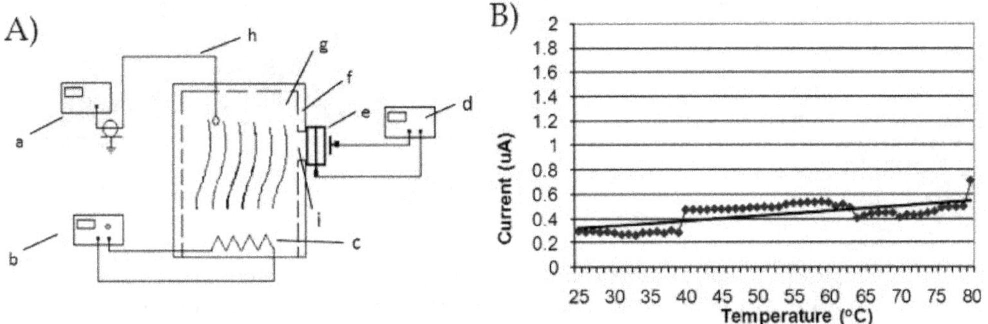

Fig. 16. CCP used as pyroelectric sensor. A) Schematic diagram, a) Digital thermometer, b) Temperature Control, c) Heater, d) Amperemeter e) CCP, f) Thermal Isolation, g) Furnace, h) Temperature Sensor, thermocouple type J, i) Furnace test window, B) Current graphic, I) Signal recorded from CCP, II) Tendency from signal recorded.

Electrical applications

For electrical application, three experimental setups were used. In the first electrical application, CCP was used as resonant element of high frequency oscillator based on CMOS inverter (Frerking, 1978), where Pt-wire electrode is connected to ground and high frequency signal of 41.66 MHz were obtained from logical inverter output is illustrated in Fig. 17(A). This signal was registered in an oscilloscope as Fig. 17(B) is displayed.

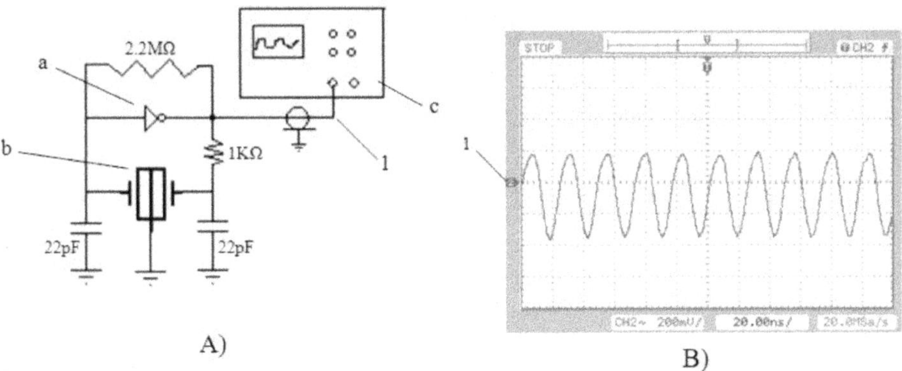

Fig. 17. CCP used as oscillator in a parallel resonator circuit. A) Schematic diagram, a) CMOS inverter, b) CCP, c) Oscilloscope, B) Graphics from schematic setup, 1) Signal obtained 41.66 MHz

When CCP is tried as important part of a high frequency oscillator circuit, it works well connecting Pt-wire electrode to ground illustrated in Fig. 17. Using the same electrical circuit but now adding a signal generator (sinusoidal, triangular wave or square wave) to Pt-wire electrode input, superposition of high and low frequency waves were obtained in Fig. 18.

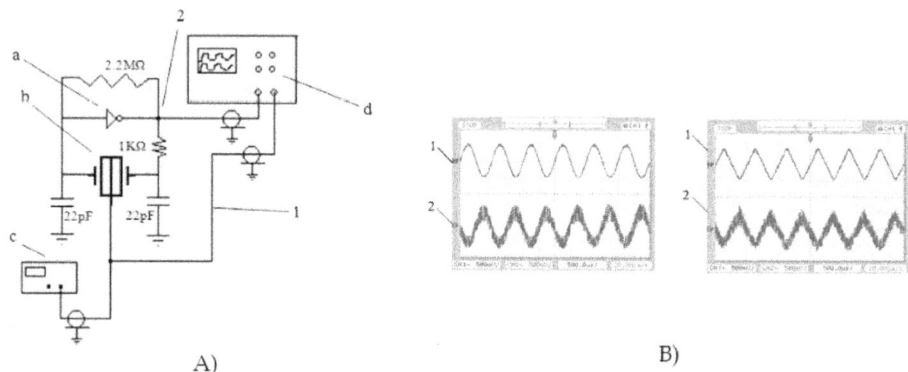

A)   B)

Fig. 18. CCP in a parallel resonator circuit enabling Pt-wire electrode as input signal. A) Schematic diagram, a) TTL inverter, b) CCP, c) Signal generator, d) Oscilloscope, B) Graphics from schematic setup, 1) 10KHz CCP input signal (sinusoidal, triangular wave, square wave) in Pt-wire electrode, 2) Wave superposition signals obtained from logical inverter output.

Electrical application, CCP as it bases of a logic gates. Fig 19(A) illustrates schematic diagram where CCP is exposed to two TTL signals, input S1 in one side face electrode was 2.5 KHz, input S2 in the other side face electrode was 10 KHz and output S3 was in Pt-wire electrode. As result of S3, it has a (offset zero) displaced giving rise to that the signal has positive and negative values (5.84 V Peak-Peak). The output S3 exhibits a clearly suppression of logical pulses (0 and 1) from S2 at the moment that S1 is in positive (digit value "1") and no suppressing pulses (0 and 1) while S1 is negative (digit value "0") as shown in Fig. 19(B).

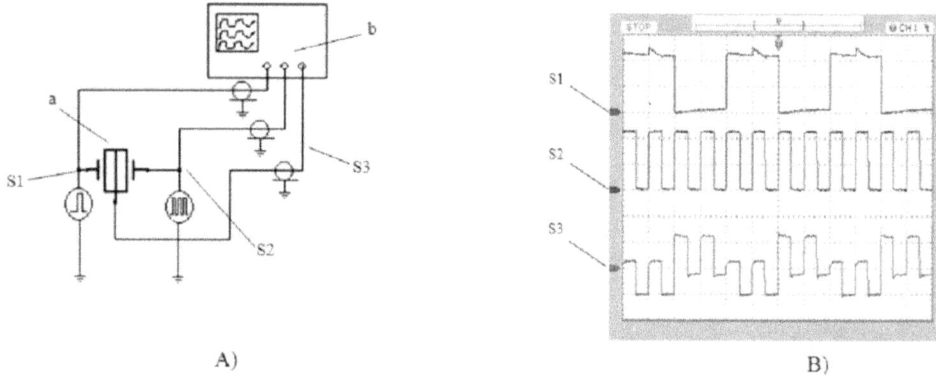

A)   B)

Fig. 19. CCP used as it bases of logic gates obtained of two digital TTL signals. A) Schematic diagram, a) CCP, b) Oscilloscope, B) Graphics from schematic setup, S1) 2.5 KHz TTL input signal on CCP electrode, S2) 10 KHz TTL input signal on electrode of CCP, S3) 5.04 V Peak-Peak output signal obtained from Pt-wire electrode.

Adding a diode on output S3 is possible to eliminate negative part from signal. Thus, to obtain an equivalent logic gate that it explains the behavior of the circuit, a truth table is realized such as Table I. In addition, getting from it a Boolean function by means of Karnaugh maps (Maini, 2007) resulting a F_1 and F_2 Boolean expressions, where F_1 implementation is a logic gate NOR with inverted input S2 and F_2 is a logic gate AND with inverted input S1, illustrated in Fig. 20.

Logical application, CCP associated to digital signals is appraised great versatility of the device, which allows glimpsing to the CCP like a device to be used similar to the logic gates demonstrated in Fig. 20.

S1	S2	S3
0	0	0
0	1	1
1	0	0
1	1	0

$$F_1 = \overline{S1 + \overline{S2}}$$

$$F_2 = \overline{S1} \cdot S2$$

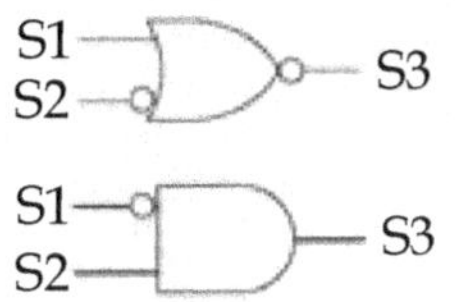

Fig. 20. Truth table for CCP obtained from waveform timing diagram as shown in Fig. 17(B), in the left Boolean expressions F1 and F2 and their implementations. The result of F1 it was a logic gate NOR with inverted input S2 and the result of F2 was a logic gate AND with inverted input S1.

5. Biomedical engineering applications

Piezoelectric Plethysmograph sensor based on Lead Lanthanum Zirconate Titanate bulk ceramic-controlled piezoelectric, implanted with Pt wire

Piezoelectric methods seem to be the most promising for skin microcirculation studies; the analysis of the skin mechanic pulse piezoelectric detection provides valuable selective information on blood flow on upper skin layers, cutting off the influence of the deeper arteries and veins (Caro, 1978). CCP offers a great variety of applications in medical physics such as the human pulse detection called piezoelectric plethysmograph sensor (PZPG).

The periodical increase of blood volume in micro-vessels due to their dilatation during the systolic raise of pressure with the following diastolic contraction over each heartbeat causes corresponding changes in the absorption of the mechanical signals which travel within the working volume. The human vascular system is elastic and multi-branched, and each branching partially reflects back the pressure wave (Janis, 2005).

The experimental setup used in order to get the patient's cardiac pulses was as follows: he was placed in a supine position and in order to record the electrical activity of his heart, the electrocardiographic (ECG) electrodes were placed on the lead I, the PZPG was placed on the index finger of the patients' left arm, as shown in Fig. 21 in order to measure the blood flow that results from the pulsations of blood occurring with each heartbeat. The ECG and PZPG signals were registered and measured their peak to peak voltage. PZPG sensor guarantees electrical isolation of the patient. Its size PZPG can be adjusted to the fingertip measurements, it can easily be extended by means of spare bands, therefore it is possible to take PZPG measurements from different locations of the body, e.g. forehead, forearm, knee, neck, in this context it is possible to use the CCP as base of frequency modulated transmitter, in this mode the modified ceramic is used as part of resonator circuit in an oscillator at the same time an electrocardiographic signal is applied to the CCP by the Pt-wire having as result a frequency modulated signal (Gutierrez & Suaste, 2009).

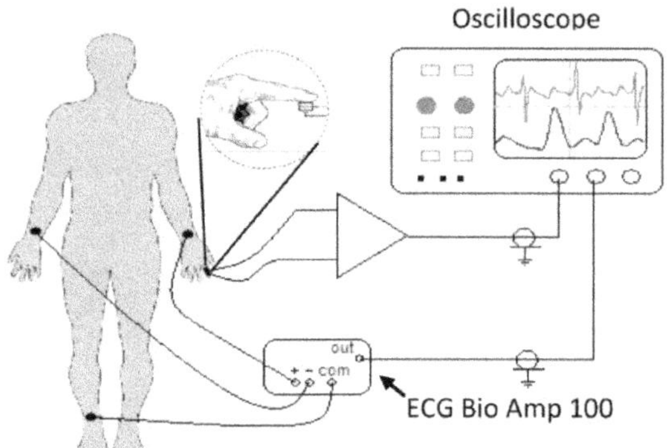

Fig. 21. Experimental setup of Piezopletysmograph in order to obtain fingertip pulses

Opacity sensor (OPS) based on photovoltaic effect of Ferroelectric PLZT ceramic with Pt wire implant

When the PLZT is illuminated after poling, voltage and current can be generated due to the separation of photo induced electron and holes caused by its internal electric field. This is considered an optical property of the material itself which has potential applications for supplying energy transfer in microelectromechanical systems and optoelectronic devices (Sturman & Fridkin, 1992). The steady current in the absence of applied voltage, called photocurrent, is considered the result of photo carriers and the asymmetric electromotive force induced by near-ultraviolet radiation (Tonooka et al., 1998). Therefore, photocurrent is a very important parameter for optical detection (Qin et al., 2007). The behavior of the photovoltaic effect in ferroelectrics is similar to that of the photovoltaic effect in semiconductor p-n junctions.

The semiconductor p-n junction is an interface; in order to extract an electrical output, it is indispensable to apply a bias voltage to it; however, this electrical stimulus is unnecessary for producing the photovoltaic effect on ferroelectrics.

There are three types of photovoltaic samples (Kobayashi et al., 2005) namely, a bulk single plate (Morikawa et al., 2000), a bulk bimorph (Brody, 1983) and film (Ichiki et al., 2006). The single plate exhibit high voltage; the bimorph has a large degree of distortion, fast mechanical response, and is suitable for mechatronics; the film shows a high current output and is a good current source for the optical sensor of the MEMS. Thus, the output power of the film is still one order of magnitude lower than that of the bulk structures reported here.

This new type sensor has some interesting advantages such as free upper face (it is not necessary to add any electrodes), it is a better voltage source than multilayer film PLZT and

the output from the photovoltaic effect of bulk PLZT is controllable by only varying the light intensity (Ichiki et al., 2005).
The thickness of each thin sample was measured on OPS and the values obtained are show in Table I. Two types of oil were used as liquid samples: soy oil and linseed oil. The thin materials and liquids were analyzed on the first setup in a one-dimensional way, other materials like plant leaves, a steel nut and a cross made with single-sided adhesive copper conducting tape 3M Code 1181 were recorded in a bi-dimensional scan.

Photovoltaic currents from translucent samples were obtained by the one-dimensional experimental setup described below. Fig. 22 shows the first setup used in order to obtain photovoltaic current from translucent samples. In this case, a fiber-coupled LASER system with a wavelength of 650 nm was used (Suaste & González, 2009; Suaste et al., 2010). This LASER was modulated to a frequency of 3 Hz with a Beckman Industrial Function Generator; the distance between sample and LASER beam was 1 cm. The electrical contacts (on the lower side face electrode and Pt-wire) of the sensor were connected to an SRS-Low Current Noise Preamplifier in order to obtain photovoltaic current. These signals were registered by an Oscilloscope Agilent which measured their peak to peak voltage.

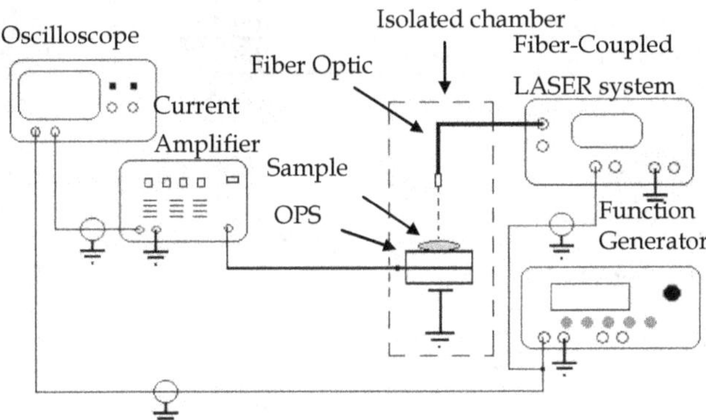

Fig. 22. One-dimensional experimental setup of OPS system.

In the one-dimensional setup, the OPS showed satisfactory results because it reacted to a wide range of detection which allowed to have higher resolution at a higher intensity of illumination. Photovoltaic current was 688 pA, in the region of highest illumination intensity, 160 mW/cm² for the OPS without sample. The graphic of OPS shows a clear difference among thin materials and liquids, Fig. 23. The OPS recording without samples determines 0% of opacity; while for the recordings with samples, this percentage increased according to the characteristics of each of them. Table I shows the percentage of opacity at 91.43 mW/cm² of illumination; this value was chosen as example of the opacity percentage of each sample.

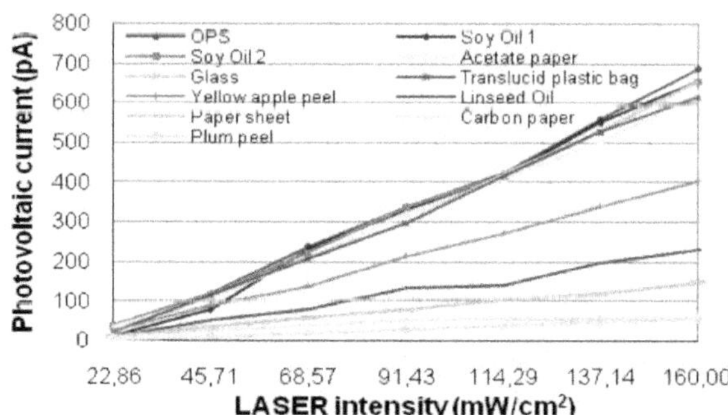

Fig. 23. The OPS detected curves for different liquids and thin solids.

Liquids and thin materials samples measured on OPS	Thickness (μm)	% of Opacity
Plum peel	76.2	91.7
Carbon paper	36.56	84.33
Paper sheet	101.6	76.61
Linseed Oil	100	60.82
Yellow apple peel	88.9	36.84
Transparent Plastic Bag	63.5	12.28
Glass	152.4	12.28
Acetate paper	109.22	5.26
Soy Oil 2	100	1.17
Soy Oil 1	100	2.92

Table 1. Opacity percentage of different liquids and thin materials at 91.43 mW/cm² of illumination using one-dimensional setup.

The second experimental setup used to obtain 2D opacity images from a plant leaf (Myrtus communis), steel nut and a copper cross is shown in Fig. 24 fiber-coupled LASER system was modulated at the same frequency of the first setup but in this case, a set of lenses which reduce the LASER beam spot to a 0.5 mm of diameter size was used. The sample was placed on the OPS and both were placed on an X-Y translation microscope stage.

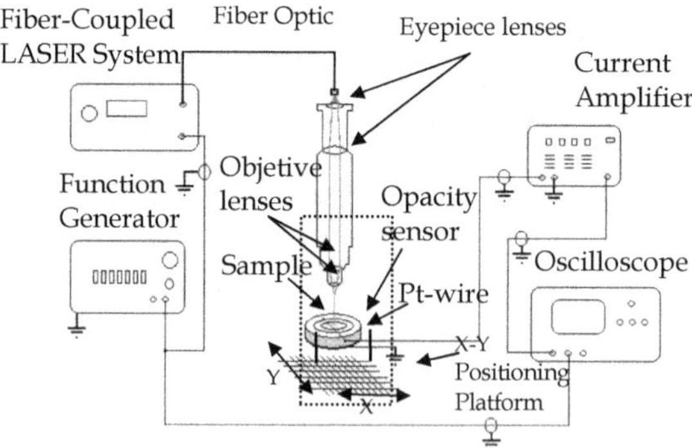

Fig. 24. Bi-dimensional experimental setup of OPS system.

The analysis of each sample was done millimeter by millimeter with a total of 100 points recording for each sample. The electrical contacts from OPS were connected to an SRS-Low Current Noise Preamplifier in order to obtain 2D scans; these signals records were registered by an Oscilloscope which measured their peak to peak voltage in order to construct a 3D graphics in which the Z-axis represents the magnitude of the photovoltaic current of each record.

In the bi-dimensional setup, firstly a 2D scan response without sample was obtained; the records were from *1* to *10* in Y-axis and from *a* to *j* in X-axis, this graphic shows a great increment in the Pt-wire zone because the ferroelectric domains are very close to each other. The maximum photovoltaic current response was 3.28 pA at 160 mW/cm² of illumination; the photovoltaic current response was lower because the LASER beam spot area was smaller. In Fig. 25 is showing the response without simple of OPS.

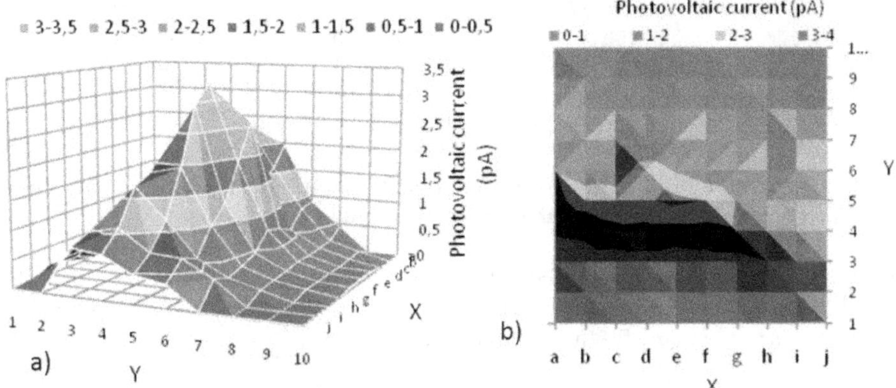

Fig. 25. OPS response without sample in a Bi-dimensional scan, a) 3D representation of OPS, b) Top view of OPS in 2D scan.

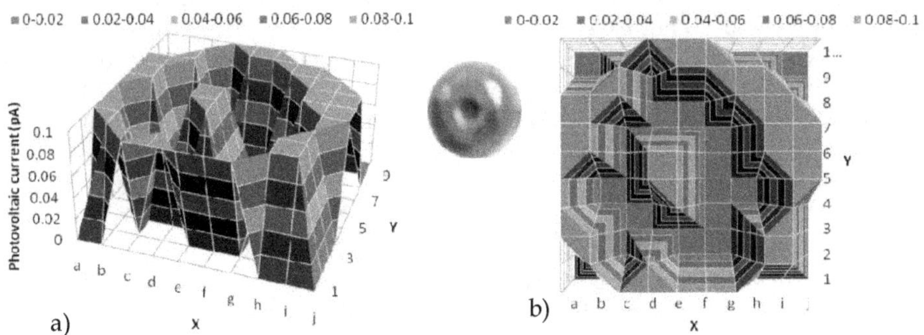

Fig. 26. OPS response with a steel nut, to the right a little picture of the steel nut, a) 3D Graphic, b) 2D scan.

Figure 26 shows the magnitude of the photovoltaic current of the steel nut registered by the OPS. The edges associated with the graphic are due to the curved form of the nut which dispersed the light and caused that the signal was detected in an area different to the one expected.

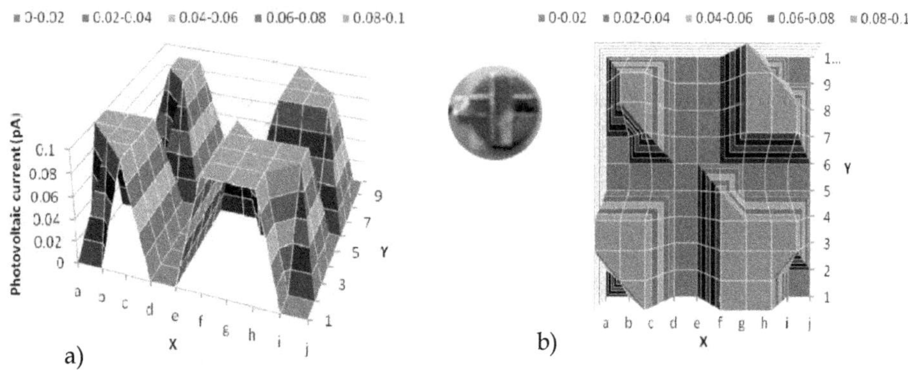

Fig. 27. OPS response with a copper cross, a) 3D graphic, b) 2D scan.

Finally, the acquisition of the signal of the copper cross, on the OPS, was obtained in order to verify each point of the sample so that its image could be constructed, Fig. 27.

These results demonstrated that the OPS allowed to perform 2D scanning and thus generated 3D graphics or images of opacity which varied according to the porcentage of opacity of each sample in the bi-dimensional setup.

Quantitative sensitivity comparison with other types of opacity/transmission would be relative due to the chemical compositions of each sensor and because finally whatever the type of sensor, they only give a percentage measure parameter. The most significant differences among ferroelectric sensors and Si based sensors are: 1) Si based sensors require electrical external supply and do not have domains; in contrast ferroelectric sensors do not require an external electrical supply; 2) the Si based sensors have greater sensitivity than ferroelectric sensors but these last offer a good response at extreme temperatures, up to 570 ºC, according to its chemical composition (ferroelectric phase transition) (Jona & Shirane, 1993).

6. Conclusions

It was shown that the PLZT bulk with Pt-wire has the following advantages: (1) the increase in surface analysis is superior because of the role of Pt-wire that works as a third electrode. (2) The OPS has a useful working range of photovoltaic current because it can capture a power of 330 mW/cm² of LASER illumination and give a photovoltaic current response of 1280 pA. Furthermore, the range of operating temperature of the OPS reached the Curie temperature of 365 ºC, while a standard silicon photo detector operating up to 125 ºC maximum; due to those parameters, the sensor proposed in this work can be used at much higher temperatures and higher illumination intensity as compared to conventional Si based

OPS. (3) The OPS is easy to manufacture and its size can be modified. The production cost is lower than that of any other semiconductor photo detector. (4) It allows to perform 2D scanning and generate 3D graphics or images of opacity depending on the percentage of opacity of the sample in the bi-dimensional setup. In general, the OPS offers good versatility as a sensor of opacity due to its ferroelectricity and could have novel applications such as before transplant 2D cornea scan in order to verify its opacity, 2D scan opacity of insects (Entomology), before transplant one-dimensional skin scan in order to analyze its color, one-dimensional scan on real time of CO_2 car emissions (Chamberlain, 2008; Dekking, 1948; Matusik et al., 2002).

Finally, the novel device CCP that has recently been developed such as is described in this chapter, offers a window of opportunity to develop, the circumstances are right for working in diverse fields of applied research.

7. References

[Anon], New portable smoke opacity meters introduced. *Diesel Progress Engines & Drives*, Vol. 60, Issue 8, pp. 34-35, 1948.

Brody, P. S. (1983). Optomechanical bimorph actuator. *Ferroelectrics*, Vol. 50(1), 27-32, 0015-0193

Caro C. (1978). *The mechanics of the circulation*, Oxford University Press, 0192633236, NY-Toronto

Chamberlain, D. (2008). The transmission opacity tester. *Paper Technology*, Vol. 49, pp. 42-4, 0 306-252X

Dekking, H. M. (1948). Opacity meter for cornea and lens. *Ophthalmologica*, Vol. 115(4), 4, 219-226

Frerking M. E., (1978). *Crystal oscillator design and temperature compensation*, Van Nostrand Reinhold Company, New York, 0-442-22459-1

González-Morán C. O., Suaste-Gómez E. (2009). Developed and experimental evidence of a ceramic-controlled piezoelectric bulk implanted with Pt wire based on PLZT, *Ferroelectrics*, Vol. 392, 98-106, 0015-0193

Gutierrez-Begovich, D.A.; Suaste-Gomez, E., Frequency modulation via a biological signal using a controlled piezoelectric ceramic, *Proceedings of Pan American Health Care Exchanges 2009*, pp.7-7, 978-1-4244-3668-2, México City, March 2009

Ichiki M. et al. (2006). Photovoltaic Effect of Crystalline-Oriented Lead Lanthanum Zirconate Titanate in Layered Film Structure, Vol. 45, No. 12, 9115-9118,

Ichiki M., Maeda R., Morikawa Y., Mabune Y. and Nakada T., (2004). Photovoltaic effect of lead lanthanum zirconate titanate in a layered film structure design, *Appl. Phys. Lett.*, Vol. 84, pp. 395-397, 0003-6951

Ichiki M., Maeda R., Morikawa Y., Mabune Y., Nakada T. and Nonaka K. (2005). Preparation and Photovoltaic Properties of Lead Lanthanum Zirconate Titanate in Design of Multilayers, *Japanese Journal Appl. Phys.*, Vol. 44, 6927-6933, 0021-4922

Jaffe B., Cook W. R. (1971). *Piezoelectric Ceramics*, Academic Press, 0-12-379550-8, London and New York

Jona F., Shirane G. (1993). *Ferroelectric crystals*, Oxford, 0-486-67386-3, New York

Kobayashi T., Ichiki M., Furue H., Maeda R., Morikawa Y. and Nonaka K. (2005) Preparation and its photovoltaic effect of ferroelectric film, *Key Engineering Materials.*, Vol. 301, 193-196, 1013-9826

Maini A. K., (2007). *Digital Electronics: Principles, Devices and Applications,* John Wiley & Sons Inc., 978-0-470-03214-5, England

Matusik, W; Pfister, H; Ngan, A, et al. (2002). Image-based 3D photography using opacity hulls, *ACM Transactions on graphics,* Vol. 21 Issue 3, 427-437, 0730-0301

Morikawa Y., Ichiki M. and Nakada T. (2000). Bimorph type optical actuator using PLZT elements, *Jnp. Soc. Mech. Eng.*, Vol. C69, 361-366

Moulson A. J., Herbert J. M. (2003). *Electroceramics,* John Wiley & Sons, 0471 49748 7, England

Plonska M., Surowiak Z. (2006). Piezoelectric properties of x/65/35 PLZT ceramics depended of the lanthanum (x) ions contents, *Molecular and Quantum Acoustics,* 27, 207-211,

Qin M., Yao K., Liang Y. C., Shannigrahi S. (2007). Thickness effects on photoinduced current in ferroelectric $(Pb_{0.97}La_{0.03})(Zr_{0.52}Ti_{0.48})O_3$ thin films, *Journal of Applied Physics,* Vol. 101, 014104-1 014104-4, 0021-8979

Spigulis J. (2005). Optical noninvasive monitoring of skin blood pulsations, *Appl. Opt.* 44, 1850-1857, 0003-6935

Sturman B. I. and Fridkin V. M. (1992). *The Photovoltaic and Photorefractive Effects in Noncentrossymetric Materials Vol. 8,* Gordon and Breach Science Publishers, 288124498X, New Jersey

Suaste-Gómez E., González-Morán C. O. (2009). Photovoltaic Effect of Lead-Free Piezoelectric Ceramics, $(Bi_{0.5}Na_{0.5})_{0.935}Ba_{0.065}TiO_3$ and $Pb_{0.88}(Ln)_{0.08}Ti_{0.98}Mn_{0.02}O_3$ (Ln = La, Eu), *Ferroelectrics,* Vol. 386, 70-76, 0015-0193

Suaste-Gómez E., González-Morán C. O. and Flores-Cuautle J. J. A. (2009). Developed and applications of a novel ceramic-controlled piezoelectric due to an implant of Pt-wire into the body of sigle disk of $BaTiO_3$ ceramic, *Proceedings of WC2009, IFMBE 25/XII,* pp. 89-92,", O. Dössel and W. C. Schlegel, 978-3-642-03892-1, Munich, Germany

Suaste-Gómez E., Flores-Cuautle J. J. A., Gónzalez-Morán C. O. (2010). Opacity Sensor Based on Photovoltaic Effect of Ferroelectric PLZT Ceramic With Pt Wire Implant, *Sensors Journal, IEEE ,* vol.10, no.6, 1056-1060, 1530-437X

Tonooka K., Poosanaas P. and Uchino K. (1998). Mechanism of the bulk photovoltaic effect of ferroelectrics, *Proceedings of SPIE Volume: 3324,* 9780819427687

Touloukian Y. S., Powell R. W., Ho C. Y., Klemens P.G. (1970). *Thermophysical Properties of Matter, Vol. 1, Thermal Conductivity: Metallic elements and alloys,* IFI/Plenum Press, 306-67021-6, New York

Wai-Kai Chen, (2005). *The electrical engineering handbook, AP Series in Engineering Series,* Elservier Academic Press, 0121709604, CA USA

12

Piezoelectric ceramic applications: micromixing in biology and medicine

Sajin Maria[1], Craciunoiu Florea[2], Moisin Ana Maria[3],
Dumitru Alina Iulia[3], Petrescu Daniel[1], Raluca Gavrila[2],
Bunea Alina Cristina[2] and Sajin Gheorghe Ioan[2]
[1]*University of Medicine "Carol Davila", Chair of Pathology, Bucharest*
[2]*National Institute for Research and Development in Microtechnologies, Bucharest,*
[3]*National Institute for Research and Development in Electrotechnics, ICPE-CA, Bucharest,*

1. Introduction

Reducing the dimensions of macroscopic biological or chemical laboratories is advantageous, the small scale allowing the integration of various processes on one chip, analogous to integrated microelectronic circuitry. The microfluidic technology based on the "lab-on-a-chip" principle allows the usage of very small biological material quantities, decreases the quantity of reactives and waste and increases the speed in obtaining the results making use of automatization, integration, modularization and parallel processing (Southern, et al., 1999; Cai, et al., 2002). Such integration is prerequisite for a fully automated data management system covering all steps of a given chemical or biological process. Finally, the miniaturization results in enhanced precision by providing more homogenous reaction conditions in a shorter time (Pollack, et al., 1999; Hames & Higgins, 1990).

Surface acoustic wave (SAW) devices as nano-pumps and micro-mixers used in microfluidic bio-medical applications are one of the uprising domains today. Surely, the conventional pumps cannot be used for the displacement of the very small liquid quantities on the substrate's surface. To move such little liquid quantities in a "lab on a chip", integrated nano-pumps on the substrate's surface, are used. Using a piezoelectric substrate as support, SAWs are generated by application of an alternative signal with a suitable frequency and amplitude to a system of interdigital transducers (IDT). SAWs are launched on the substrate's surface normally on the IDT lines and move like the water waves on the surface of a lake. This movement is used to move small drops of liquid on the substrate's surface.

Concerning the use of SAWs in micromixing applications, although the wavelengths are of the order of micrometers and wave amplitudes are of the order of nanometers, they are sufficient to produce fluidic nanocurrents inside the liquid droplets. These nanocurrents have chaotic movements that superimpose allowing mixing of their content. One of the applications of this microfluidic applications are "in situ" hybridization stations. In these stations, standard micro-meshes of DNA, proteins or other biological fluids are put in contact with a piezoceramic substrate supporting a SAW. Microagitation of the fluids during the incubation dramatically shortens the hybridization time (Toegl et al., 2003).

However, biological fluids (biological media, cell suspensions, various solutions and some chemicals) may be very aggressive to the materials used in biomedical engineering. It is a less studied aspect of biocompatibility.

In this respect, two aspects were studied: (i) the influence of a piezoceramic substrate on a biological medium represented by a cell population and (ii) the influence of this biological medium (a cell suspension) and processing chemicals on this kind of substrate (possible surface damaging, roughness or porosity increase).

2. Elements concerning fabrication technology of biocompatible piezoelectric ceramics

The known materials with piezoelectric properties are potassium sodium tartrate ($NaKC_2H_4O_6.4H_2O$) also known as Rochelle salt, barium titanate ($BaTiO_3$), lead titanate ($PbTiO_3$), lead zirconate titanate ($PbZr_{1-x}Ti_xO_3$), zinc oxide (ZnO), aluminium nitrate (AlN), etc. One of the most used piezoelectric ceramic is the lead zirconate titanate (PZT) (Jaffe et al. 1971). It presents excellent piezoelectric properties, high Curie temperature and high spontaneous polarization, properties that made it widely used in electrical applications. By using bulk PZT substrates, the working frequency is extended to about 20 MHz, limited by the thickness and porosity of the active piezoelectric element.

The modified PZT ceramic, doped with different elements, can be prepared through a classical technology (Setter, 2002) following a flow shortly presented below.

The active piezoelectric elements used as substrates in SAW micromixing components were elaborated in this technology in which, in the first stage, the ceramic powder is obtained and then, it is further processed into a bulk element through shaping, pressing and sintering processes. These two stages are complementary and constitute the technological flow in the conventional ceramic processing.

The processing developed in these two main stages is given in the table below:

No.	Stages	Processing	Intermediary products
1.	Elaboration of the ceramic powder	dosage	Ceramic powder
		homogenization	
		milling	
		calcination	
		milling	
		properties testing	
2.	Elaboration of bulk ceramic	shaping	Ceramic plates – active piezoelectric elements
		pressing	
		sintering	
		cutting/ lapping/polishing	
		electrical poling	
		morphological/structural investigation	
		properties testing	

Table 1. The main two stages in the conventional ceramic technology

Each main stage ends with a specific control of the powder respectively of the bulk material. The main properties of the ceramics, as requested by the application, are: a prevailing perovskit type crystalline phase, high purity and homogeneity of the ceramic material, micronic dimension of the ceramic grains, high densification (that means the lowest possible porosity), fine superficial roughness of the ceramic plates and, last but not least, biocompatibility with the biologic medium to be handled.

These characteristics are achieved starting with the powder processing in the first stage of the technological flow. It is necessary to obtain fine granulated ceramic powder because, mainly the sintering process, contributes to the growth of the ceramic grains.

The apparent density of the bulk is determined both by the granulation of the ceramic powder and by the pressing conditions. For a granulation controlled ceramic powder, significant differences occur between the normal uniaxial pressing process, followed by sintering, and the hot pressing or isostatic pressing process. The roughness of the bulk surfaces, very important for the quality of the interdigital structures, is dependent on the ceramic granulation. The roughness of the surfaces is obtained by mechanical processing (grinding, lapping, polishing) and sometimes by chemical corrosion.

In order to get products with the requested characteristics, the development of the elementary processing has to be controlled and optimized. A schematic presentation of the technology used for the preparation of PZT ceramics is given in Fig.1. The following important facts can be extracted:

- the purity, of the raw materials (1);
- the efficiency of the mixing processes and the possible impurification (2);
- the reactivity of the materials mixing and the reduction of the PbO loss (3);
- the optimizing of milling by granulation control (4);
- the sintering conditions (9);
- the poling technique (11).

After homogenizing, the mixture is calcinated as a powder, with the aim:

a) of removing the residual water from the carbonates and the volatile impurities;

b) of obtaining the chemical reactions between the components (oxides and/ or carbonates) and to get the solid solution;

c) of diminishing the volume reduction, due to the reactions;

The calcination temperature must be high enough to facilitate the reactions and, in the same time, sufficiently low to enable the subsequent milling process. For the PZT system the temperature has to also be low to avoid the PbO evaporation. The temperature, duration and firing diagrams of the calcination have to be experimentally established for each situation.

The milling contributes to the material homogenization and to the compositional differences which appear after calcination. A milling to $1 \div 10$ μm grain dimension is recommended. Too large grains induce porosity and too fine grains determine colloidal properties of the powder which disturb the technological process.

The sintering conditions of the piezoelectric ceramics are determined experimentally, too. For PZT ceramics, sintering temperatures between 1100^0C and 1300^0C are recommended, and special methods have to be used to maintain the partial pressure of PbO.

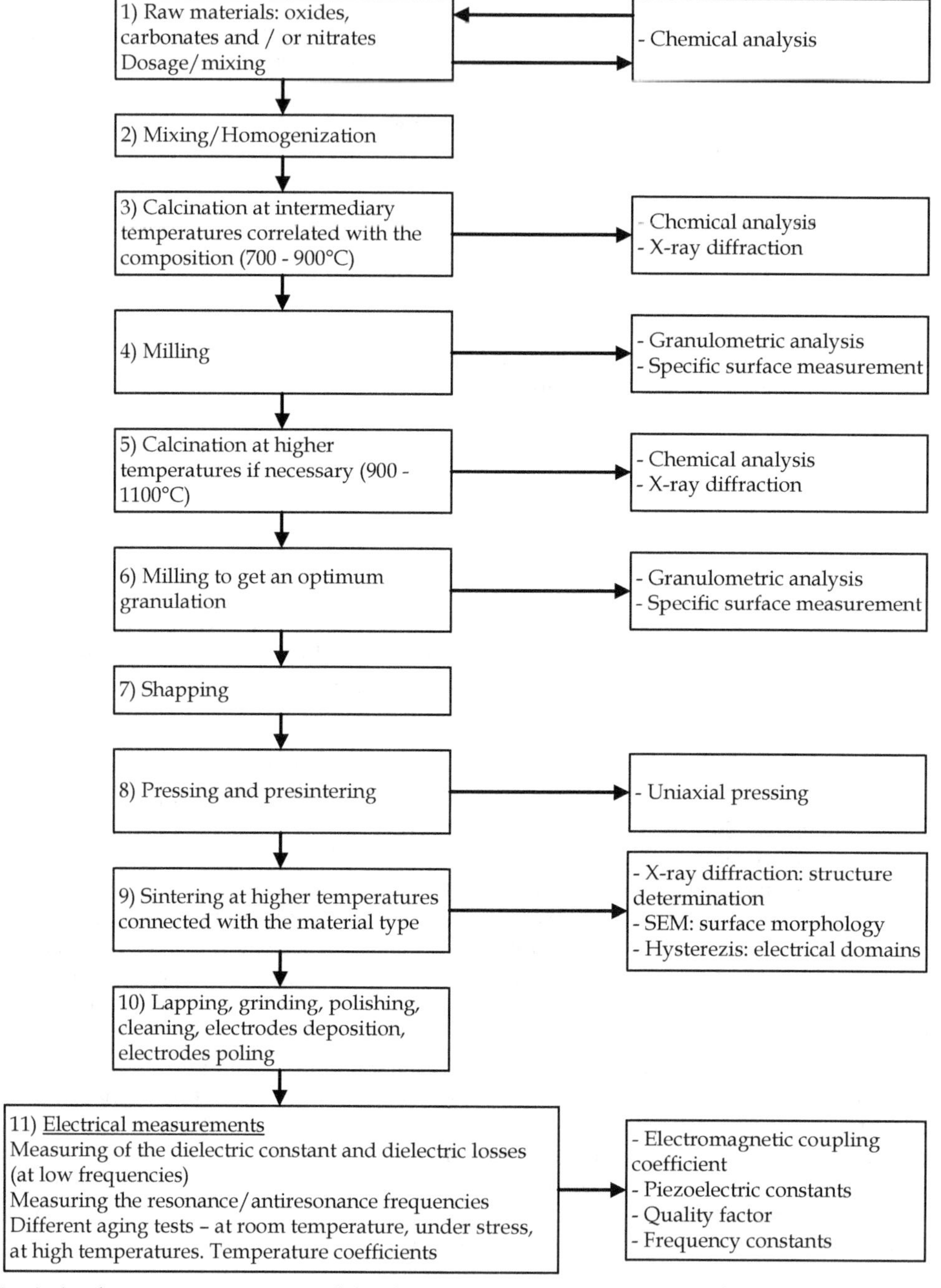

Fig. 1. A schematic presentation of the classical ceramic technology

The active piezoelectric ceramic elements have different shapes and dimensions, and also smooth surfaces. High precision lapping machines are used with different abrasive materials such as aluminium oxide, silicium carbide or diamond powder.

The piezoceramic active elements are obtained by electrical poling, this process being specific for each application. It is important to get high remnant polarization values, thus obtaining high piezoelectric properties. There are different poling modes:

- transversal, for a rectangular plate, with $P \perp V$,
- radial, for a disc element, with $P \perp V$,
- longitudinal, with $P \parallel V$ for a cylindrical element,
- thickness, $P \parallel V$ for a disc element,
- shearing, for a rectangular plate element.

where P is the electrical poling vector and V, the vibration vector. The corresponding coupling coefficients are k_{31}, k_p, k_{33}, k_t si respectively k_{15}.

For each composition it is necessary to determine the optimum poling parameters, avoiding the breakdown of the ceramic material determined by the porosity and different defects of the ceramic element (cracks, dislocations, vacancies or other defects at microscopic scale). The dielectric strength also depends on the thickness and shape of the ceramic samples.

The induced polarization into the ceramic sample depends on the poling duration, electrical field and working temperature. The applied electrical field, together with an increased temperature, accelerate the mobility and ordering of the dipoles. After poling, the measured piezoelectric properties give, indirectly, the value of the induced polarization. The most important parameter for the present application is the electromagnetic coupling coefficient, directly correlated to the remnant polarization. The evaluation of the polarization degree of the samples is made by measuring the fundamental resonance and antiresonance frequencies, and also the first harmonic, as well as the electrical admittance characteristics in correlation with the frequency. The resonance frequency corresponds to the maximum value of the admittance, the antiresonance frequency to its minimum value. By using these values one can compute the electromechanical coupling coefficients, the Poisson coefficient a.s.o.

The research focused on the technological variants able to produce submicronic and nanostructured piezoelectric materials. Experiments were carried out on different compositional variants and the results can be extended to other piezoelectric systems which are interesting for applications.

For the micromixing application, a piezoceramic powder with submicronic granulation was obtained, by using advanced milling processing (Kong et al., 2002). The obtained grain dimensions were submicronic, in the range (500 – 1000 nm), the small dimensions determining a densification of the ceramic bulk, the reduction of the surface roughness of the active ceramic elements and the altering of some properties, mainly the dielectric losses. A granulometric analysis of the ceramic powder was performed. Some results obtained on PZT-F are given in the histograms in Fig.2. The milling duration was respectively 1, 2 and 3 hours and ceramic grain dimensions under 0.5 μm were obtained.

By increasing the milling duration one also observes the appearance of fine powder agglomerations. Therefore it is necessary to find an optimum for the milling conditions in order to reduce these agglomerations. The values obtained for the specific surface are given in Table 2.

This ceramic powder was used to get bulk active piezoceramic elements. Different thermal treatments were studied to diminish the size increase of the grains in order to finally get uniform and fine grained materials.

Milling	Attritor Mill/ 1h	Attritor Mill/ 2h	Attritor Mill/ 3h
Area of the specific surface [m²/cc]	23.44	20.88	17.49

Table 2. The area of the specific surface for a PZT-F powder for different milling durations.

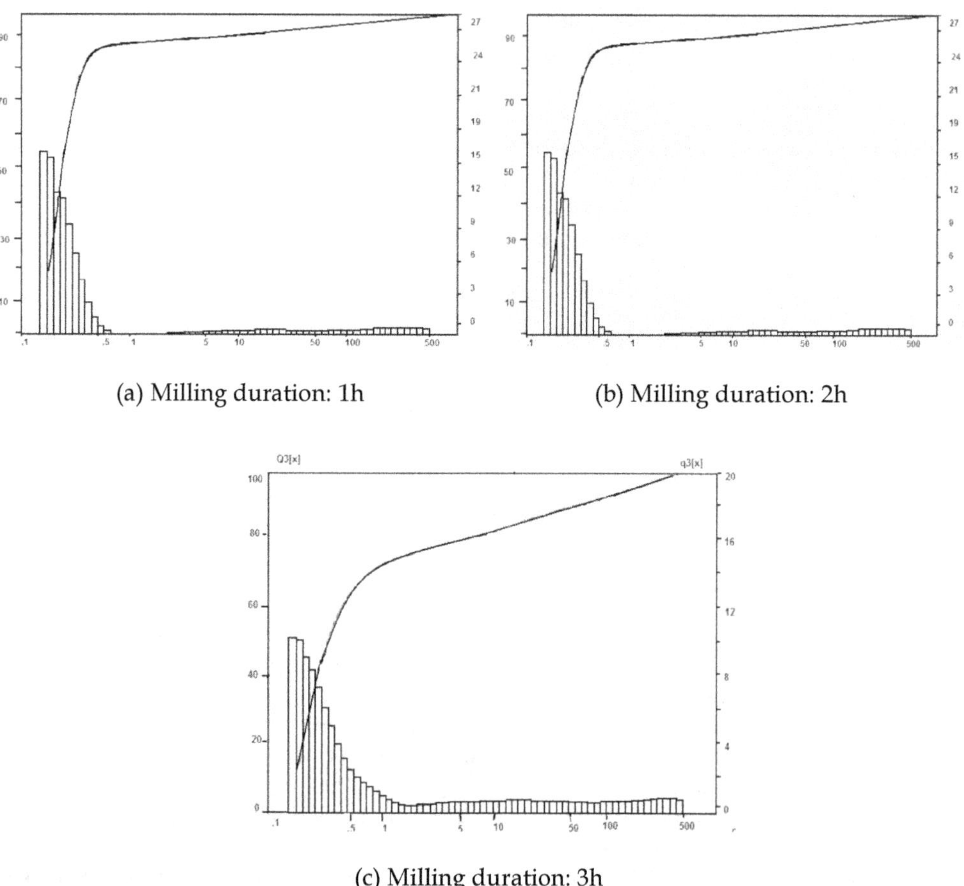

(a) Milling duration: 1h

(b) Milling duration: 2h

(c) Milling duration: 3h

Fig. 2. The distribution of the dimensions of the ceramic grains after milling in an Attritor mill.

The influence of the thermal treatment on the grain dimensions was evidenced by microscopic analysis. Thus, in Fig.3 the results obtained for a ceramic material of the PMN-PT system, sintered three hours at 1000⁰C, 1050⁰C and respectively 1100 ⁰C, are presented (Blank D.H.A. et al., 2005; Moisin et al., 2006).

(a)   (b)

(c)

Fig. 3. SEM images obtained for PMN-PT sintered at 1000⁰C/3h (a), 1050⁰C/3h (b), 1100⁰C/3h (c)

During the sintering process, the crystalline structure is fulfilled, in this example a prevailing perovskit type structure was obtained. In the same time an increase of the dimension of the ceramic grains takes place. In the images obtained through scanning electronic microscopy (SEM), at the same magnification (5.000 X) one can observe the appearance of ceramic grains having a mean dimension smaller than 1 μm.

3. SAW resonators for use in micromixing applications

3.1. Interdigital transducers (IDT) for SAW generation

In this chapter, only some elements concerning the interdigital transducers as elements of SAW resonators will be presented. More complete data may be found in literature as, for instance, in (Morgan, 2007).

In order to convert an electrical signal in a surface acoustic wave structures consisting of alternating conducting segments deposited on the surface of a piezoelectric material are used. They are named InterDigital Transducers (IDT) and can be seen in Fig.4.

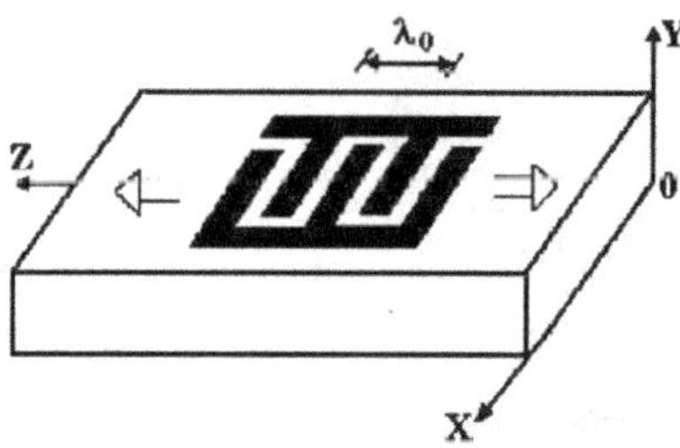

Fig. 4. IDT structure on a piezoceramic substrate

By applying an electrical sinusoidal signal to the interdigital transducer set as in Fig.4, a distribution of electric field components, as in Fig.5, results in the subjacent piezoelectric material.

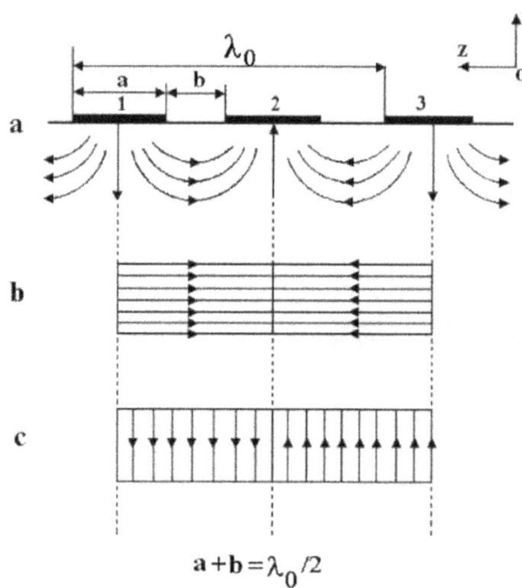

Fig. 5. Electric field distribution induced by an interdigital transducer in a piezoelectric material:
a)   Realistic aspect of electric field lines;
b)   Longitudinal approximation;
c)   Transversal approximation.

Due to the piezoelectricity of the substrate, mechanical deformations are generated which generate Rayleigh surface waves. These waves propagate along the Oz axis. The surface waves produced by the neighbouring transducer are added in-phase if the periodicity of the interdigital structure is equal to half the wavelength $\lambda_0/2$ at the frequency f_0. This means:

$$a + b = \frac{\lambda_0}{2} \qquad (1)$$

The wave generated by the semi-section between segments 1 and 2 (see the figure) which propagates in the negative direction of the Oz axis will be proportional with $\sin\dfrac{2\pi}{\lambda_0}(z + vt)$, where z is the distance on the Oz axis, v is the velocity of the wave and t is the time travelled.

If the condition (1) is fulfilled, this wave arrives under the semi-section between segments 2 and 3 with a phase shift expressed by:

$$-\frac{2\pi}{\lambda_0}(a + b) = -\pi \tag{2}$$

As a result, this wave will be added to the wave generated by segments 2 and 3 which have a $\pi/2$ phase shift. The electric – elastic (and reverse) conversion induced by the interdigital structure is a selective conversion and the transfer characteristic for a periodic transducer with equal segments (uniform covering) is proportional to:

$$\left[\frac{\sin\left(N\pi\dfrac{f - f_0}{f_0}\right)}{N\pi\dfrac{f - f_0}{f_0}}\right] \tag{3}$$

where N is the number of the sections of the transducer and f_0 is the synchronic frequency. The synchronic frequency determines the periodicity of the IDT structure. For example, for a frequency f_0 = 300 MHz on a lithium substrate (v = 3488 m/s) results $\lambda_0 = \dfrac{v}{f_0} = 11.6\ \mu m$

that means a conductor width of 2.9 μm.

Conversion attenuation electric – elastic (and reverse) is defined as:

$$a_c (dB) = -10\log\frac{P_1}{P_2} \tag{4}$$

where P_1 is the active power at the electrical input of the transducer and P_2 is the active power at the elastic output.

3.2. IDT fabrication and results concerning SAW resonant frequency

The SAW micro-mixer structures were processed (Sajin et al., 2005) on a piezoceramic substituted $Pb(TiZr)O_3$ material known as PZT ceramic. The exact formula of this substituted ceramic was $Pb_{0.92}Nd_{0.08}[(Ti_{0.53}Zr_{0.47})]O_3$. The ceramic wafer dimensions were: diameter D = 40 mm and thickness T = 0.5 mm.

For the micromixing applications, the resonator structure was designed for a frequency f_0 = 10 MHz. The wavelength, in this case, is λ = 48 μm allowing IDT lines with a width w = 12 μm. The distance between two IDT lines was, also, 12 μm.

Two such structures obtained by a suitable photolithographic process on the same piezoceramic wafer are shown in Fig.6 (a) where (a) is the metallization not removed from the ceramic disk and (b) are the interdigital transducers (IDT) detailed in Fig.6 (b). The micromixing zone is (c) and the ceramic surface of this zone will be the subject of further biocompatibility experiments.

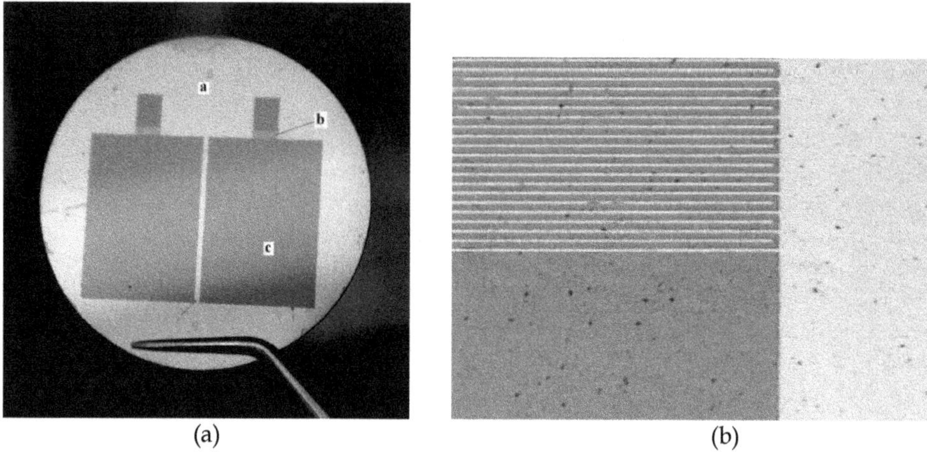

          (a)   (b)

Fig. 6. SAW micro-mixer structure.
(a) – Two structures on a ceramic wafer; (b) – IDT structure; (c) – The micromixing zone.

The IDT system launches surface acoustic waves (SAW) in region (c) of the structure which is the active mixing surface. There are two identical structures on the shown ceramic wafer. In order to be used as micromixers, these structures will be separated by an adequate diamond cutting operation. Subsequently, each of them will be mounted in a suitable mechanical and electrical assembly in order to be used as a mixing device.
In order to verify the resonator functionality, the resonance frequency is measured. In this respect, the active structure of the micromixing device comprising the IDT and the mixing surface (see Fig.7) is cut out and mounted in a special test fixture especially conceived for this measurement. The measuring setup is presented in Fig.8 while the SAW resonator mounted on the test fixture is shown in Fig.7.

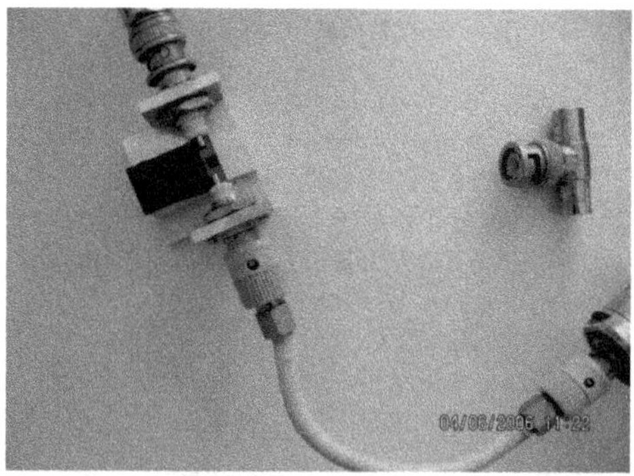

Fig. 7. The active structure of the micromixing device mounted on the test fixture

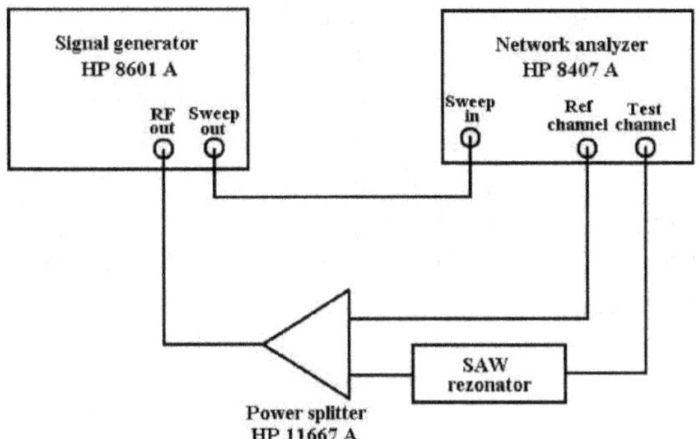

Fig. 8. Measuring setup for the SAW resonator

The measurement results on the SAW launching capabilities comprising resonant frequency and S_{21} parameter for 2 devices are presented in Fig.9 (a) - (b).

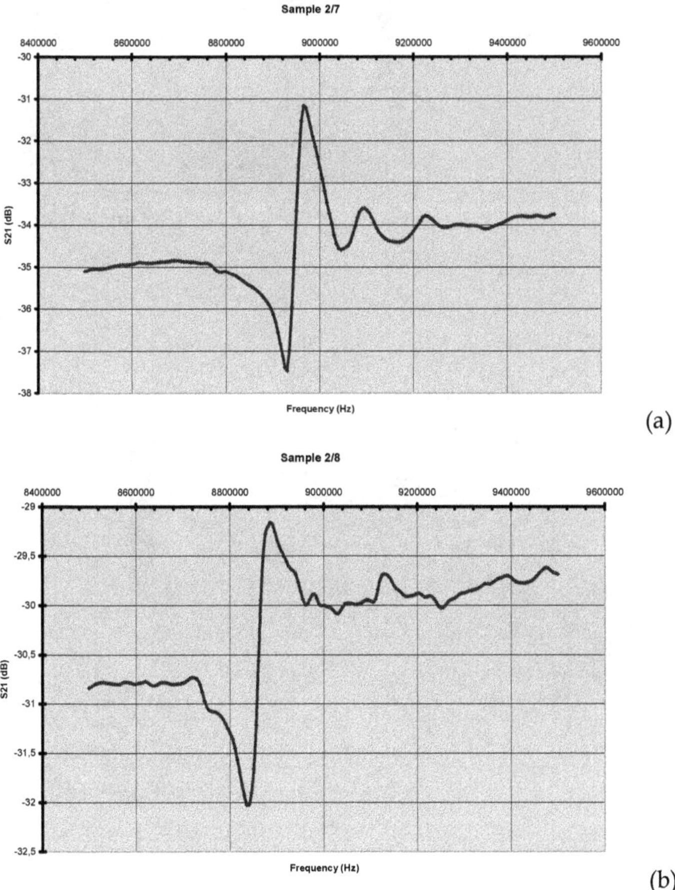

(a)

(b)

Fig. 9. (a) and (b). Resonant frequencies for two SAW structures to be used in micromixers.

The resonant frequency is about. 8.85 MHz. The frequency dispersion is due to limited precision of the metallic lines forming the IDT. Indeed, as it was previously stated, the roughness of the ceramic substrate is of prime importance, as a rough surface limits the possibility to obtain the necessary narrow and long metallic lines. Two different illustrative cases are exemplified in Figs. 10 – 11 and Figs. 12 – 13 respectively. The atomic force microscopy (AFM) images in Fig. 10, with scan area sizes varying from 20 µm to 5 µm, depict the surface roughness of one of the substrates. Pits of about 6-8 µm in size could be noticed in the substrate topography in this case.

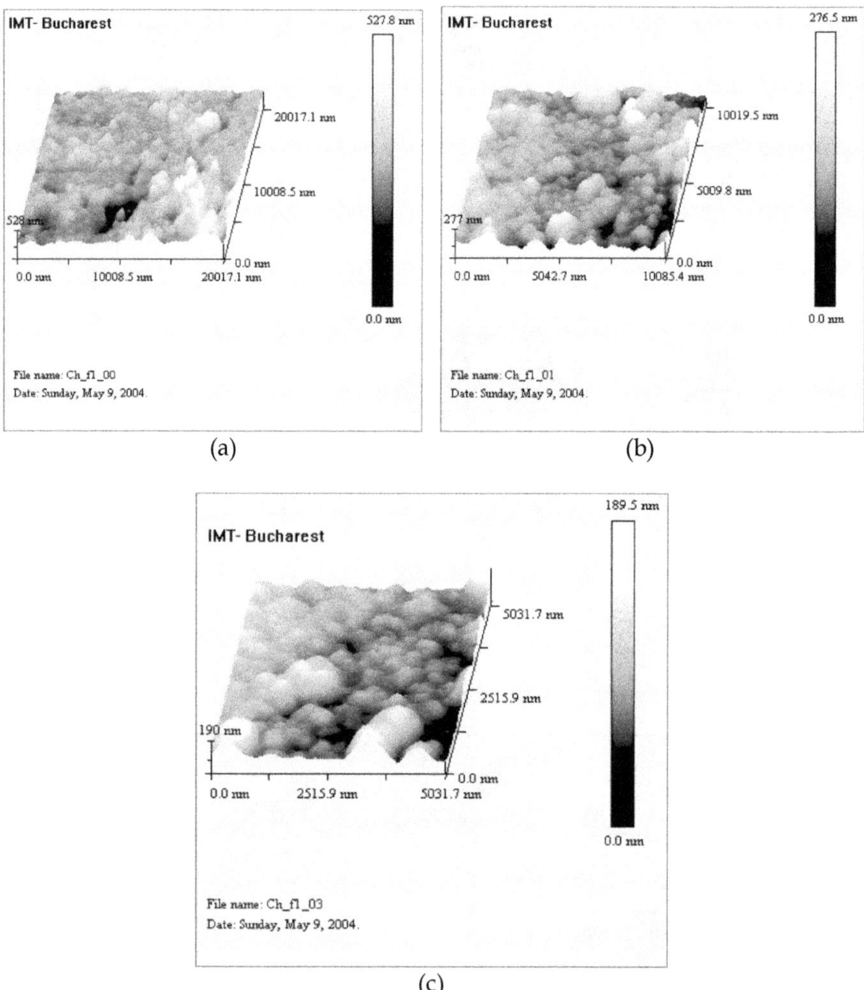

Fig. 10. Surface roughness for one of the used piezoelectric ceramics. Scan area size is respectively:

(a)   20 µm
(b)   10 µm
(c)   5 µm

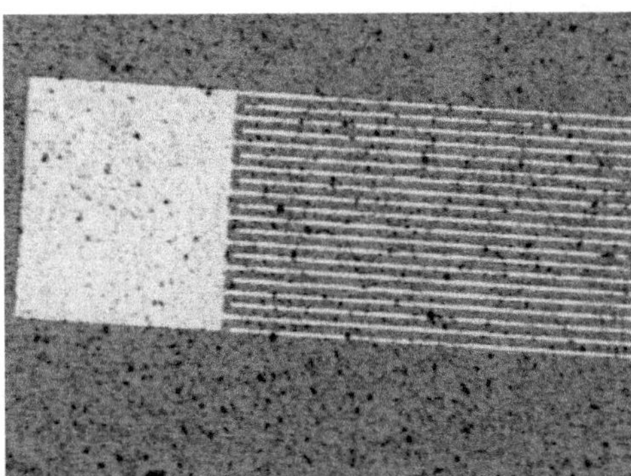

Fig. 11. Interdigital transducers obtained on the ceramic surface with the roughness like in Fig.10.

In this case the ceramic substrate was too rough to allow obtaining metallic lines accurate enough for IDT applications. This is visible in the IDT structures presented in Fig.11. For comparative purposes, remember that the widths of the metallic lines as well as the gap between two lines are 12 μm.

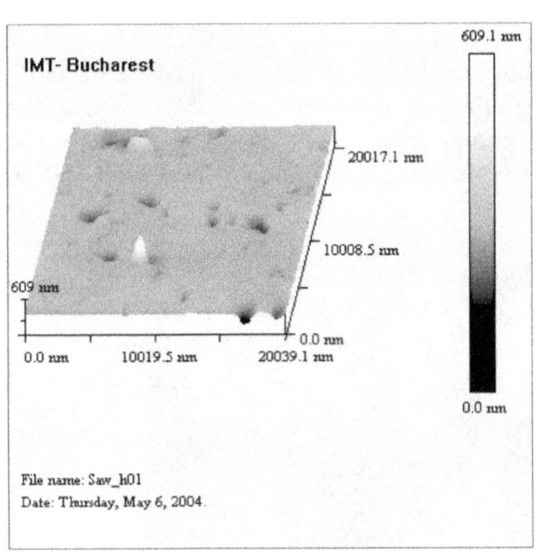

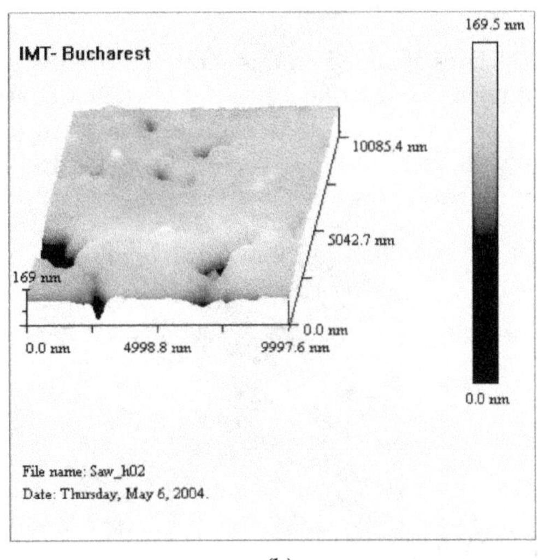

(a)   (b)

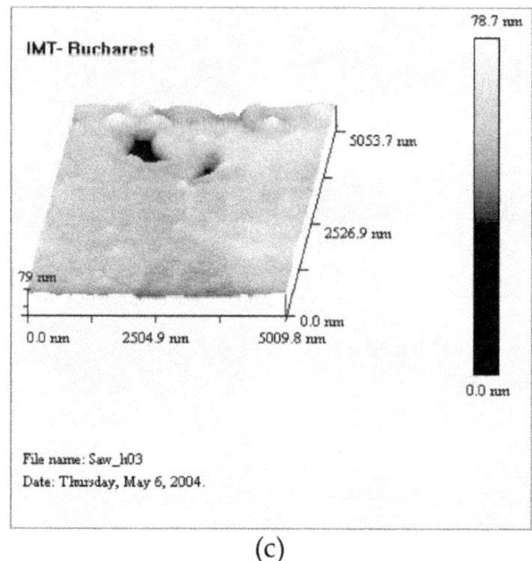

(c)

Fig. 12. Surface roughness of the second piezoelectric ceramic. Resolutions were, respectively:
(a)   20 µm
(b)   10 µm
(c)   5 µm

IDT lines obtained on this ceramic surface are shown in Fig.13. In this case, the better quality of the ceramic surface allowed a much accurate definition of the IDT lines obtained with the same technology as in the previous case (for comparison see Fig.11). The width of the metallic lines in Fig.13 (measured by optical microscopy) is $\cong$ 15 µm and the space between two adjacent lines is 12 – 13 µm.

Fig. 13. IDT lines on piezoelectric ceramic with the surface roughness shown in Fig. 12.

The measurements of the surface roughness in both cases (Fig.10 and Fig.12) were made using an atomic force microscope (AFM).

3.3. Construction and use of the micromixing device

Finally, in order to handle it in the micromixing device, the piezoelectric ceramic wafer carrying the SAW resonator is glued on a laboratory glass slide. Inside the box of the micromixing device there is a space shaped to support a glass slide with the standard dimmensions 76×26 mm² used in the laboratory medical techniques. Athwart of this slide another slide is placed carrying the piezoelectric ceramic wafer. The whole micromixing device is presented in Fig.14 (a) … (c) where Fig.14 (a) shows the glass slide carrying the SAW resonator, Fig. 14 (b) shows the micromixer with the active mixing element (ceramic wafer carrying the SAW resonator) in working position and Fig.14 (c) shows the assembled micromixer.

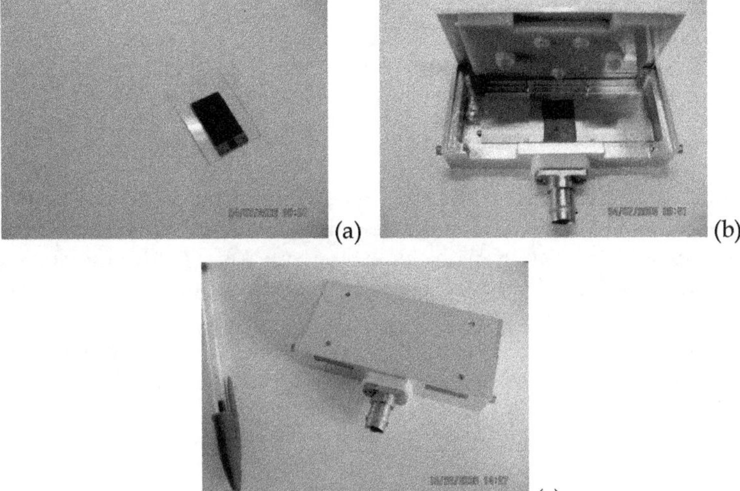

Fig. 14.   (a) - The glass slide carrying the SAW resonator;
          (b) - The micromixer with the active mixing element in working position;
          (c) - The assembled micromixer

The alternative current is supplied to the mixing structure via a BNC connector and a system of elastic contacts inside the mixing box.

3.4. Experimental results

A micromixing experiment involves the following stages:
- Small drops of the substances to be mixed are placed on the glass slide in a chessboard arrangement. These droplets are placed only on the surface of the slide which will be covered by the mixing surface of the SAW resonator – zone (c) in Fig.6.
- The glass slide is inserted in the micromixer box with the droplets upward.
- The glass slide having the SAW resonator attached is placed on the glass slide supporting the substance droplets in such a manner that the piezoelectric wafer is in contact with the droplets of substances to be mixed. Although the mixing box is provided with guiding elements for a perfect placement of the SAW resonator, the operation will be undergone with care.
- The distance between the glass slide and the SAW wafer mixing surface is 50 μm, obtained by careful mechanical processing of the mixing box. This distance can be changed between 75 μm and 20 μm.
- The lid of the mixing box is closed and the alternative signal with a frequency of 8.85 MHz is applied to the SAW interdigital transducers through a coaxial cable fitted with a BNC connector. Application of the alternative signal to the IDT launches the surface acoustic waves that propagate on the piezoelectric ceramic wafer surface. These surface acoustic waves are efficiently absorbed by the fluids placed on the glass slide inducing the mixing process.

4. Study of the Bio-compatibility Between Piezoceramic Material and Biologic Cell Suspension

In an attempt to use such a ceramic as substrate for a microfluidic surface acoustic waves (SAW) micro-mixer, the compatibility between a PZT substituted piezoceramic and a biological cell suspension was studied (Sajin et al., 2006). Two aspects were studied: (i) the influence of this kind of substrate on the development of a cell population and (ii) the influence of the cell suspension on the piezoceramic substrate.

4.1. Methods

First, the ceramic wafer was mirror polished on one face (the active face). Then, the wafer was successively washed with tap water, deionized water and sterile water, in order to remove all the remaining dust from the polishing process. Finally, the surface of such a prepared wafer was analyzed by optical microscopy and by atomic force microscopy (AFM) in order to estimate the roughness.

As biological medium a suspension of DC3F cells were used (pulmonary fibroblasts of Chinese hamsters) in D-MEM buffered with PBS, in a concentration of 5×10^5 cells placed in three 50 mm diameter Petri dishes.

The experimental ceramic wafer was put in a Petri dish containing the cell suspension. Two other Petri dishes containing the same biological medium but without a ceramic wafer were kept as controls. All Petri dishes were maintained at 37^0C in 5% CO_2 atmosphere and were observed daily for 4 days, in order to acknowledge the cell population development. Fig.15 shows the piezoelectric ceramic wafer in the Petri dish containing the biologic medium.

Fig. 15. Petri dish containing the ceramic wafer immersed in the cellular culture.

4.2. Influence of the piezoelectric ceramic substrate on biological fluids

The Figs.16 - 19 showing the development of cell cultures in experimental and in control Petri dishes were taken on the 3rd day of the experiment.

In Fig.16 (a) and (b) one may see, at the optical microscope, that in the control Petri dishes the cells are attached on the dish bottom and walls. They form a continuous cellular layer and the color of the culture medium is orange.

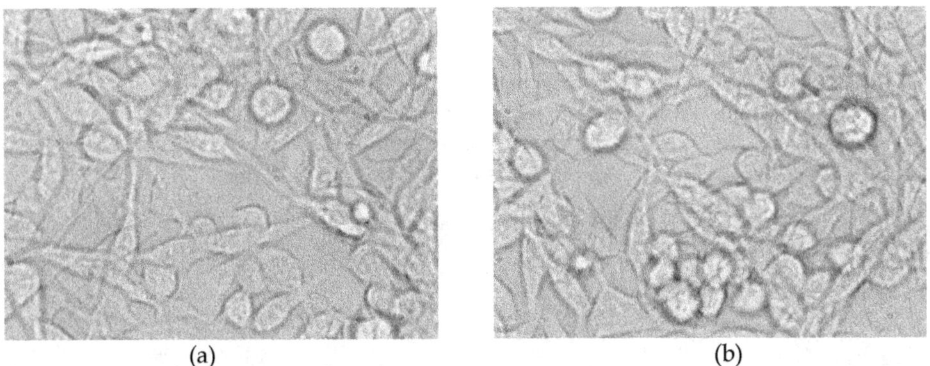

(a)   (b)

Fig. 16. (a) and (b). Continuous cellular layer in two locations in the control Petri dishes on the 3rd day of the experiment. Phase contrast microscopy 100×.

Fig. 17. (a), (b), (c) and (d) show the situation in the experimental Petri dish containing the ceramic wafer at the same time, on the 3rd day.

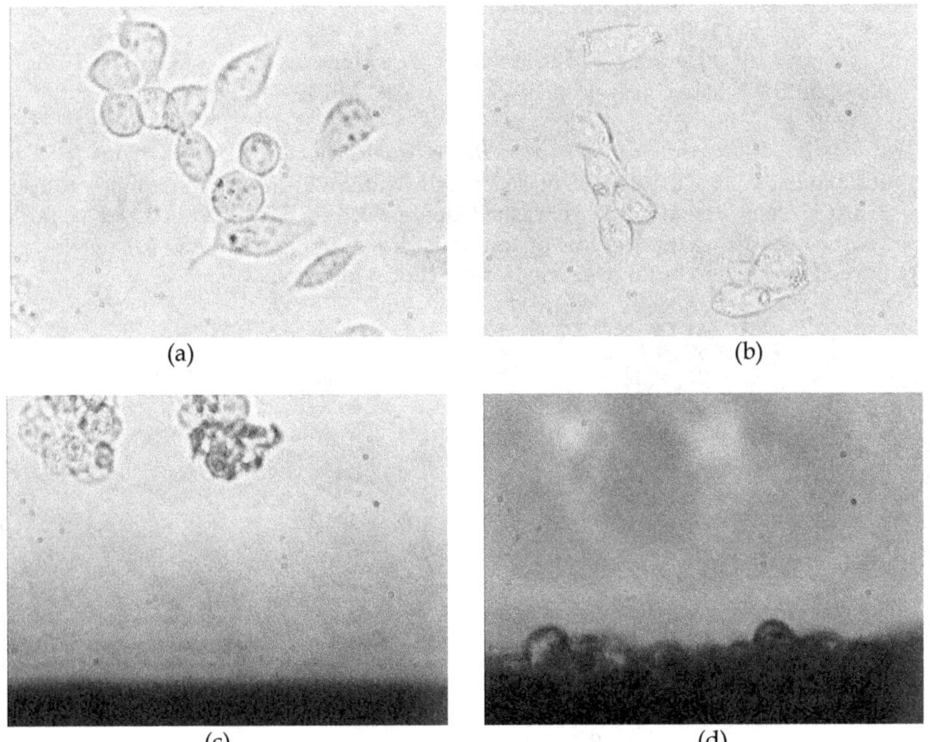

(a)   (b)

(c)   (d)

Fig. 17. (a) and (b). Groups of viable cells floating in suspension in two locations of the reference Petri dishes on the 3rd day of the experiment. Phase contrast microscopy 100×.
(c) and (d). Clusters of dead cells in two locations on the Petri dish containing the piezoceramic wafer, on the 3rd day. Phase contrast microscopy 40×.

One may see, in Figs.17 (a) and (b) groups of viable cells floating in suspension. In Fig.17 (c) groups of dead cells are visible floating in suspension near the ceramic wafer and in Fig.17 (d) a cluster of dead cells are attached to the edge of the PZT piezoceramic wafer. Live cells are not yet attached to the ceramic wafer and the culture medium is pink.

In the 4th day the experiment was stopped and the ceramic wafer was washed in HBSS and fixed in glutaraldehide 25%. The result is visible in Fig.18 (a) and (b).

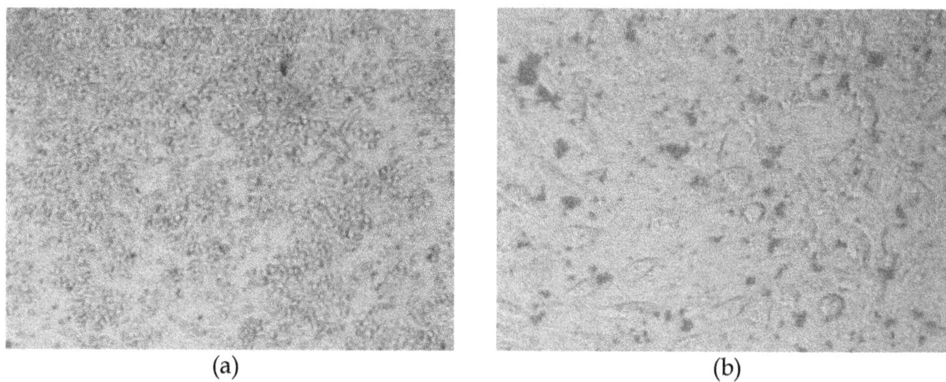

(a)   (b)

Fig. 18. (a) and (b). Ceramic substrate after washing in HBSS and the fixation in glutaraldehide 25%. Phase contrast microscopy (a) 20× ; (b) 40×.

One may see cellular groups attached to the piezoelectric wafer but without forming a continuous layer. Finally, the surface of the ceramic wafer was treated for 4 min with trypsin (5 ml / 37^{0}C). Then, the wafer was washed down with HBSS and deionized water. The surface of the ceramic wafer after this operation is shown in Fig.19 (a) and (b). One may see, again, the cellular groups but they do not form a continuous layer.

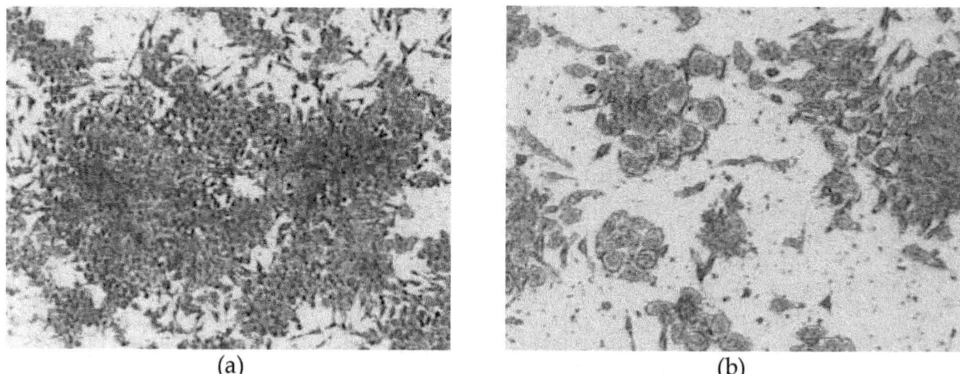

(a)   (b)

Fig. 19. (a) and (b). Aspect of the piezoceramic wafer surface after trypsin treatment. Phase contrast microscopy: (a) – 20× ;   (b) – 40×

At the same time, the two control Petri dishes show an abundant cellular development.

4.3. Influence of biological media on the piezoelectric ceramic substrate

The initial texture of the mirror polished piezoelectric ceramic was investigated by an Atomic Force Microscope (AFM). The surface finish of the initial ceramic substrate is shown in Fig. 20 (a) and Fig 20 (b), in 2D and 3D rendering, for two scan area sizes: 20×20 µm² and 2×2 µm² respectively.

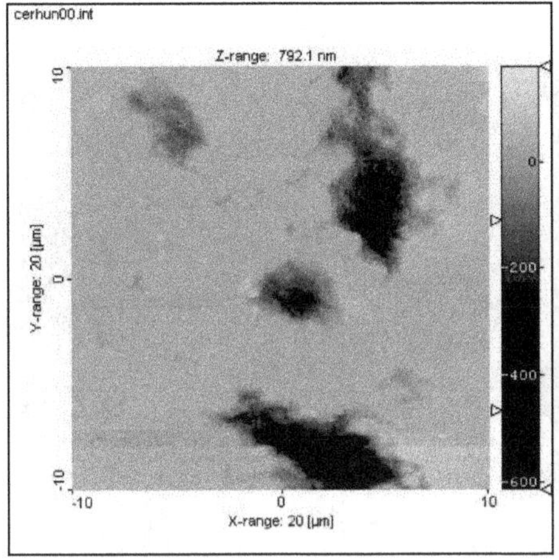

Fig. 20. (a) – 2D

Fig. 20. (a) – 3D

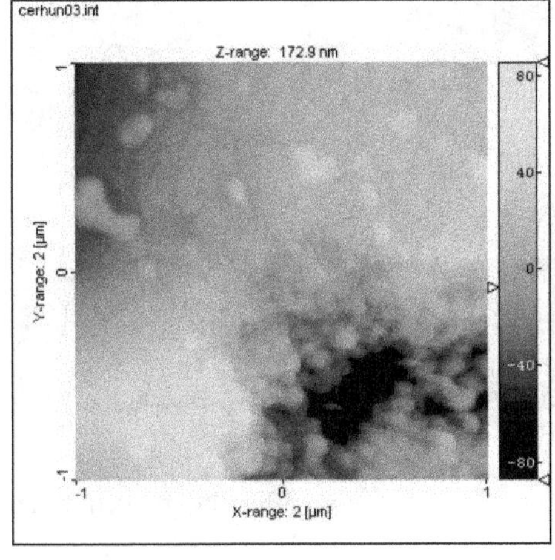

Fig. 20. (b) – 2D

Fig. 20. (b) – 3D

Fig.20 (a) and (b). AFM recordings of the piezoceramic wafer surface roughness prior to the experiments, in 2D and 3D views. Scan area: (a) – 20×20 µm²; (b) – 2×2 µm².

One can see in Fig.20 (a) that the surface is relatively smooth, yet featuring several holes of about 800 nm depths. One of these holes, together with its surrounding relatively flat area, is presented with higher resolution in Fig.20 (b). More measurements performed at different locations of the sample surface essentially revealed similar features as those shown in Fig.20. After the experiment, the ceramic wafer was cleaned and a new set of AFM measurements was performed. The results are shown in Fig.21 (a) and (b) for 20×20 μm^2 and respectively 5×5 μm^2 scan area, both in 2D and 3D views.

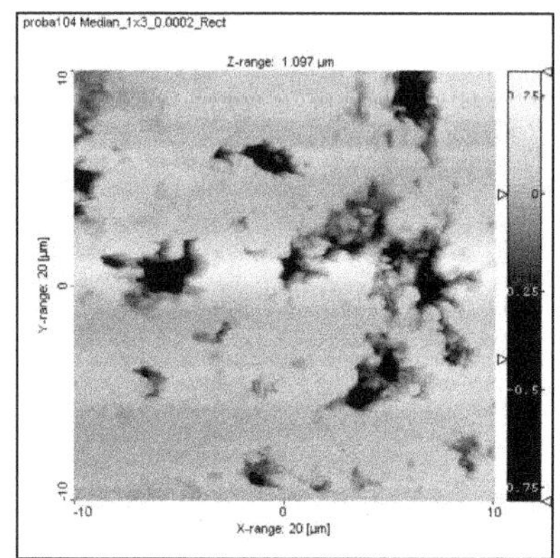

Fig. 21. (a) – 2D

Fig. 21. (a) – 3D

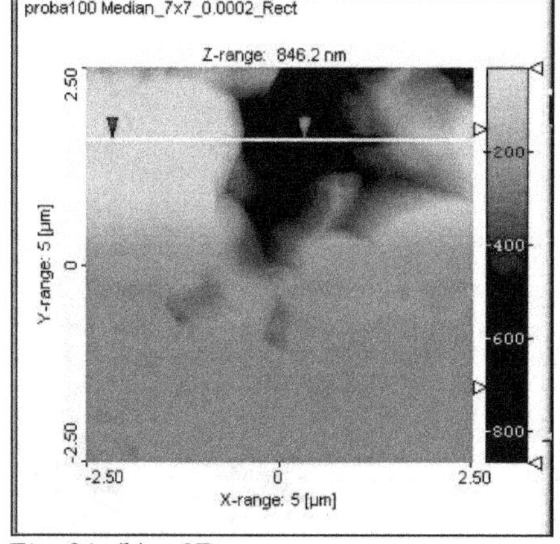

Fig. 21. (b) – 2D

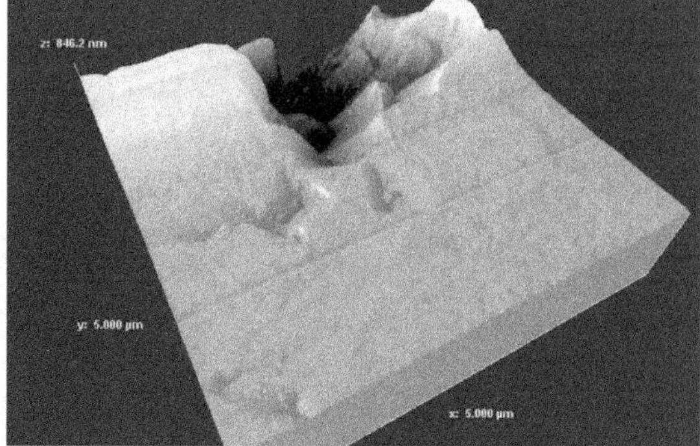

Fig. 21. (b) – 3D

Fig.21. AFM recordings of the piezoceramic wafer surface roughness after the experiments, in 2D and 3D views. Scan area: (a) – 20×20 μm^2;   (b) – 5×5 μm^2

As in the previous case, several additional scans were run, in different areas of the wafer, in order to check the uniformity of the surface morphology. Analogous to the previous case, at all locations of the ceramic surface, the morphological features were similar to those displayed in Fig. 21.
AFM measurements allowed to ascertain that the roughness of the ceramic wafer surface was essentially the same as it was prior to the experiments.

A hole of about 750 nm depth could be seen in Fig.21 (b), comparable in size with the holes in Fig.20 (a). A profile of this hole is shown in Fig. 22, proving that the depth of the cavities featured in the initial surface topography (about 750 nm in this case) did not increase as a result of the action of cellular suspension and chemicals used in experiments.

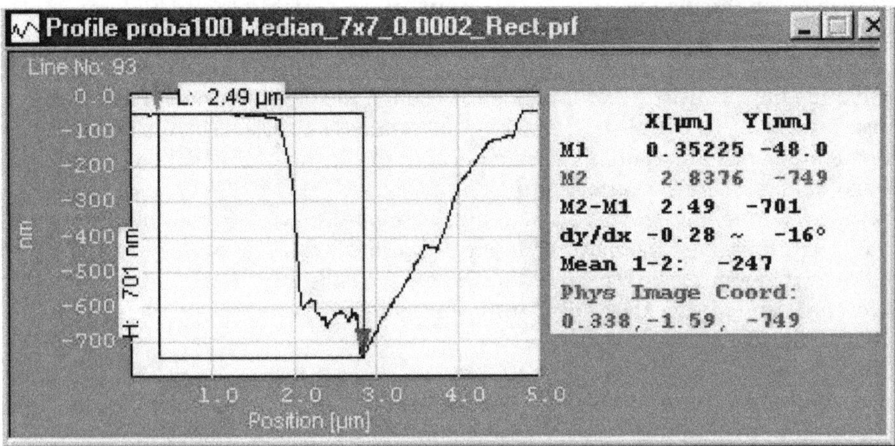

Fig. 22. Depth profile of the hole in Fig.21 (b) – 3D.

Therefore, the AFM investigations revealed that the texture of the ceramic wafer surface preserved its main features during the experiments.
As far as the cleaning of the piezoceramic wafer after the experiment is concerned, it was a more difficult task than expected. In the photo displayed in Fig.23, one may see the experimental ceramic wafer (left side) compared to a genuine one, of the same type and polished in the same manner (right side).

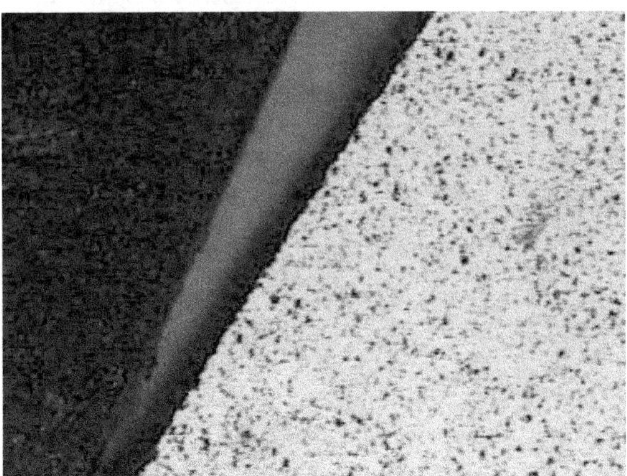

Fig. 23. Comparison between experimental ceramic wafer and a genuine one.

The photo was taken after a ceramic surface washing stage in a stream of deionized water. The adherence of the cellular remains on the ceramic surface is remarkable. Even after using a very strong mixture of oxidants there were still some cellular remains on the ceramic surface. The cellular remains could be totally removed only by re-polishing the ceramic wafer surface, but this operation could drastically damage the SAW interdigital transducers.

4.4. Discussion on the biocompatibility between ceramic substrate and biologic medium

In order to obtain a microfluidic micro-mixer for applications in medicine and biology, experiments were made on biocompatibility between piezoelectric ceramic and the aggressive medium consisting of a biological cellular suspension and the chemicals used. The objectives were to establish if this kind of ceramic substrate has an influence on the development of a cell population and reciprocal, if the action of the biological medium and chemicals could damage the surface of the ceramic wafer.

For the first objective, the growth of the cellular population was affected by the presence of the ceramic substrate. As one may see in Fig.17, on the 3rd day of experiment the cells had difficulties to fix themselves and grow on the ceramic substrate. Even on the 4th day, when the experiment ended, the cells failed to form a continuous layer over the whole ceramic surface (see Figs.18 and 19).

Regarding the second objective, the PZT ceramic surface was not affected. The substrate roughness and the surface porosity were not modified following the contact with the cell suspension and chemicals. As a negative aspect, the cellular remains (dead cells) on the wafer surface are rather difficult to clean out.

A method of solving this last problem may consist in coating the entire mixing surface with another material which is easy to clean, but this solution may reduce the mixing efficiency. In addition, the question of the biocompatibility is transferred from the ceramic-biological medium system to the ceramic protection layer-biological medium system.

5. References

Blank D.H.A., Moisin A.M., Addemir O., Andronescu E., Sajin G., Enescu V., Balazs L., Sztaniszlav A., Ferenc J., Sugurovas V., (2005). Ceramic Substrate with Controlled Piezoelectric Properties for Surface Acoustic Wave Applications. *NATO SfP Project 974130 Final Report, 2005.*

Cai, W.W., Mao, J.H., Chow, C.W., Damani, S., Balmain, A., Bradley, A., (2002). Genome-wide detection of chromosomal imbalances in tumors using BAC microarrays, *Nature Biotech.* 20(2002), pp. 393-396, ISSN: 1087-0156.

Hames, H.D., Higgins, S.J., (1990). *Nucleic acid hybridization - a practical approach,* IRL Press, Oxford, Washington DC. ISBN: 9780947946616.

Jaffe, B., Cook, W.R., Jaffe, H., (1971). *Piezoelectric Ceramics,* Academic Press, London, New York, 1971. DOI: 10.1016/0022-460X(72)90684-0.

Kong, L.B., Ma, J., Zhu, W., Tan, O.K., (2002). Lead zirconate titanate ceramics achieved by reaction sintering of PbO and high-energy ball milled $(ZrTi)O_2$ nanosized powders *Journal of Alloys and Compounds* 236, 2002, p.242, ISSN: 0925-8388.

Moisin, A.M, Dumitru, A.I., Pasuk, I., Stoian, G., (2006). Structural investigation of PMN-PT system, *Journal of Optoelectronics and Advanced Materials* 8 (2), 2006, p.555, ISSN: 1454-4164.

Morgan D., (2007). *Surface Acoustic Wave Filters,* Elsevier, ISBN: 978-0-1237-2537-0.

Pollack, J.R., Perou, C., Alizahdeh, A., Eisen, M., Pergamenschikov, A., Wiliams, C., Jeffrey, S., Botstein, D., Brown, P., (1999). Genome-wide analysis of DNA copy-number changes using cDNA microarrays, *Nat. Genetics 23(1999)*, pp.41-46, ISSN: 1061-4036.

Sajin, G., Moisin, A.M., Craciunoiu, F., Dumitru A.I., (2005). SAW Resonators on ceramics with controlled piezoelectric properties, *Proceedings of the 28th International Semiconductor Conference, CAS*, Sinaia, Romania, 03 - 04 Oct., 2005, p. 95 - 98. ISBN 0-7803-9214-0.

Sajin G., Petrescu D., Sajin M., Craciunoiu F., Gavrila R., (2006). Testing the biocompatibility of the piezoelectric ceramics and biological media. *Proceedings of the World Congress on Medical Physics and Biomedical Engineering "Imaging the Future Medicine"*, 27 August - 01 Sept. 2006, Seoul, Korea, pp. 3179 - 3182, ISBN 3-540-36839-6.

Setter, N., editor, (2002). *Piezoelectric Materials in Devices*, EPFL, Lausanne, ISBN: 2-9700346-0-3

Southern, E., Mir, K., Shchepinov, M., (1999). Molecular interactions on microarrays, *Nat. Genetics (Suppl.)*, Vol. 21(1999), pp. 5-9, ISSN: 1061-4036.

Toegl, A., Kirchner, R., Gauer, C., Wixforth, A., (2003). Enhancing Results of Microarray Hybridizations Through Microagitation, *Journal of Biomolecular Techniques*, 14, Sept.2003, pp.197–204.

13

Piezoelectric-ceramic-based microgrippers in micromanipulation

Xin Ye and Zhi-jing Zhang
Beijing Institute of Technology
P.R. China

1. Introduction

Microgrippers are widely applied in micromanipulation systems covering manufactural, biological and medical fields, and are especially important in microassembly process. There are several reasons. Firstly, as the size of microminiature parts is in millimetre or micron, the strength and stiffness compared to those of macro components is much smaller and thus microminiature parts are more easily damaged. Secondly, since geometric feature size for most of microminiature parts ranges between 0.01mm and 10mm, has not yet emerged a microgripper with which the scope to clipping all parts can be achieved. Thirdly, adhesion phenomena in microassembly process are a difficult problem that limits the development of microgrippers. At the beginning of the century, scholars have included the microgripper technology into research projects and made a lot of initial results. According to the driving force, microgrippers can be divided into piezoelectric-ceramic-based (Heyliger & Brooks, 1995), electrostatic-based (Majumder et al., 2001), electromagnetic-force-based (Quandt, 1997; Ashley, 1998), electric-based, vacuum, adhesive-material-based, shape-memory-alloy-based clampers, and so on. Li Q. and Li Y. (Li Q. & Li Y., 2004) summed up a detailed overview of microgripper profile and types.

Compared with other kinds of microgrippers, piezoelectric-ceramic-based microgripper is the most widely used one. Piezoelectric ceramics, owing to the unique properties such as direct and reverse piezoelectric effects, are utilized to generate controllable displacement in microgrippers far and wide.

2. Two categories of piezoelectric-ceramic-based microgrippers

Generally, one microgripper consists of two parts, clamper and actuator. A clamper usually contains two or more arms, and is driven by the actuator to grab tiny objects. Therefore, according to two different functions of piezoelectric ceramics, piezoelectric-ceramic-based microgrippers can be classified to two categories. The piezoelectric ceramic works as clamper or actuator. Although piezoelectric ceramics are adopted widely through both categories of microgrippers, the two mechanisms are constitutionally different.

When piezoelectric ceramic works as clamper, piezoelectric bimorph or heterogeneous bimorph is adopted in most cases. A piezoelectric bimorph consists of two piezoelectric elements stacked up. The heterogeneous bimorph consists of a piezoelectric element on the top of a non-piezoelectric element. Both elements of a piezoelectric bimorph serve two functions, electric and elastic, while in a heterogeneous bimorph one element serves only an elastic function, the other serves both functions. The microgrippers with piezoelectric bimorph or heterogeneous bimorph can achieve greater deflections and clamping force compared with those of the latter category. In addition, it is compatible with IC process. However, the driving voltage is high and cracking of the bimorph is severe. We developed a piezoelectric bimorph microgripper. The bimorph piezoelectric ceramics were chosen as the clampers. Theoretically, two pieces of piezoelectric ceramic were stuck together in order to form a laminated beam. Polarization of the two pieces was in the same direction, both along the thickness direction. However, during actual preparation, a substrate layer was added between the two piezoelectric layers. The structure is illustrated in Fig. 1. One end of the laminated beam was fixed up to form a cantilever. After electric field was applied, one piece of piezoelectric ceramics stretches while the other shrinks. Deformation direction can be changed by exchanging positive and negative connections. The free end of the laminated beam can be bent free to output force and displacement.

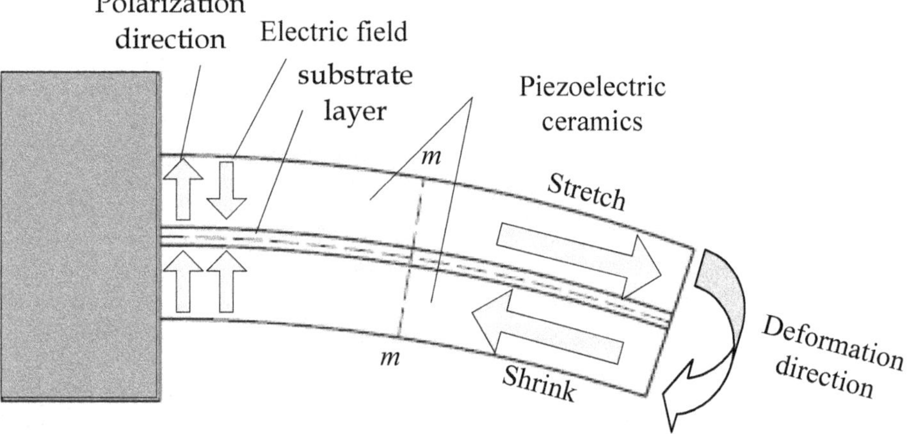

Fig. 1. The structure and working principle of parallel bimorph piezoelectric ceramic laminated beam.

When piezoelectric ceramic works as actuator, the clamper is usually a displacement amplification mechanism based on flexure hinges. This kind of microgripper has great advantages such as precise controllable output, zero friction, and zero clearance. However, this kind of microgrippers can not be made small enough for the minimum size is to a large extent limited by displacement amplification mechanisms and piezoelectric ceramics.

3. Microforce detection technology used in microgrippers

Since the manipulated objects are tiny and fragile, it is crucial important to detect the microforce during the micromanipulation process. Generally, microforce detection technology develops towards two directions, contact and non-contact detection styles. In the process of contact detection, hypersensitized force sensors are bonded with the clamper of microgrippers and utilized to obtain voltage information which is deduced to microforce after rectification, filtration, and amplification. During non-contact detection process, the clamping part's tiny deformation is detected by visual servo systems and then the microforce is deduced. In addition, forces are usually converted into light, sound, electricity, or heat. Accordingly, new experimental principle and technology, such as Optical Force Sensing method, are invented. Meanwhile, as human eyes cannot distinguish objects in micromanipulation clearly, scanning electron microscope (SEM), scanning tunneling microscope (STM), and atomic force microscopy (AFM) are utilized in the non-contact detection process.

3.1 Contact microforce detection

Strain gauge pressure sensor is one of the most important part in contact microforce detection devices. The main component is the resistance strain gauges. Resistance strain gauge is a piece of a sensitive device that an electrical signal can be converted into the measured strain. There are two categories of resistance strain gauges–metal strain gauges and semiconductor strain gauges (Smits & Choi, 1990).

The working principle of metal strain gauges is deformation after external force leads to changing resistance. Metal resistance strain gauge is divided into wire and foil strain gauges.

Semiconductor strain gauges work under external forces, apart from distortion takes places, its resistance will change. Horizontal effect, creep and hysteresis of the semiconductor strain gauge is quite small. And the frequency response range is wide for the information can be measured from the static response to high-frequency dynamic strain. With the development of integrated semiconductor manufacturing process, a variety of small and ultra-small semiconductor strain sensors can be made combining this technology with semiconductor strain gauges. Thus, the measurement system can be greatly simplified. The semiconductor strain gauge is more sensitive than the metal strain gauge, but its non-linear error is also bigger. Therefore, a semiconductor strain gauge with invariable electric current source has chosen as the main component in the micro force sensing system (Liu & Xu, 1990).

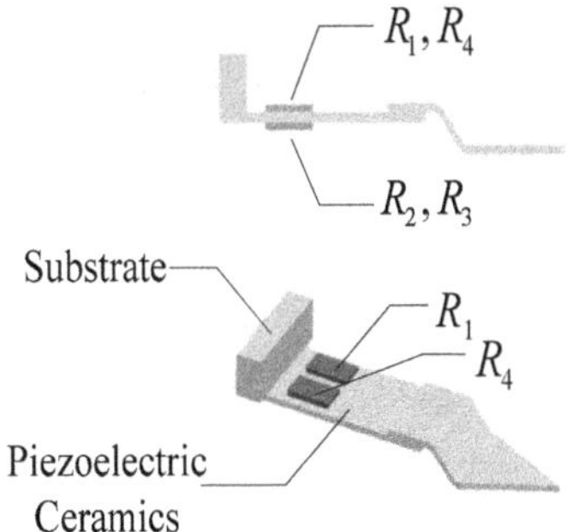

Fig. 2. Affixed position of semiconductor strain gages

Fix the parallel bimorph piezoelectric ceramic to the main body of the micro-gripper. Here, the piezoelectric ceramic can be used as a cantilever. Fig. 2 Shows the location of the semiconductor on the parallel bimorph piezoelectric ceramic.

The cantilever performs parallel bimorph bending purely when it's only effected by the moment M_e, the relationship between the end displacement w and M_e is illustrated as follows.

$$w = \frac{M_e l^2}{2EI} \qquad (1)$$

Here, E is the equivalent Young's modulus.
So,

$$M_e = \frac{2EIw}{l^2} \qquad (2)$$

The relationship between the stress σ and the strain ε is shown as follows.

$$\sigma = \varepsilon E \qquad (3)$$

The equation to calculate the stress on the pure bending cantilever is as follows.

$$\sigma = \frac{M_e y}{I} \qquad (4)$$

Here, y is the coordinate of the thickness direction; I is the moment of inertia of the section of the cantilever. And, y equals half of the value of h, which is 0.3 mm. By integrating equations (2), (3) and (4), the following equation can be got.

$$\varepsilon = \frac{hw}{l^2} \qquad (5)$$

A full-bridge circuit is applied to the invariable electric current source. The relationship between the input current of the circuit I_0 and the output voltage V_0 is shown as follows.

$$V_0 = R_0 K_0 I_0 e \qquad (6)$$

By integrating the equations (5) and (6), the relationship between the displacement w and the output voltage of the full-bridge circuit V_0 can be illustrated as follows.

$$w = \frac{V_0 l^2}{R_0 K_0 I_0 h} \qquad (7)$$

The output voltage of the full-bridge strain sensing circuit is given as follows.

$$U_0 = K \varepsilon U_i \qquad (8)$$

Here, U_i is the sensitivity of the full-bridge strain sensing circuit, which stays at 3V all along the experiment; U_0 is the output voltage of the full-bridge circuit; K is the parameter of the strain sensitivity; ε is the strain.

According to the model of the force effected on the cantilever, the relationship between the gripping force and the moment is given as follows.

$$F_g = M_e / l_g \tag{9}$$

Here, l_g is the length of the strain gauge.

3.2 Non-contact microforce detection

As we don't involve any non-contact microforce detection technology, an optical force sensing system developed by (Yu & Bradley, 2000), illustrated in Fig. 3., is introduced. Contact force is detected by a protuberance on a cantilever tip. The displacement of the cantilever is obtained by that of orthogonal diode laser point. Thus, contact force can be deduced by the displacement multiplied by cantilever's elastic coefficient.

Microgripper with two fingers, illustrated in Fig. 4, is improved by the non-contact microforce detector. An improved cantilever without protuberance, used as one of the fingers, is connected with the other finger with protuberance by a piezoelectric actuator. The open-close action of the microgripper is realized by piezoelectric actuator extension and contraction. The contact force aroused by microgripper and gripping objects causes the displacement of cantilever. And as the elastic coefficient of the cantilever, contact force can be deduced by displacement detected by the optical force sensing system.

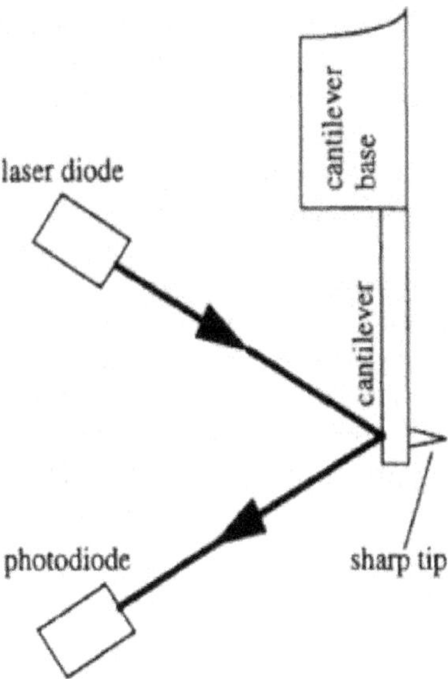

Fig. 3. Basic structure of optical force sensing system

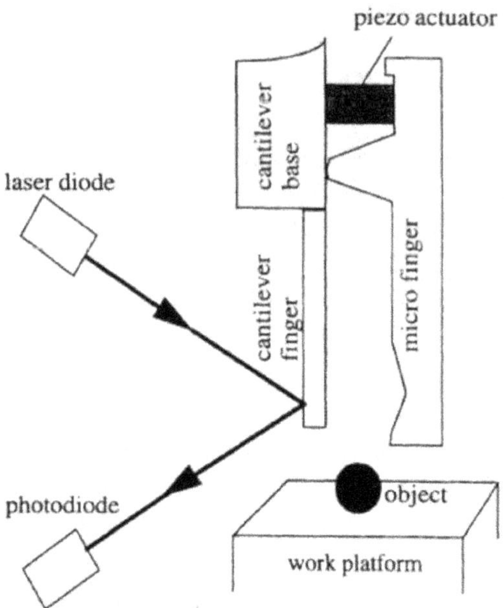

Fig. 4. Principle of microgripper with built-in optical force sensor

4. A bimorph piezoelectric ceramic microgripper integrating micro-force detecting and feedback

A combined micro-gripper offers new methods to settle the gripping and assembly problems of miniature structures and systems. Both of the large gripping extension and the gripping accuracy can be acquired by combining micro-gripper and a small linear motor with work distance of 15mm. However, two difficulties exist in combined micro-gripping technologies. First, though the gripping of some big structures in the demanded scales can be easily achieved by the combined micro-gripper, it's hard to avoid the shed of parts during the motion. Second, gripping forces range from micro-Newtons to Newtons, so it's difficult to ensure the precision of micro force sensing.

This chapter focuses on the gripping of miniature mechanical structures and systems, shows a combined two-chip piezoelectric ceramic microgripper integrated micro-force sensing and feedback function. This gripper uses the bend of two-chip piezoelectric ceramic and the motion of a linear electromotor to grip miniature structures and systems. A reinforcement machine is designed for the two-chip piezoelectric ceramic, which can avoid the shed of parts during the motion. The gripping stability and micro-force sensing and feedback are detailedly investigated theoretically and experimentally.

4.1 A combined two-chip piezoelectric ceramic micro-gripper
Abundant knowledge of the miniature parts to be gripped is required as preparation of the design and manufacturing of grippers. 3D miniature structures have peculiar demands for

their grippers. First, the gripping extension of the micro-gripper must be large enough. Miniature structures can be as big as 10mm in geometry, so a larger gripping extension than generic micro-grippers is needed. Second, a high gripping precision is needed so that miniature structures as small as 0.01 mm in geometry can be successfully gripped. Third, a high gripping stability is requested to ensure a stable gripping of big parts during the motion.

The combined two-chip piezoelectric ceramic microgripper integrated micro-force sensing and feedback function is designed to meet the specific requirements of 3D miniature structures, which is shown in Fig. 5. The basic frame and function are as follows.

(a) The micro-gripper of the combined two-chip piezoelectric ceramic microgripper contains 5 portions — two-chip piezoelectric ceramic, fixed brace, reinforcement for the piezoelectric ceramic, linear fine-motion platform and pedestal. The linear fine-motion platform and the fixed brace are installed on the pedestal, while the two-chip piezoelectric ceramic and the reinforcement for the piezoelectric ceramic are installed on a folded plank which is connected onto the linear fine-motion platform.

(b) The implementing part of the micro-gripper is the two-finger mechanical gripper with one finger fixed and the other moving. The top end of the fixed brace is made of rigid alloy steel, and the top end of the moving finger is made up of less rigid two-chip piezoelectric ceramic. The tiny bending of the two-chip piezoelectric ceramic and the fixed brace work together to grip a complex 3D miniature structure.

(c) The linear fine-motion platform can produce large displacement, which can make up for the defect of small displacement produced by the tiny bending of two-chip piezoelectric ceramic. The two-chip piezoelectric ceramic's bending can only bring small displacement; what's more, the gripping force will counteract some displacement. So it's hard to meet the requirement of displacement during the assembly just by using two-chip piezoelectric ceramic. It's necessary to develop a linear coarse-motion module with large displacement output as a makeup for the fine-motion module.

(d) The reinforcement for the piezoelectric ceramic is made up of a lever machine with two reinforcing patches rotating around an axis, as shown in Fig. 6. The stretch and closing of the two reinforcing patches are controlled by an electromagnet installed on the top ends of them. When the electromagnet is cut, the spring releases so that the two reinforcing patches close. Here, the reinforcement for the piezoelectric ceramic can ensure that the two-chip piezoelectric ceramic's fixedness. When the electromagnet is switched on, the spring shrinks so that the two reinforcing patches stretch. Here, the two-chip piezoelectric ceramic can bend as long as the current is not cut. Through the stretch and closing of the electromagnet, the two patches of the reinforcement is able to clamp two sides of the piezoelectric ceramic. Therefore, the gripped part will not shed during the gripping and transfer for the reason of the inadequate stiffness.

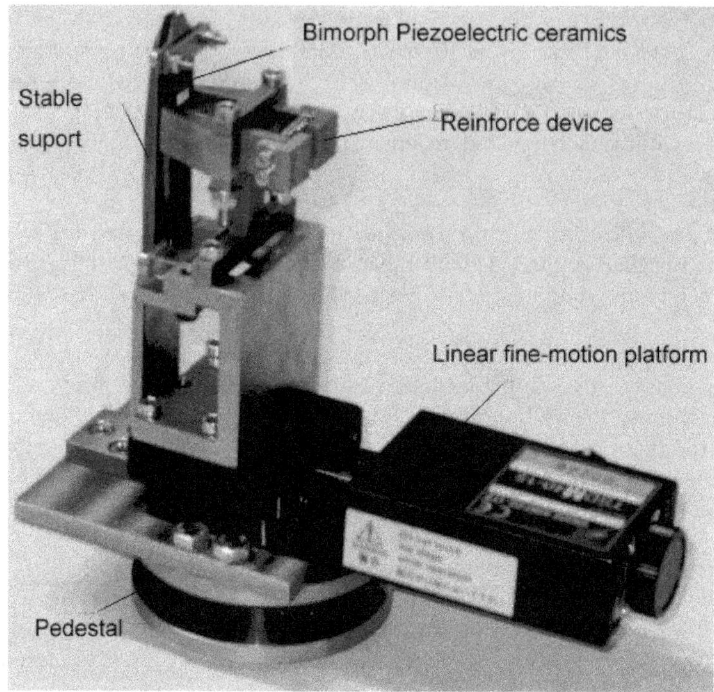

Fig. 5. Microgripper based on macle pizeelectric ceramics

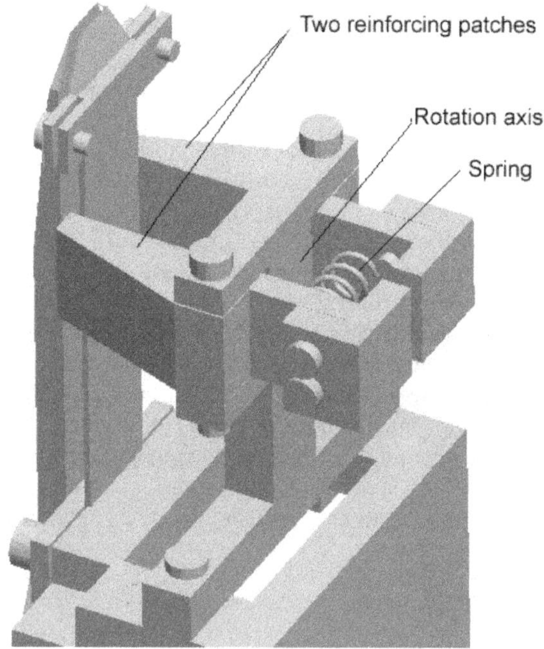

Fig. 6. Reinforcing device for piezoelectric ceramics

4.2 Micro force sensing and feedback

In this stage, researches mainly focus on two aspects–the strain sensing circuit as well as the amplifying and filter circuit. The function of the strain sensing circuit is to get the accurate voltage or current which can reflect the change in the resistance of the strain gauge. But because the parts to be gripped are very small, the bending of the two-chip piezoelectric ceramic is so tiny that the output of the strain sensing circuit is also tiny. It's necessary to apply an amplifying circuit to amplify to micro signal and a filter circuit to filter noises and other disturbing signals.

A full-bridge circuit with an invariable electric current source as its power supply is used in the strain sensing circuit. A reverse amplifier using negative feedback technology is adopted in the amplifying circuit, in which the gain is only decided by the resistance ratio. The filter circuit is designed into a twice Sallen-Key low-pass filter, shown as Fig. 7 (a) and (b).
An A/D acquisition card at the end of the sensing and filter circuits can gather the signals sensed and filtered by the previous circuits and change the analog data into digital data. Then the digital information is sent back to the control system, carrying out the feedback of the signal and providing the basis for the adjustment of the micro-gripper. This is the closed-loop control.

Fig. 7. (a) Amplification and filtering circuit

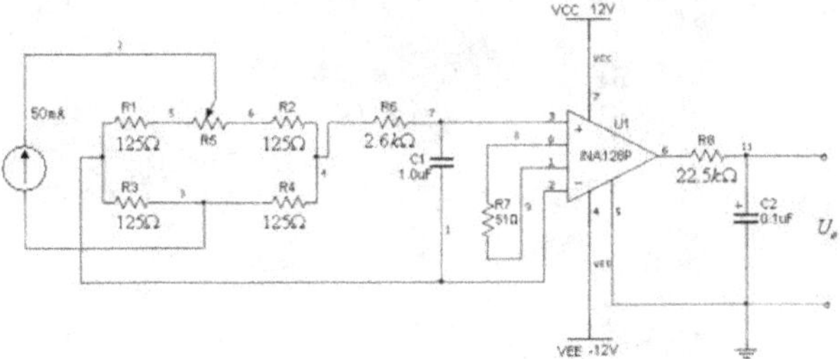

Fig. 7. (b) Amplification and filtering circuit diagram

4.3 Examples of the micro-gripping

The two-chip piezoelectric ceramic plays an important role in the combined two-chip piezoelectric ceramic micro-gripper integrated micro-force sensing and feedback function. The embodiment of its importance is shown as follows. (a) The shape of the two-chip piezoelectric ceramic is a flat cuboid, which is easy to stick the strain gauge on it. (b) The characteristic of the two-chip piezoelectric ceramic that it can bend when supplied with electricity provides a method to sensing the strain by the strain gauge.

When the two-chip piezoelectric ceramic is supplied with electricity and bending, the amplifying and filter circuits sense the output voltage of the strain sensing circuit, converse it into the corresponding end displacement, and test the capability of the two-chip piezoelectric ceramic and the relationship between the output voltage of the strain sensing circuit and the end displacement of the two-chip piezoelectric ceramic.

High precision optics displacement meter, as shown in Fig. 8, is applied to measure the true end displacement of the two-chip piezoelectric ceramic, which will be the benchmark of the end displacement.

The optics displacement meter can transform the actual displacement-the relative displacement of the certain point on the measured part-to the displacement of the reference point in a digital image, and the actual bending information will be processed by the digital image processing. The optics displacement meter won't do any harm to the measured part or import any added measuring error. Besides, its measuring precision is high. So the results got by the optics displacement meter can be considered as the ideal displacement of the two-chip piezoelectric ceramic. By comparing the end displacement got by the micro force sensing circuit and the ideal displacement got by the optics displacement meter, the measuring precision of the micro force sensing circuit can be tested.

Fig. 8. Piezoelectric ceramics' displacement detected by optical displacement detection device

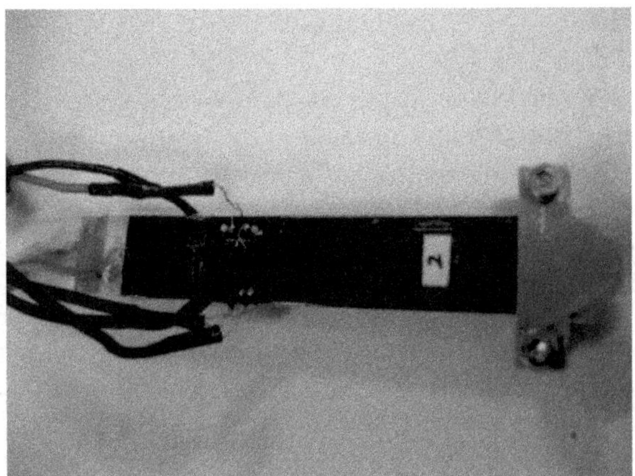

Fig. 9. Macle piezoelectric ceramics affixed with strain gages

A full-bridge circuit is applied, with two of the total four strain gauges sticked on where the strain is the maximal positive and the other two on where the strain is the maximal negative. This arrangement can get the highest sensitivity. Switch on the driving circuit of the two-chip piezoelectric ceramic. The driving voltages are cataloged into a series, and the output voltages are recorded correspondingly to the driving voltage series, shown as: $U_1, U_2, \cdots, U_n$. These output voltages are the V_0 in the equation (6) and by putting them in equation (7) the displacement got by the measuring circuits $w_1, w_2, \cdots, w_n$ can be calculated. A fine linear electrical source with invariable voltage and current is used as the driving power supply. The driving voltages range from 0 V to 150 V, with each one higher than the previous one by 10 V, which is 0, 10, 20, 30, ..., 150 V, and then change degressively from 150 V to 0 V. Gather the output voltages of the amplifying and filter circuits and converse them into the end displacement information of the two-chip piezoelectric ceramic. Meanwhile, the actual displacements are recorded by the optics displacement meter. The results are shown as Fig. 10.

Fig. 10. (a) shows the relationship between the driving circuit of the two-chip piezoelectric ceramic and the input and output voltages of the measuring circuits. When the driving voltages change from 0 V to 150 V and then change backwards, the output voltages change from -6V to 6V and then go backwards to -6V. But when the input voltages curve declines, the output voltages curve declines too but its shape is not the same as the shape of the output voltages curve when the input voltages curve rises. A difference appears. Fig. 10. (b) shows the comparison of the actual displacement got by the optics displacement meter and the displacement got by the measuring circuit, where the actual displacement shows a typical hysteresis curve marked with ▲ which reflects the characteristic of the piezoelectric ceramic and the displacement got by the measuring circuits marked with □ is too small to be counted compared to the actual displacement. Fig. 10. (c) and (d) shows the displacement got by the measuring circuits and the actual displacement respectively in proper coordinates, where their shapes are almost the same although the displacement got by the measuring circuit is two orders of magnitude smaller than the actual displacement. And then we can get the conclusions.

(a) The relationship between the end displacement of the two-chip piezoelectric ceramic and the driving voltage is not strictly linear, while the hysteresis is evident. (b) The displacement got by the measuring circuits shows the hysteresis, but its curve is S-shaped which is not accordant with the relationship between the end displacement of the two-chip piezoelectric ceramic and the driving voltage. This illuminates an aberrance of the signal. (c) The displacement got by the measuring circuits is two orders of magnitude smaller than the actual displacement got by the optics displacement meter, so it can almost be ignored. This illuminates that noises and other disturbing signals influence the measuring circuit largely, but the displacement got by the measuring circuits can already reflect the changing trend of the displacement.

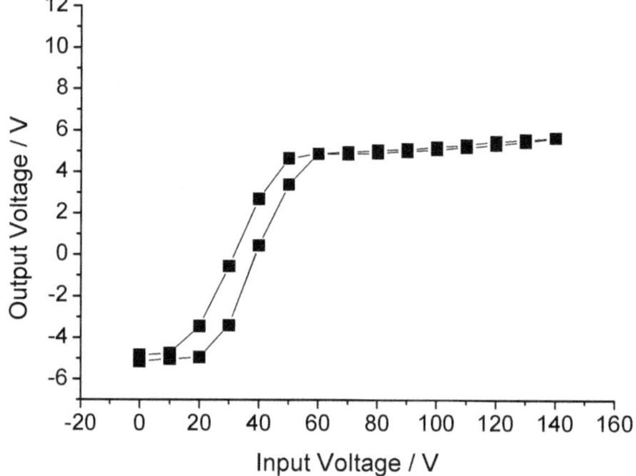

(a) Input and output voltage of detection circuit

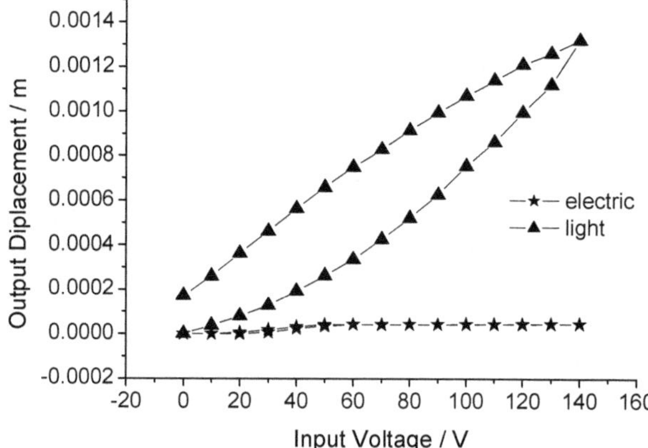

(b) Comparison of displacement separately detected by optical displacement device and amplification filtering circuit

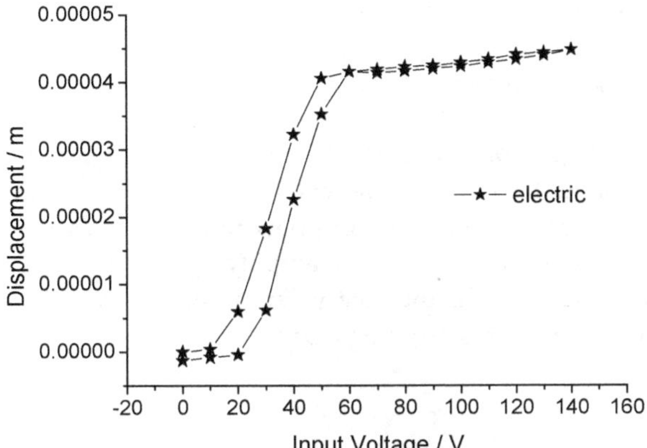

(c) Displacement obtained from amplification filtering circuit

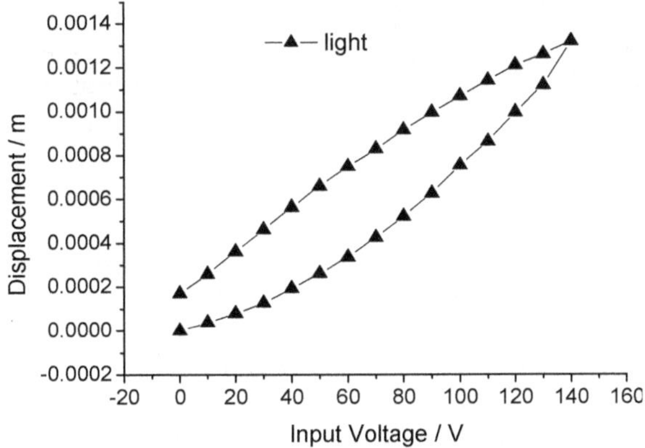

(d) Displacement obtained from optical displacement device
Fig. 10. Detection result

4.4 Micro force sensing experiments

This experiment is based on the output voltage got by the amplifying and filter circuit when the two-chip piezoelectric ceramic is bending. By comparing the output voltages at the same driving voltage with a gripped part on the micro-gripper and without a gripped part, the critical driving voltage and the average output voltage when gripping a part can be found, and then they can be conversed into corresponding micro gripping force.

Switch on the driving circuit of the two-chip piezoelectric ceramic and the amplifying and filter circuit, catalog the driving voltages in a series, and record output voltages of the measuring circuit respectively as follows: $U_1, U_2, \cdots, U_n$. When the gripper doesn't contact the part to be gripped, the output voltage increases as the driving voltage increases; when the gripper contacts the part, the bending of the two-chip piezoelectric ceramic can't be controlled by the driving voltage and stays invariable. Then the output voltage of the full-bridge circuit doesn't change either. So it's considered that the part has been gripped when the output voltage doesn't change. By using the voltage here to solve equation (5), (6), (8) and (9), the micro gripping force can be got.

Change the driving voltage from 0V to 140V with each 10V higher than the previous one-that is 0V, 10V, 20V, 30V...140V, gather the voltage information got by the amplifying and filter circuit, and converse the voltage into gripping force. The measuring result is shown as Fig. 11. Fig. 11. shows the output curve of the amplifying and filter circuits. The results data are divided into two groups–one is the data when no part is gripped, and the other one is the data when the micro-gripper is gripping a part.

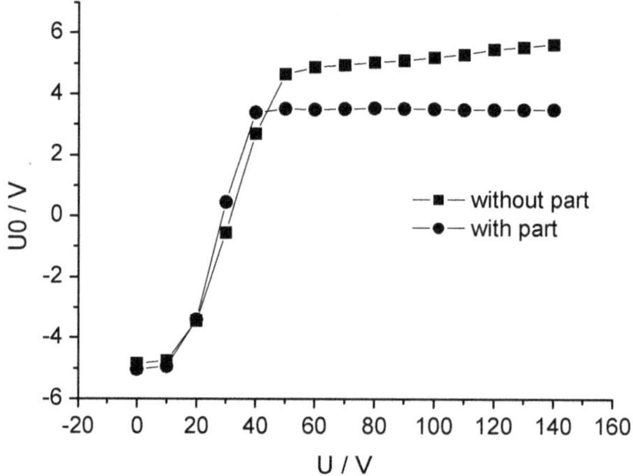

Fig. 11. Voltage result from micro force detection experiment

It can be concluded from Figure 8 that when no part is gripped the output voltage of the amplifying and filter circuit increases as the driving voltage increases and when the micro-gripper is gripping a part, the output voltage increases as the driving voltage increases from 0V to 40V, which is the same as when no part is gripped, but the output voltage stays invariable as the driving voltage continues to increase from 40V to 140V. So the 40V driving voltage is the critical voltage as the micro-gripper is gripping a part, which means that driving voltage upwards 40V will cause no more bending of the two-chip piezoelectric ceramic and no more changes of the measuring circuit. Here, the average output voltage is 5.213mV, and according to this the gripping force can be calculated to be 5.374mV.

5. Conclusion

In this chapter, we introduced the application of piezoelectric ceramics in microgrippers of micromanipulation area, especially microassembly. Two categories of piezoelectric-ceramic-based microgrippers were presented. The basic structures of bimorph piezoelectric ceramic clampers were established and the corresponding mathematical equations were deduced. After simply introducing the amplification mechanisms with flexure hinges, the reverse piezoelectric effect used by actuators of microgrippers were explained. Then, the microforce detection technology used in microgrippers was presented. The resistance strain gage was the essential and effectual constituent of contact detection style. The principles of metal resistance and semiconductor strain gauge were explained. Optical force sensing method, as a typical example of non-contact microforece detection method, was explained in detail. Finally, a practical example, a bimorph piezoelectric ceramic microgripper integrating microforce detecting and feedback, was developed. Experimental results indicated that the microgripper is possessed with not only clamping properties but also microforce detection function with milli-Newton scale accuracy.

6. References

Ashley, S. (1998). Magnetrostrictive actuators. *Mechanical Engineering*, Vol. 6, page numbers (69-70)

Heyliger, P. & Brooks, S. (1995). Free vibration of piezoelectric laminates in cylind rical bending. *International Journal of Solids Structures*, Vol. 32, No. 20, page numbers (2495-2960)

Li, Q. & Li, Y. (2004). *Basic Technology on Microassembly and Micromanipulation*, Tsinghua University Press, ISBN 7-302-08355-X, Beijing

Liu, M. & Xu Y. (1990). *Piezo- ferroelectric material and devices*, Huazhong University of Science and Technology Press, ISBN 7-5609-0491-2,Wuhan

Majumder, S.; McGruer, E. & Adams G. (2001). Study of contacts in an electrostatically actuated microswitch. *Sensors and Actuators A: Physical*, Vol. 93, No. 1, page numbers (19-26).

Quandt, E. (1997). Giant magnetostrictive thin film materials and applications. *Journal of Alloys and Compounds*, Vol. 258, page numbers(126-132)

Smits, G. & Choi, S. (1990). The constituent equations of piezoelectric heterogeneous bimorphs, *Proceedings of IEEE Ultrasonics Symposium*, pp. 1275-1278, ISBN, Honolulu, HI, Dec. 1990, IEEE, Piscataway, NJ, USA

Yu, Z. & Bradley N. (2000). Integrating optical force sensing with visual servoing for microassembly. *Journal of Intelligent and Robotic Systems*, Vol. 28, page numbers(259-276)

14

Application of piezoelectric ceramics in pulsed power technology and engineering

Sergey I. Shkuratov[1], Evgueni F. Talantsev[2] and Jason Baird[1,3]
[1]Loki Incorporated, Rolla, MO 65409,
[2]Pulsed Power LLC, Lubbock, TX 79416,
[3]Department of Mining and Nuclear Engineering, Missouri University of Science and
Technology, Rolla, MO 65409-0450,
U.S.A.

1. Introduction

Poled ferroelectrics are capable of storing electromagnetic energy in a form of bonded electric charge. The electric charge density of poled lead zirconate titanate (PZT) ceramic samples can exceed 30 $\mu C/cm^2$ and the electromagnetic energy density stored in the samples can reach 4 J/cm^3. Releasing the electric charge stored in ferroelectrics during a microsecond time interval can generate pulses of high voltage, high current and high power. The generation of pulsed voltage and pulsed currents by shock-compressed lead zirconate titanate and barium titanate ferroelectric ceramics was reported for the first time at the end of the 1950s (Neilson, 1957). Studies of the physical and electrical properties of shock-compressed ferroelectric and piezoelectric materials with light gas guns and explosively accelerated pellets, which initiated planar shock waves in the investigated samples have been performed since the 1960s and continue until present time (Reynolds & Seay, 1961; Reynolds & Seay, 1962; Halpin, 1966; Cutchen, 1966; Halpin, 1968; Lysne, 1973; Lysne & Percival, 1975; Lysne, 1975; Lysne & Percival, 1976; Bauer & Vollrath, 1976a; Bauer & Vollrath, 1976b; Mineev & Ivanov, 1976; Duvall & Graham, 1977; Lysne, 1977; Novitskii, et al., 1978; Dick & Vorthman, 1978; Dungan & Storz, 1985; Novitskii & Sadunov, 1985; Setchell, 2003; Setchell, 2005; Setchell et al., 2006; Setchell, 2007).

Since 1990s we have been performing design work, experimental studies (Shkuratov et al., 2004; Shkuratov et al., 2006a; Shkuratov et al., 2006c; Shkuratov et al., 2007a; Shkuratov et al., 2007b; Shkuratov et al., 2007c; Shkuratov et al., 2007d; Shkuratov et al., 2008a; Shkuratov et al., 2008b; Shkuratov et al., 2010) and theoretical investigations (Tkach et al., 2002, Shkuratov et al., 2007d; Shkuratov et al., 2008a) of miniature pulsed power generators based on shock depolarization of ferroelectric materials. Development of these autonomous prime power sources is important for the success of modern research and development projects (Altgilbers et al., 2009; Altgilbers et al., 2010).

Results of systematic studies of miniature ferroelectric generators (FEGs) are presented in this chapter. We describe the design of a miniature FEG, consider principles of its operation, give examples of performance of the generators in the high resistance and charging modes, and operation of the FEG equipped with a power conditioning stage.

2. Explosively Driven Ferroelectric Generator Design

A schematic diagram of our miniature shock wave FEGs that was used in experimental studies described in this chapter (Shkuratov et al., 2004; Shkuratov et al., 2006a; Shkuratov et al., 2006c; Shkuratov et al., 2008a) is shown in Fig. 1. It contained a ferroelectric element, an explosive chamber, a metallic impactor (flyer plate) and output terminals. The explosive part of the FEG contained a plastic cylindrical detonator support that holds an RP-501 detonator (supplied by RISI (Online A)) and a high explosive charge. The air gap between the flyer plate and the ferroelectric element front contact plate (the acceleration path) was 5 mm.

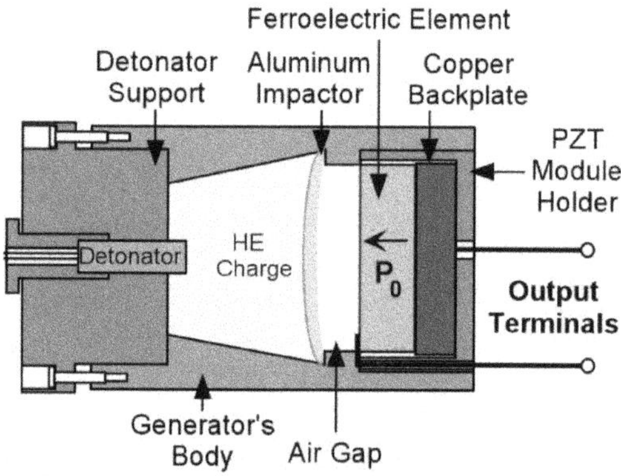

Fig. 1. A schematic diagram of a miniature explosively driven shock wave ferroelectric generator (FEG).

The metallic impactor was responsible for initiation of a shock wave in the ferroelectric element. A flyer plate accelerated to high velocity by the detonation of a high explosive charge impacted the face plate of the ferroelectric element and initiated a shock wave in the body of the ferroelectrics. The direction of propagation of shock wave in the ferroelectric element was parallel to the polarization vector P_0 (Fig. 1). This type of generator is referred to as a longitudinal FEG. The overall dimensions of the FEGs did not exceed 50 mm.

Desensitized RDX (essentially the same as Composition 4, or C-4) high explosives were used in all FEGs described in this chapter. The detonation velocity of desensitized RDX is 8.04 km/s and the dynamic pressure at the shock front reaches 36.7 GPa. The mass of the C-4 charge was varied from 12 to 17 g, depending on the charge holder configurations.

To provide a planar impact in the ferroelectric element, we designed the impactor (Fig. 1) as a curved plate that deformed into a quasi-flat structure under blast loads. To derive the correct curvature and thickness of the flyer plate material, the arrival times and locations of the explosive shock was calculated based on the principles of geometric optics and impedance mismatch as applied to detonation waves. Then, the shock transit time through the flyer plate material was calculated in order to calculate shock time of arrival along concentric circular loci on the target side of the flyer plate. The combination of shock arrival times and locations on the plate surface allowed for calculation of the plate curvature needed to produce an essentially flat flyer plate upon impact with the PZT target.

The cylindrical body of the FEG was made of material with good electrical insulating characteristics to avoid electric breakdown during the operation of the generator. We performed a series of experimental studies to determine the effects of generator body material on the explosive and electrical operation of the FEG (Shkuratov et al., 2004; Shkuratov et al., 2006c). Three types of plastics were used for the generator bodies: polyethylene, polyvinylchloride (PVC) and polycarbonate. The experimental results showed that each of these materials worked well.

In addition, we performed a series of experimental investigations to determine the effect of generator body wall thickness on the electrical operation of the FEG (Shkuratov et al., 2004; Shkuratov et al., 2006c). As a result of these studies it was found, that there is no significant difference in the generation of high-voltage pulses by FEGs having body wall thicknesses in the range from 1 mm to 30 mm. Based on this result, we reduced the total volume of the FEGs for four times. Physics behind this result is related to the fact that the generator's body does not effect on the explosive and electrical operation of the FEG, it only holds the ferroelectric element and explosive part together until detonation of the high explosives. The generator's body can be a lightweight and thin plastic shell (less than a millimeter larger than the diameter of the ferroelectric element).

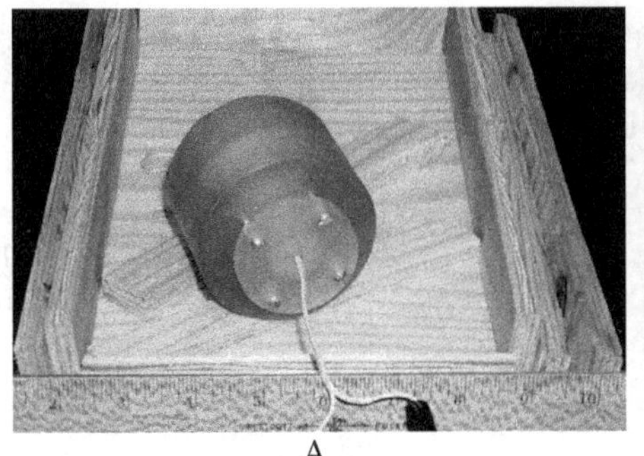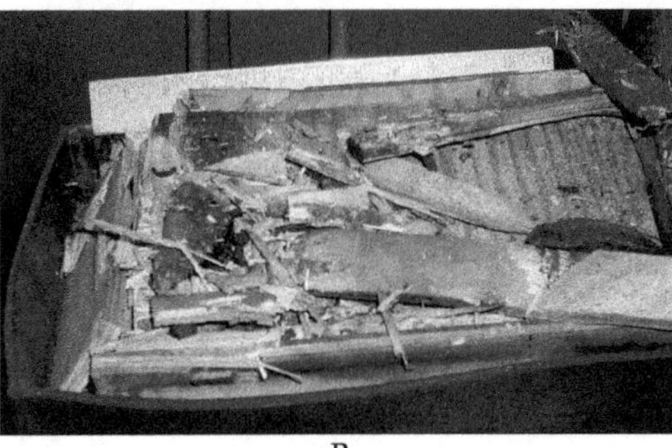

A   B

Fig. 2. Shock wave ferroelectric generator before (A) and after (B) explosive and electrical operation.

The ferroelectric element (Fig. 1) was bonded to a copper back-plate that was 1 mm larger in diameter than the ferroelectric element. The back-plate thickness was 5 mm. A silver-loaded epoxy was used to bond the ferroelectric disk to the back-plate. The copper back-plate provided mechanical impedance matching for minimizing reflection of the stress wave when it reached the back face of the ferroelectric energy-carrying element. The silver-loaded epoxy provided an electrical contact and reduced the capacitive reactance of the bond to a negligible value. The ferroelectric element, along with the copper back-plate, was centered in a cylindrical plastic holder (Fig. 1).

Two output electric terminals of the FEG were connected to two contact leads that were connected to the metallic contact plates deposited on the surface of the ferroelectric disk. One contact lead was bonded to the front plate of the ferroelectric disk (which was subjected to flyer plate impact) with silver-loaded epoxy. Another lead was bonded in the same manner to the copper back-plate. Figure 2 shows an FEG before and after explosive detonation and electrical operation.

3. Generation of Pulsed High Voltage by Longitudinally-Shock-Compressed PZT 52/48 Ferroelectrics

There are well known difficulties in the generation of pulsed power in a high resistance load with explosively driven electrical generators (including all types of magnetic flux compression generators, magneto-hydrodynamic generators, etc.) (Altgilbers et al., 2010). We demonstrated in a series of systematic experimental studies that explosively driven longitudinal FEGs are an exception to this list (Shkuratov et al., 2004; Shkuratov et al., 2006a; Shkuratov et al., 2006c; Shkuratov et al., 2008a; Shkuratov et al., 2008b). FEGs are explosively driven pulsed power sources that effectively generate high voltage across high-resistance loads.

In these studies we used lead zirconate titanate $Pb(Zr_{0.52}Ti_{0.48})O_3$ (PZT 52/48) poled piezoelectric ceramics (trade mark EC-64, supplied by ITT Corp. (Online B)). PZT 52/48 is a ferroelectric material widely used in modern mechanical and electrical systems due to its excellent piezoelectric properties. This material is in mass production and commercially available in a variety of shapes, sizes and trademarks. It is used in form of ceramics since 1960s (Jaffe et al., 1971) and recently in form of thin films (Shur et al., 1997, Kuznetsov et al., 2006).

The parameters of PZT 52/48 are as follows: the density is $7.5 \cdot 10^3$ kg/m³, the dielectric constant is $\varepsilon = 1300$, the Curie temperature is 593 K, the Young's modulus is $7.8 \cdot 10^{10}$ N/m², the piezoelectric constant is $d_{33} = 295 \cdot 10^{-12}$ C/N, and the piezoelectric constant is $g_{33} = 25 \cdot 10^{-3}$ m²/C (Online B).

FEG ferroelectric elements investigated in (Shkuratov et al., 2004; Shkuratov et al., 2006a; Shkuratov et al., 2006c; Shkuratov et al., 2007a; Shkuratov et al., 2007b; Shkuratov et al., 2007c; Shkuratov et al., 2008a; Shkuratov et al., 2008b) were PZT 52/48 disks polarized along the cylindrical axis. Silver contact plates (electrodes) were deposited on both ends of the PZT disks by electron beam deposition, and each disk was poled by the manufacturer to its remnant polarization value. The diameters of the ferroelectric disks were 25.0, 25.4, 26.1, and 27.0 mm, and their thicknesses varied from 0.65 mm to 6.5 mm (Table 1).

Diameter, D (mm)	Thickness, h (mm)
26.1	0.65
27.0	2.10
25.0	2.50
25.4	5.10
25.0	6.50

Table 1. Sizes of investigated PZT 52/48 ferroelectric elements of the FEGs.

To investigate FEG pulsed power generation in the high-resistance mode, we used a Tektronix P6015A high-voltage probe as a load. The impedance of the probe was 10^8 Ω and its capacitance was 3 pF, so the current in the electrical circuit of the FEG was negligible (less than $5 \cdot 10^{-4}$ A at 50 kV output voltage).

The experimental setup and circuit diagram of the system for investigation of the operation of the FEG in the high resistance mode are shown in Fig. 3. The high-voltage output terminal of the FEG (positive plate of the ferroelectric element) was connected directly to the high-voltage probe, and the negative (front) plate of the PZT disk was grounded. Note that high-voltage diodes or high-voltage rectifiers were not used in these experiments.

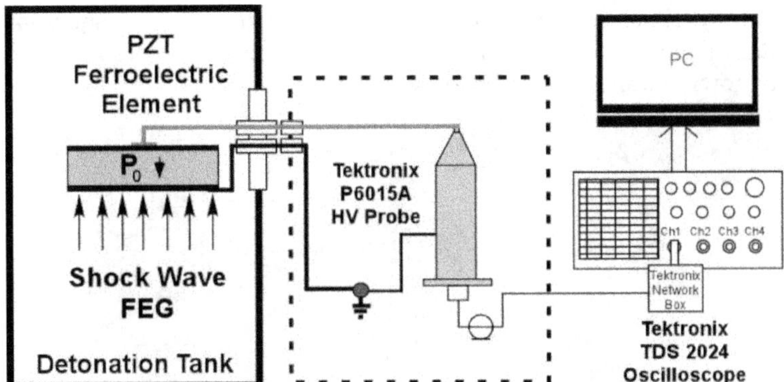

Fig. 3. Schematic diagrams of the measuring system for investigating the operation of FEG in the high-resistance mode.

The operation of the FEG was as following. After detonation of high explosives, an accelerated flyer plate impacted the ferroelectric element and initiated a shock wave. The shock wave propagated through the ferroelectric element and depolarized it. The depolarization process released the induced electric charge on the metallic contact plates of the ferroelectric element and a voltage pulse (electromotive force pulse) appeared on the output terminals of the generator.

The waveform of a typical electromotive force (e.m.f.) pulse produced by an FEG containing a PZT 52/48 ferroelectric element of $D = 25.0$ mm/$h = 6.5$ mm is shown in Fig. 4. The amplitude of the e.m.f. pulse reached $E_g(t)_{max} = 21.4$ kV and the full width at the half maximum (FWHM) was 1.1 µs.

In Fig. 4, the increase in the e.m.f. pulse from zero to its peak value is the direct result of the depolarization of the ferroelectric element due to explosive detonation shock wave action. Shock wave depolarization induced an electric charge that was released at the contact plates of the ferroelectric disk. Because of the high resistance and low capacitance of the load in this mode of operation, the released electric charge was charging the ferroelectric element (it is initially a capacitor) to a high voltage. The rise-time of the e.m.f. pulse is related to the shock front propagation time through the ferroelectric disk thickness.

Since there was no electric charge transfer from the ferroelectric element to the external circuit in this mode of operation of the FEG (Fig. 3), one might expect that the waveform of the e.m.f. produced by the FEG would be a square pulse of $E_g(t)_{max}$ amplitude, with a flat top and lasting several or more microseconds (until the mechanical destruction of the ferroelectric element occurred). It follows from the experiments, however, that after reaching its maximum value the e.m.f. pulse does not hold its maximum value, but decreases rapidly (see Fig. 4).

To understand this phenomenon it is necessary to take into a consideration that the shock wave propagating through the ferroelectric element had very complex characteristics (Reynolds & Seay, 1961; Reynolds & Seay, 1962; Halpin, 1966; Setchell, 2003). The shock represents the superposition of a number of elastic and inelastic acoustic and shock waves (Setchell, 2003) that depolarize the ferroelectric element and simultaneously change its physical properties significantly. Apparently, the rapid decrease of the e.m.f. pulse after it reached its peak value, $E_g(t)_{max}$, was the result of a significant increase in the electrical conductivity of the shock-compressed ceramic material (behind the shock as it traveled through the disk) and a corresponding leakage current in the element, or of internal electrical breakdown within the ceramic disk.

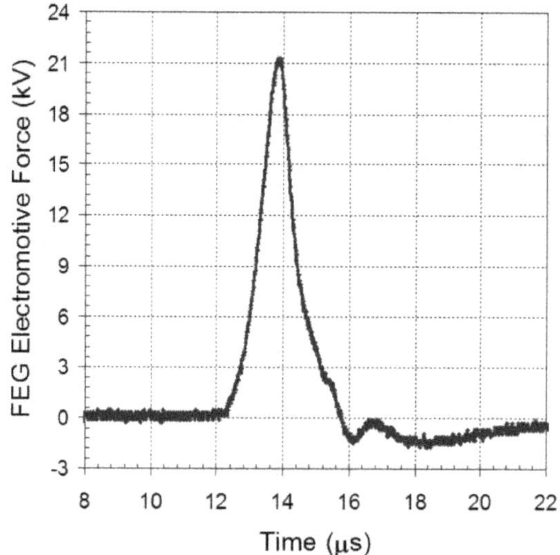

Fig. 4. A typical waveform of high voltage pulse produced by the FEG containing PZT 52/48 element with D = 25.0 mm/h = 6.5 mm.

In spite of complex and not completely understood physical processes in shocked ferroelectrics, we have found out that there are two linear relationships between PZT 52/48 element thickness and parameters of the high voltage pulse produced by the FEG. They are presented in Figure 5.

The first linear relationship is between PZT element thickness, h, and the amplitude of the e.m.f pulse, $E_g(t)_{max}$. The second linear relationship is between h and FWHM of the e.m.f. pulse. The experiments documented in Fig. 5 show that the e.m.f. pulse amplitude and FWHM are highly reproducible. An increase in PZT 52/48 element thickness leads to an increase of both the e.m.f. pulse amplitude, $E_g(t)_{max}$, and the pulse width.

For a PZT 52/48 ferroelectric element with h = 0.65 mm, $E_g(t)_{max}$ = 3.6 ± 0.23 kV with FWHM of 0.2 ± 0.04 μs. An increase in the ferroelectric element thickness to 2.5 mm almost triples the e.m.f. pulse amplitude, to 8.9 ± 0.07 kV. The FWHM increases more than 3 times to 0.62 ± 0.02 μs. For a ferroelectric element having h = 5.1 mm, $E_g(t)_{max}$ = 17.03 ± 0.25 kV with a FWHM of 0.9 ± 0.03 μs.

It follows from the experimental data shown in Fig. 5 that the amplitude of the e.m.f. pulse is directly proportional to the PZT 52/48 element thickness, with a coefficient of proportionality equal to 3.4 ± 0.5 kV/mm. Apparently, this electric field strength (3.4 kV/mm) is the internal electrical breakdown field for PZT 52/48 ferroelectric material. The

FWHM is directly proportional to the thickness of the PZT 52/48 ferroelectric element, with a coefficient of proportionality equal to 0.21 ± 0.02 µs/mm.

Therefore, we experimentally demonstrated that miniature explosively driven ferroelectric generators are capable of producing high-voltage pulses of several tens of kilovolts for a few microseconds in a high-resistance load.

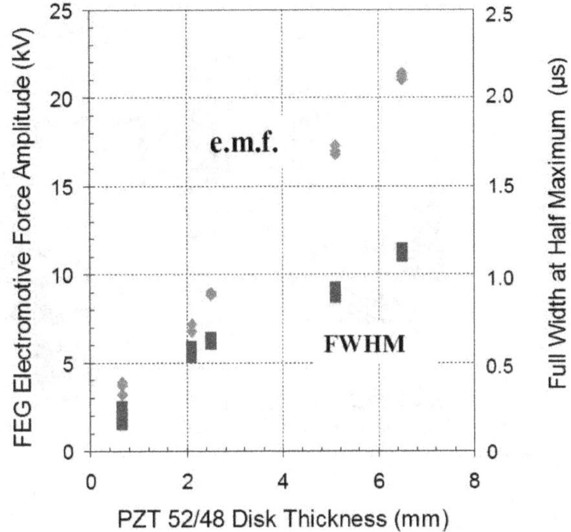

Fig. 5. From (Shkuratov et al., 2004; Shkuratov et al., 2006a; Shkuratov et al., 2006c; Shkuratov et al., 2007a; Shkuratov et al., 2008a; Shkuratov et al., 2008b): experimentally obtained dependence of e.m.f. pulse amplitudes (diamonds) and FWHM (squares) produced by FEGs versus thicknesses of PZT 52/48 ferroelectric elements.

4. Depolarization of PZT 52/48: Longitudinal Explosive Shock Versus Quasi-Static Thermal Heating

The quantity of electric charge released in the electrical circuit of an FEG during its explosive operation determines the amount of electrical energy produced by the generator in its load circuit. Therefore, the efficiency of the device depends on the degree of the depolarization of the ferroelectric element under the action of a shock wave. In this section, we describe the results of comparative systematic studies of the depolarization of PZT 52/48 ceramic by means of two techniques – thermal heating, and longitudinal compression by shock waves from the detonation of an explosive charge.

A schematic diagram of the experimental setup we developed for thermal depolarization of ferroelectric samples is shown in Fig. 6 (Shkuratov et al., 2010). In this setup, we used an automatically controlled Thermolyne 47900 furnace, and during each experiment placed a ferroelectric sample in a thermal bath (a beaker filled with ultra-fine sand) close to a K-type thermocouple. We connected the thermocouple to a SPER Scientific thermometer (Model 800005), and the signal cables from the ferroelectric sample to a Keithly 2400 pico-ampere meter.

The heating rate did not did not exceed 1 K/min for all studied samples. Detailed description of technique of thermal depolarization experiments is given in (Shkuratov et al., 2010).

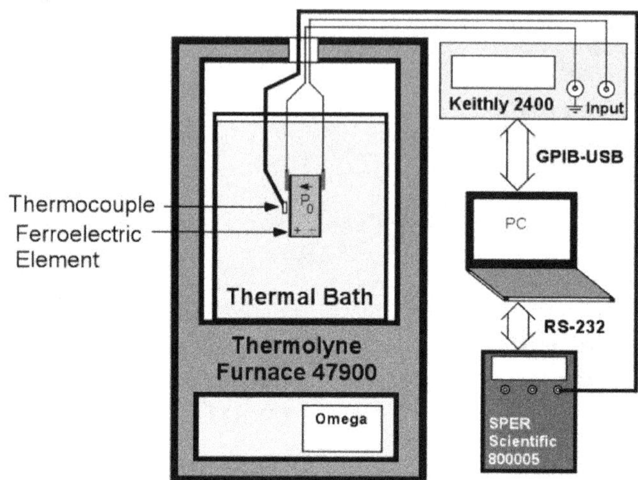

Fig. 6. A schematic diagram of the experimental setup used for measurement of thermal depolarization of ferroelectric samples.

A diagram of the measuring circuit we developed for explosive shock depolarization experiments is shown in Fig. 7 (Shkuratov et al., 2006c; Shkuratov et al., 2007b; Shkuratov et al., 2010). A schematic of shock wave FEG used in these experiments is shown in Fig. 1. The load loop (Fig. 7) was made of a copper strip 12.0 mm wide and 1.0 mm thick. The load loop (Fig. 7) resistance and inductance were $R_L(100\ \text{kHz}) = 0.57\ \Omega$ and $L_L(100\ \text{kHz}) = 0.98\ \mu\text{H}$, respectively.

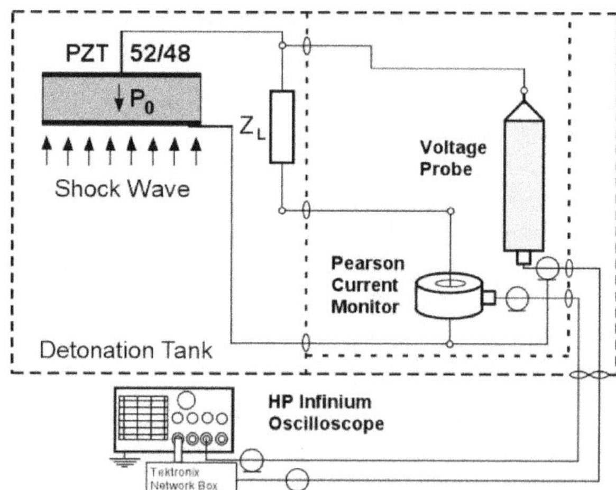

Fig. 7. A schematic diagram of the experimental setup for investigation of longitudinal shock depolarization of the PZT 52/48 ceramic samples.

Typical plots of the thermally induced current, $I_h(T)$, for PZT 52/48 ceramic disk of 26.1 mm diameter are shown in Fig. 8. The $I_h(T)$ is not a monotonic function of temperature, and it has a few well-resolved peaks. The relative amplitudes of the peaks and their positions varied with disk thickness (Shkuratov et al., 2010). Each sample we investigated, no matter the size (see Table 1), had the most pronounced peak of $I_h(T)$ at $T = 405\pm3$ K. This temperature is in good agreement with the temperature at which the samples were poled by the manufacturer. Commercial PZT 52/48 samples are poled by ITT Corp at temperatures ranging from 400 K to 410 K (Online B). Heating the ferroelectric samples higher than 420 K can lead to an increase in leakage currents to a level that could result in thermal runaway and electrical breakdown during the poling procedure (Moulson & Herbert, 2003).

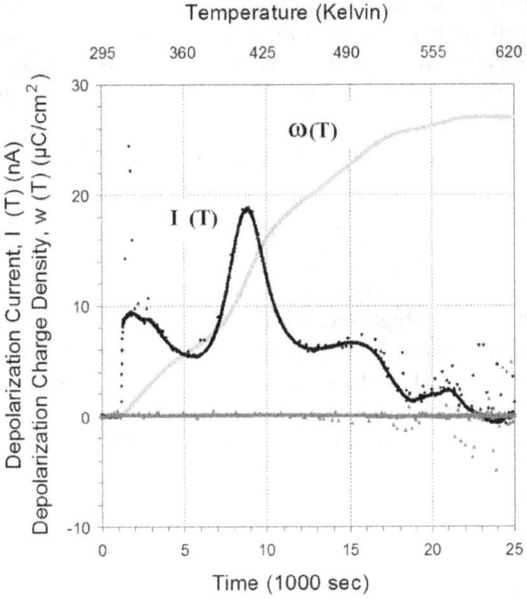

Fig. 8. Typical depolarization curve for PZT 52/48 disks of diameter $D = 26.1$ mm/ thickness $h = 0.65$ mm. Initial depolarization current (first heating cycle) is black curve. The electric charge released in the circuit is light gray curve. Second heating cycle depolarization current is dark gray curve.

It follows from our experimental results that a second heating cycle of the PZT 52/48 samples caused no current flow in the circuit (Fig. 8). In this case, the current recorded from each investigated sample was practically zero.

This is direct experimental evidence that during the first heating cycle (Fig. 8), PZT 52/48 undergoes a structural phase transition into the cubic phase on the PZT composition-temperature phase diagram (Jaffe et al., 1971). The samples are completely depolarized during the first heating cycle, and all compensating electric charge density stored in the samples' electrodes is released to the external circuit.

The depolarization charge density, ω, is the integral of the experimentally measured depolarization current divided by the electrode area, A

$$\omega(T(t)) = \frac{1}{A} \cdot \int_0^{t_F} I_h(T(t)) \cdot dt \tag{1}$$

where t_F is the time of the depolarization process. The experimental data for all sample sizes we studied are summarized in Table II. The thermally-induced depolarization charge density was consistent for all sample sizes, ω = 27.7 μC/cm^2.

The flyer plate of the FEG longitudinally impacted the ferroelectric body so that the shock wave traveled in a direction parallel to the ferroelectric polarization vector P_0 (Fig. 7). Prior to the flyer plate impact, the electric field in the ferroelectric sample was equal to zero because the surface charge density (the bonded charge) compensated the polarization of the sample, P_0, that was obtained during the poling procedure. When a shock wave depolarized a ferroelectric disk, a pulsed e.m.f. appeared on the metallic electrodes of the ferroelectric element. When the electrical circuit of the generator was closed (a load is connected to the output terminals of the FEG) the e.m.f. caused a pulsed electric current, $I(t)$, in the circuit. Integration of the $I(t)$ waveform from 0 to t gives the momentary value of the electric charge, $\Delta Q(t)$, released to the electrical circuit due to depolarization of the PZT

$$\Delta Q(t) = \int_0^{t_F} I(t) \cdot dt \tag{2}$$

It should be noted that t_F in the shock experiments ranges from 10^{-6} to 10^{-5} s. It is about 9-10 orders of magnitude lower than t_F in the thermal depolarization experiments (the time for the thermally induced charge to be released).

Typical waveforms of the shock induced depolarization current produced by a PZT 52/48 disk of D = 26.1 mm / h = 0.65 mm, and the electric charge released from the sample are presented in Fig. 9.

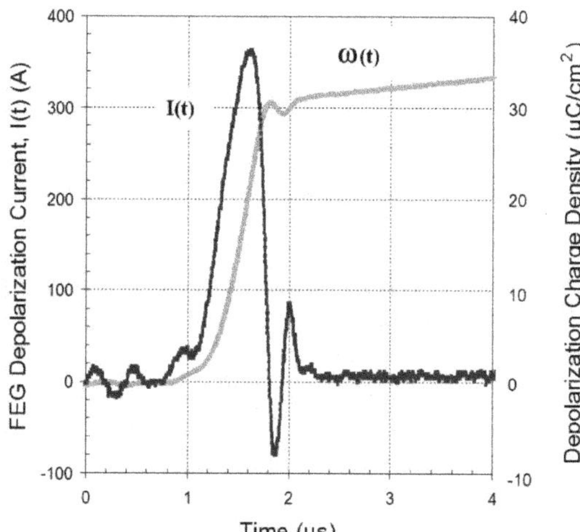

Fig. 9. Waveform of the pulsed depolarization current, $I(t)$, and the depolarization charge density, $\omega(t)$, released due to longitudinal shock compression of PZT 52/48 disk with D = 26.1 mm / h = 0.65 mm.

The peak depolarization current generated by longitudinally-shock-compressed PZT element was $I(t)_{max}$ = 360 A, with FWHM of 0.5 µs. The electric charge density released from the sample due to the shock compression was ω = 30.6 µC/cm². The depolarization charge density for all PZT 52/48 samples we investigated are presented in Table 2.

Diameter of the PZT 52/48 disks (mm)	26.1	27.0	25.4
Thickness of the PZT 52/48 disks (mm)	0.65	2.1	5.1
Thermal depolarization charge density (µC/cm²)	27.9±1.8	27.8±1.5	27.9±1.6
Adiabatic (shock induced) high-pressure depolarization charge density (µC/cm²)	29.7±2.4	27.5±2.2	26.7±2.2

Table 2. Sizes of PZT 52/48 disks and electric charge released by disks due to thermal depolarization and adiabatic (longitudinal-shock-wave) high-pressure depolarization.

It follows from our experimental results (Shkuratov et al., 2010) that increasing the thickness of the PZT disk led to decreased shock depolarization current amplitude, and increased current pulse FWHM. The latter is apparently the result of the longer shock wave front propagation time in the thicker samples. Figure 10 presents the amplitude of the depolarization current generated by longitudinally-shock-compressed PZT 52/48 disks, versus the thickness of the disks. It follows from our experimental results that the current amplitude is inversely proportional to the disk thickness.

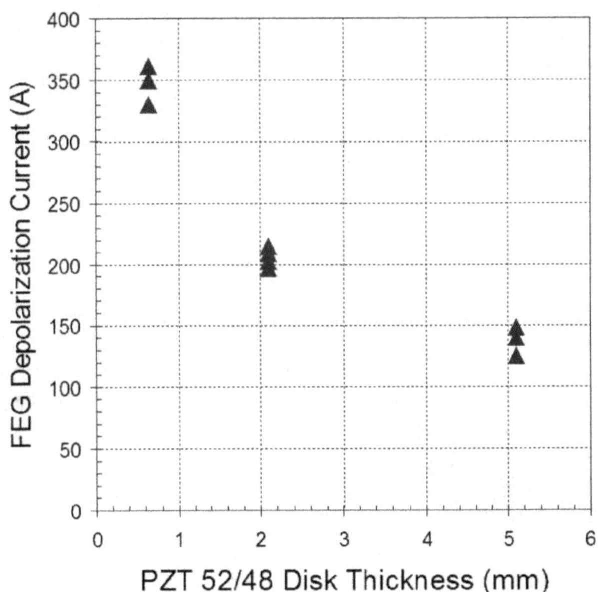

Fig. 10. Amplitude of shock induced depolarization current versus thickness of PZT 52/48 disks.

Experimental data for both the thermal and adiabatic (shock-induced) depolarization of all PZT 52/48 samples we studied are shown in Fig. 11. The shock pressure in the bulk PZT 52/48 ceramic elements was determined in (Shkuratov et al., 2010), P_{SW} = 1.5 ± 0.2 GPa.

It is clear from results presented in Fig. 11 that the depolarization charge released due to longitudinal shock wave compression of PZT 52/48 disks is almost equal than that released due to thermal heating. This is direct evidence of almost complete depolarization of the PZT 52/48 elements due to their compression by longitudinal shock waves within the FEGs.

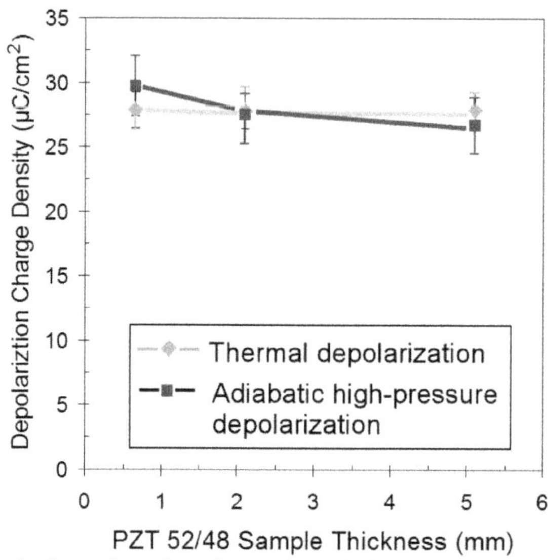

Fig. 11. Experimental data for the thermal and longitudinal explosive shock wave depolarization of PZT 52/48 samples.

It follows from the results we obtained (Fig. 11) that despite the small size of ferroelectric generators and their correspondingly imperfectly-shaped shock wave fronts, these devices provide almost complete shock-wave depolarization of the PZT 52/48 ferroceramic elements.

One of the effects we observed was a monotonic decrease of the electric charge released from shock-compressed PZT 52/48 samples with increasing sample thickness (Fig. 11). A possible explanation of this effect can be based on the shock wave splitting phenomena observed earlier by Reynolds and Seay (Reynolds & Seay, 1961; Reynolds & Seay, 1962). They showed that a longitudinal shock wave in PZT 52/48 splits into two shock waves; the split shocks result in decreased depolarization charge. It is possible that such shock splitting has only a minor effect in thin samples. An increase of the thickness of the samples leads to an increase in shock wave travel distance, and to a more significant separation of the shock waves. Other possibilities may include the dispersion of the shock as it crushes the sample while transiting the its length, or shock attenuation as the rarefaction(s) behind the shock catch(es) the shock, or some combination of these processes.

In Section 3 of this Chapter we have demonstrated that the output voltage produced by the FEG is directly proportional to the thickness of PZT 52/48 element up to $h = 6.5$ mm. The effect of reduction of the electric charge transfer from longitudinally-shock-compressed PZT 52/48 elements with increasing the element's thickness we detected may pose a problem in the development of FEGs utilizing ferroelectric elements thicker than 10 mm for producing ultrahigh output voltage.

5. Pulse Charging of Capacitor Bank by Explosively Driven FEG

Various types of pulsed power systems built around capacitive energy storage devices are widely used in modern technology for production of charged-particle beams, generation of pulsed high power microwaves and others pulsed power technologies (Mesyats, 2005). In these generators, electric energy is provided to the capacitive energy storage from high-voltage power sources powered from a conventional 110/220 V - 50/60 Hz supply line. The operation theory of these generators is well-developed (Mesyats, 2005).

However, certain special applications require that the pulsed power system be autonomous. Another necessary condition is compactness of the device as a whole. We proposed and experimentally studied an autonomous two-stage pulsed power system based on a shock-wave ferroelectric generator as a charging source for a capacitive energy storage (Shkuratov et al., 2007c; Shkuratov et al., 2007d; Shkuratov et al., 2008a; Shkuratov et al., 2008b). We have experimentally demonstrated that a miniature explosively driven FEG can successfully charge capacitor banks of different capacitances. We have developed a model and made a successful digital simulation of the operation of the FEG-Capacitor bank system. We present some results of these studies in this section.

5.1. FEG-Capacitor Bank System

A diagram of the experimental setup we developed for investigations of the FEG-Capacitor bank system is shown in Fig. 12 (Shkuratov et al., 2008a; Shkuratov et al., 2008b).

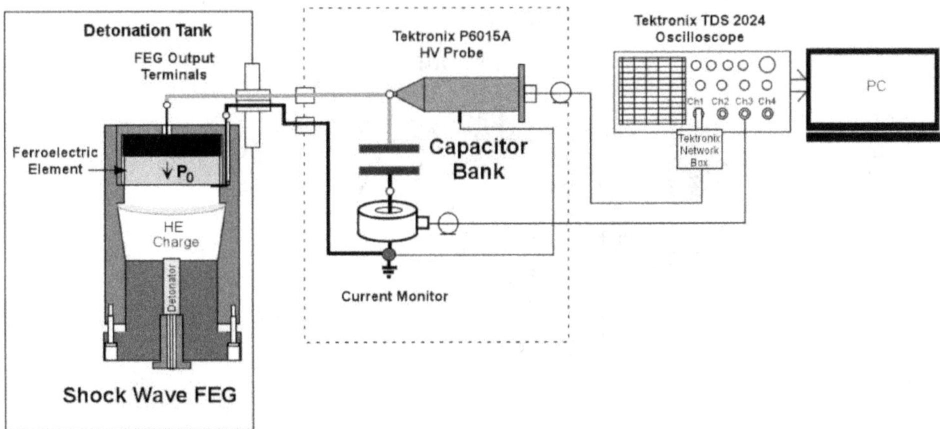

Fig. 12. Schematic diagrams of the measuring system for investigating the operation of the FEG-Capacitor bank system.

The high voltage output of the FEG was connected to the high-voltage terminal of the capacitor bank and to the Tektronix P6015A high-voltage probe (Fig. 12). The negative plate of PZT disk was connected to the ground terminal of the capacitor bank through a commercial current monitor. We did not use high-voltage diodes or rectifiers in these experiments.

The first series of FEG-Capacitor bank experiments was performed with PZT 52/48 elements having diameter $D = 26.1$ mm and thickness $h = 0.65$ mm. The capacitance of the capacitor bank was $C_L = 18$ nF. It is more than two times higher than initial capacitance of the FEG, $C_G = 7.1 \pm 0.1$ nF.

Figure 13(A) shows a typical waveform of the high voltage produced by an FEG across an 18 nF capacitor bank. It is not a single pulse, but a series of oscillations. The frequency of oscillations is about 1.0 MHz. The peak voltage amplitude of the first pulse was $U(t)_{max} = 2.16$ kV, the FWHM of the first pulse was 0.54 µs, and $\tau = 0.34$ µs.

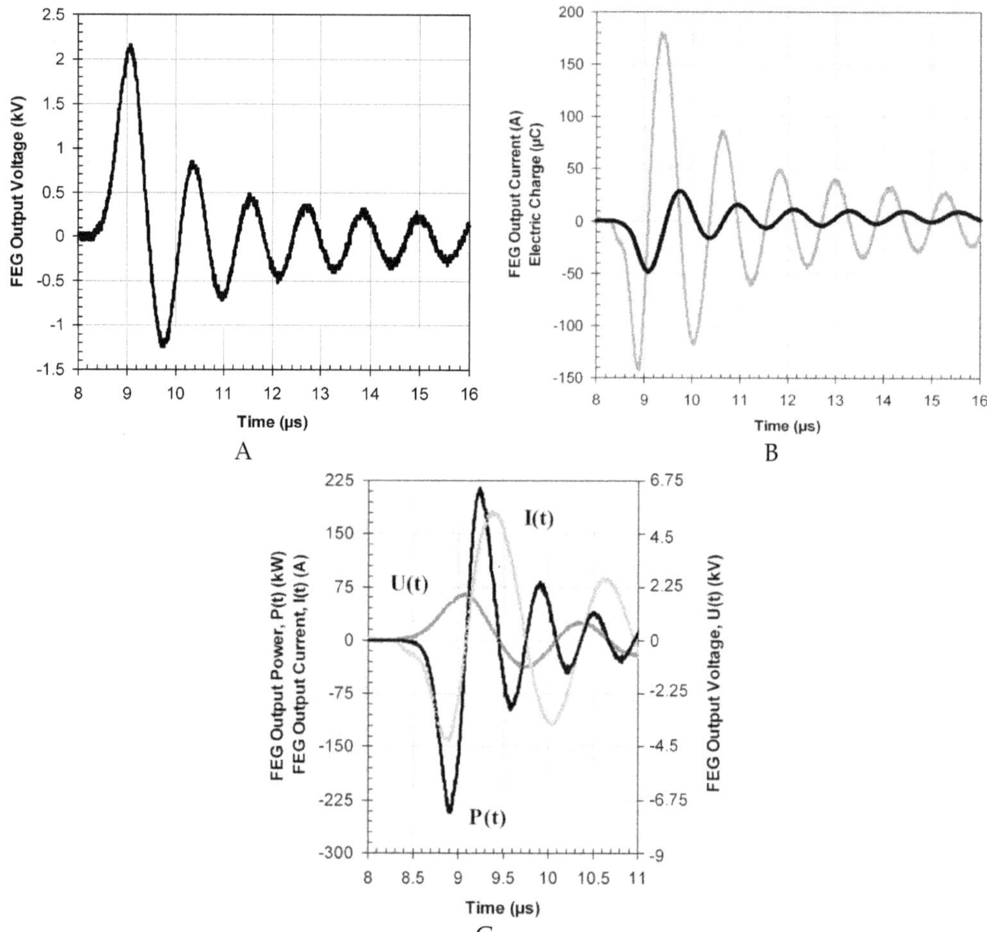

Fig. 13. A – waveform of the output voltage; B – waveform of current (gray) and circulation of electric charge (black); C – waveforms of voltage (dark gray), current (light gray) and power (black) produced by the FEG. PZT 52/48 disk with $D = 26.1$ mm / $h = 0.65$ mm. 18 nF capacitor bank.

The peak energy delivered to the capacitor bank in this experiment [the first pulse in Fig. 13(A)] reached $W(t)_{max} = C_0 \, U(t)_{max}^2/2 = 42$ mJ. The average amplitude of the first high voltage pulse produced by the FEG across the capacitor bank in this series of experiments was $U(t)_{max\ ave}$ = 2.07 ± 0.22 kV. The average peak energy delivered to an 18 nF capacitor bank reached $W(t)_{max\ ave}$ = 39 ± 0.4 mJ.

Figure 13(B) shows the waveform of the charging current, $I(t)$, produced by the FEG in the circuit and circulation of electric charge. The peak amplitude of the first current pulse was $I_1(t)_{max}$ = 140 A, the FWHM was 0.3 μs and the rise-time τ = 0.52 μs. The peak amplitude of the second current pulse was higher than the first one and reached $I_2(t)_{max}$ = 180 A, with FWHM = 0.45 μs and τ = 0.31 μs. Integration of the charging current, $I(t)$, waveform from 0 to t gives the momentary value of the electric charge, $\Delta Q(t)$, transferred to the external electrical circuit during explosive operation of the FEG (Eqn. 2).

As it follows from the experiment [Fig. 13(B)], the electric charge transferred from a PZT module during explosive operation of the FEG to the capacitor bank, ΔQ_{max} = 50 μC, is 33.3% of the initial charge stored in the ferroelectric element due to its remnant polarization, Q_0 = 150 μC (see Table 2 in Section 4 of this Chapter).

Waveforms of the output high voltage $U(t)$, current $I(t)$, and power $P(t)$ pulses produced by an FEG across 18 nF capacitor bank are shown in Fig. 13(C). The power dissipated in the load, $P(t)$ was determined as the product of the instantaneous value of the output voltage $U(t)$ by the instantaneous current in the circuit, $I(t)$: $P(t) = I(t) \cdot U(t)$. The peak output power reached $P(t)_{max}$ = 0.24 MW.

It should be noted that peak power produced by the FEG is 5 to 7 times higher than that produced by recently developed miniature prime power sources based on shock wave demagnetization of $Nd_2Fe_{14}B$ ferromagnets, shock wave ferromagnetic generators (FMGs) (Shkuratov et al., 2002a; Shkuratov et al., 2002b; Shkuratov et al., 2002c; Shkuratov et al., 2003a; Shkuratov et al., 2003b; Shkuratov et al., 2006b). At the same time the FMG produces much longer power pulse than the FEG (FWHM = 0.3 μs for FEG and FWHM = 8.0 μs for FMG).

The next series of experiments was performed with double less capacitance of the bank C_L = 9 nF. The output voltage oscillated as it did in the experiments with an 18 nF capacitor bank (Fig. 13). The frequency of oscillations was slightly higher, ~ 1.1-1.2 MHz in compare with C_L = 18 nF. The average amplitude of the first high voltage pulse produced by the FEG across a 9 nF capacitor bank was $U(t)_{max\ ave}$ = 2.41 ± 0.33 kV. The average peak energy delivered to a 9 nF capacitor bank reached $W(t)_{max\ ave}$ = 26 ± 0.5 mJ.

It follows from experiments with 9 nF and 18 nF capacitor banks that increasing the capacitor bank capacitance leads to increasing the energy transferred from the PZT module to the capacitor bank. This result was confirmed in the third series of FEG-Capacitor bank experiments.

The third series of experiments was performed with capacitance of the bank C_L = 36 nF. Figure 14(A) shows a typical waveform of the high voltage produced by an FEG across a 36 nF capacitor bank. Results of these experiments were different from the results obtained with the 18 nF and 9 nF capacitor banks. The FEG produced a series of oscillations, but the amplitude of the first half-wave is significantly higher than the next one. Oscillations were damping quickly. Amplitude of the first half-wave of output voltage was $U(t)_{max}$ = 1.82 kV with FWHM = 0.85 μs, and τ = 0.93 μs.

The energy delivered to a 36 nF capacitor bank was $W(t)_{max}$ = 60 mJ and the specific energy density of the PZT element was 171 mJ/cm^3. The average amplitude of the first half-wave of high voltage across a 36 nF capacitor bank was $U(t)_{max\ ave}$ = 1.75 ± 0.14 kV.

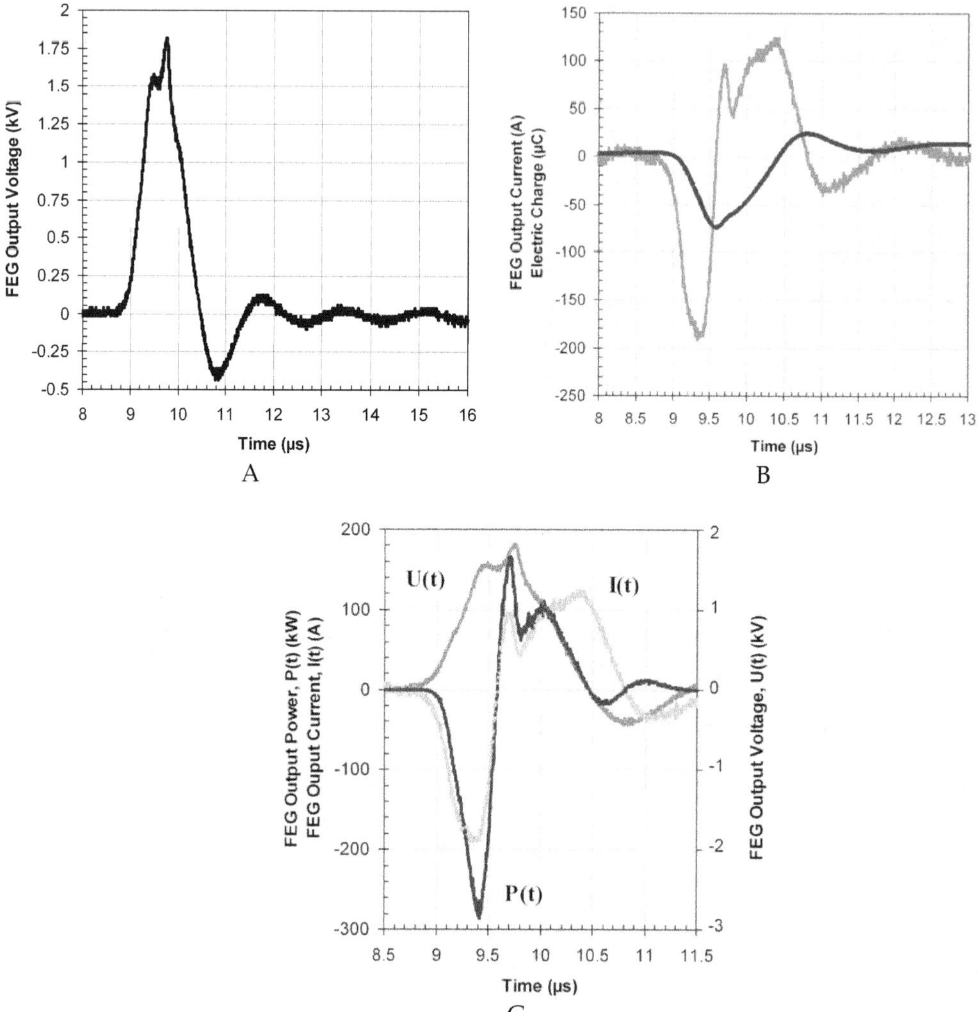

Fig. 14. A – waveform of the output voltage; B – waveform of current (gray) and circulation of electric charge (black); C – waveforms of voltage (dark gray), current (light gray) and power (black) produced by the FEG. PZT 52/48 disk with D = 26.2 mm/h = 0.65 mm. 36 nF capacitor bank.

The average peak energy delivered to a 36 nF capacitor bank reached $W(t)_{max\ ave}$ = 55 ± 0.3 mJ. Fig. 14(B) shows the waveform of the output current, $I(t)$, produced by the FEG in the circuit and circulation of electric charge. The total charge delivered from the PZT energy-carrying element to the 36 nF capacitor bank in this experiment was ΔQ_{max} = 73 µC, which is 49% of the initial charge.

The peak power produced by the FEG was $P(t)_{max}$ = 0.29 MW (Fig. 14). It is about 8 times higher than that produced by miniature FMG charging the same capacitor bank (Shkuratov et al., 2006a). Table 3 summarizes results of FEG-Capacitor bank experiments for all three capacitance of the bank. It follows from our experimental results that the capacitance of the capacitor bank effects significantly on the character of processes in the FEG-Capacitor bank circuit.

Capacitance (nF)	Voltage Amplitude (kV)	Transferred Energy (mJ)
9 nF	2.41 ± 0.33 kV	26 ± 0.5
18 nF	2.07 ± 0.22 kV	39 ± 0.4
36 nF	1.75 ± 0.14 kV	55 ± 0.3

Table 3. Amplitude of maximum output voltage, $U(t)_{max}$, generated in FEG-Capacitor bank system and energy transferred from the FEG module to the capacitor bank as a function of capacitance of the bank. PZT 52/48 elements with diameter $D = 26.1$ mm and thickness $h = 0.65$ mm.

We performed systematic experimental studies of FEG-Capacitor bank systems with FEGs containing PZT 52/48 elements of different sizes (Shkuratov et al., 2007c; Shkuratov et al., 2008a; Shkuratov et al., 2008b). Figure 15(A) summarizes experimentally obtained high-voltage pulse amplitudes produced by FEGs of three types versus the capacitance of the capacitor bank.

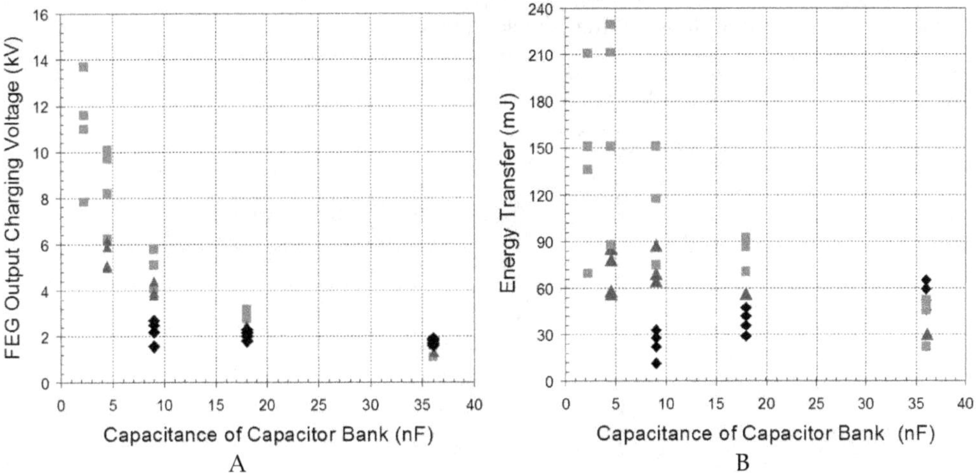

A

B

Fig. 15. Performance of FEG-Capacitor bank system. A - amplitude of the voltage pulse produced by FEGs across capacitor banks of different capacitance. B - energy delivered from FEGs to capacitor banks of different capacitance. PZT 52/48 elements of $D = 26.1$ mm/$h = 0.65$ mm (diamonds), $D = 27.0$ mm/$h = 2.1$ mm (triangles) and $D = 25.4$ mm/$h = 5.1$ mm (squares).

It follows from the experimental results that increasing the capacitance of the bank leads to a gradual decrease in the high-voltage produced by the FEGs containing all three types of PZT 52/48 ferroelectric elements ($D = 26.1$ mm/$h = 0.65$ mm, $D = 27.0$ mm/$h = 2.1$ mm and $D = 25.4$ mm/$h = 5.1$ mm).

The highest voltage, $U_{charge\ max}$ = 13.7 kV, was obtained across a 2.25 nF capacitor bank from the FEG containing PZT element of D = 25.4 mm/h = 5.1 mm. The lowest voltage, $U_{charge\ min}$ = 1.2 kV, was obtained across a 36 nF capacitor bank from the same type of FEG [Fig. 15(A)]. Figure 15(B) summarizes experimental results obtained in (Shkuratov et al., 2007c; Shkuratov et al., 2008a; Shkuratov et al., 2008b) for the energy delivered from FEGs to capacitor banks of different capacitances. It can be clearly seen the maximum energy transfer from the FEG containing PZT 52/48 elements of D = 25.4 mm/h = 5.1 mm to the capacitor bank at bank capacitance of 4.5 nF [Fig. 15(B)]. Increasing the capacitance of the bank from 4.5 to 36 nF results in significant decreases in the energy transfer from 177±27 mJ to 47±12 mJ. Decreasing the capacitance from 4.5 to 2.25 nF leads to a decrease in the energy transferred to 156±25 mJ.

A similar effect was observed with FEGs containing PZT 52/48 elements of D = 27.0 mm/h = 2.1 mm [Fig. 15(B)]. In that case, the maximum energy transfer took place with capacitance of the bank of 9 nF.

5.2. Theoretical Model of FEG-Capacitor Bank System

To explain oscillatory behavior of output signals produced in the FEG-Capacitor bank system (Fig. 13) we developed a computer model of the system (Shkuratov et al., 2007d; Shkuratov et al., 2008a). A schematic diagram illustrating the depolarization of a ferroelectric module under longitudinal shock wave impact is shown in Fig. 16.

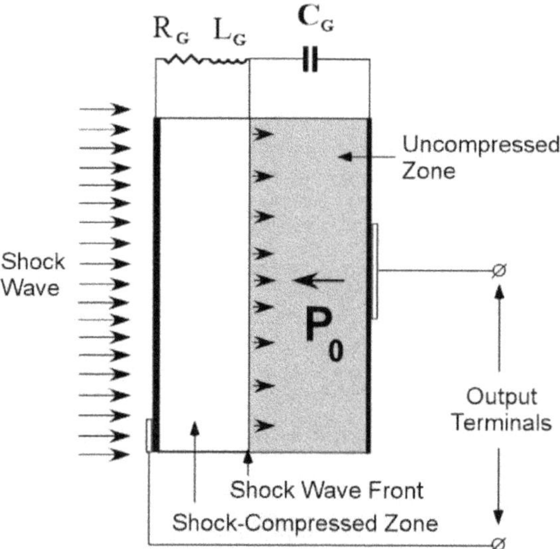

Fig. 16. Schematic diagram illustrated depolarization of PZT element under longitudinal shock wave impact.

When longitudinal shock wave passes through the polled ferroelectric element, its volume is divided into two parts or two zones, the shock-compressed zone (through which the shock wave has already passed), and the uncompressed zone (through which the shock wave has not passed). The difference in these two zones is in the value of polarization (the compressed zone is depolarized), electrical conductivity and other physical properties (Reynolds & Seay, 1961; Reynolds & Seay, 1962; Halpin, 1966; Setchell, 2003).

The equivalent circuit of the FEG-Capacitor bank system we employed in the simulation is shown in Fig. 17 (Shkuratov et al., 2008a). The shock-compressed part of the ferroelectric element is represented in the equivalent circuit as inductance, L_1, and resistance, R_1.

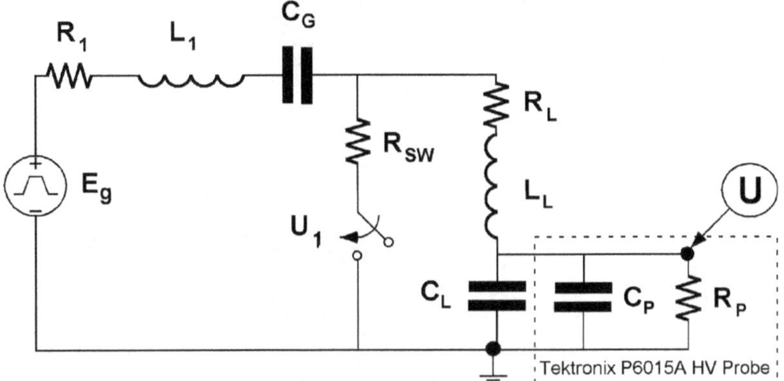

Fig. 17. The equivalent circuit employed for digital simulation of the FEG-Capacitor bank system (see the text).

The uncompressed part is represented as C_G. L_1 and R_1 are connected in series to C_G. The inductance, resistance and capacitance of the load (capacitor bank) and connecting cables are represented in the equivalent circuit as L_L, R_L, and C_L, respectively. The capacitance and resistance of Tektronix P6015A high-voltage probe are represented in the circuit as $C_P = 3$ pF and $R_P = 100$ MΩ, respectively. The internal electrical breakdown in the PZT element is simulated with switch U_1 having resistance R_{sw}. It closes when the voltage across the capacitor bank reaches the maximum value.

Result of simulation of the operation of FEG-Capacitor bank system containing PZT 52/48 element with $D = 26.1$ mm/$h = 0.65$ mm and $C_L = 18$ nF is shown in Fig. 18(A). The voltage across the bank oscillates as it did in the experiment (Fig. 13). The parameters of the system were as follows: $C_G = 7$ nF, $L_1 = 5$ μH, $R_1 = 0.2$ Ω, $C_L = 18$ nF, $L_L = 2$ μH, $R_L = 2$ Ω, $R_{sw} = 0.3$ Ω.

Result of simulation of the operation of FEG-Capacitor bank system containing PZT 52/48 element with $D = 26.1$ mm/$h = 0.65$ mm and $C_L = 36$ nF is shown in Fig. 18(B). The parameters of the system C_G, L_1, R_1, L_L, R_L were equal to those for the case with capacitor bank of 18 nF [Fig. 18(A)]. The different parameters in comparison with the case of 18 nF bank are: capacitance of the bank $C_L = 36$ nF and resistance of the PZT element after internal electrical breakdown $R_{sw} = 4.3$ Ω. The output voltage of the FEG is 20% lower than in case of an 18 nF bank [Fig. 18(B)] and it is practically a single pulse as it was in the experiment (Fig. 14).

Based on results of digital simulation we can make the conclusion that a key parameter responsible for the oscillatory mode of operation of FEG-Capacitor bank system is the resistance of PZT element after electrical breakdown. It means that the intensity of the internal electrical breakdown in the shock-compressed PZT module has significant effect on the processes in the FEG-Capacitor bank system.

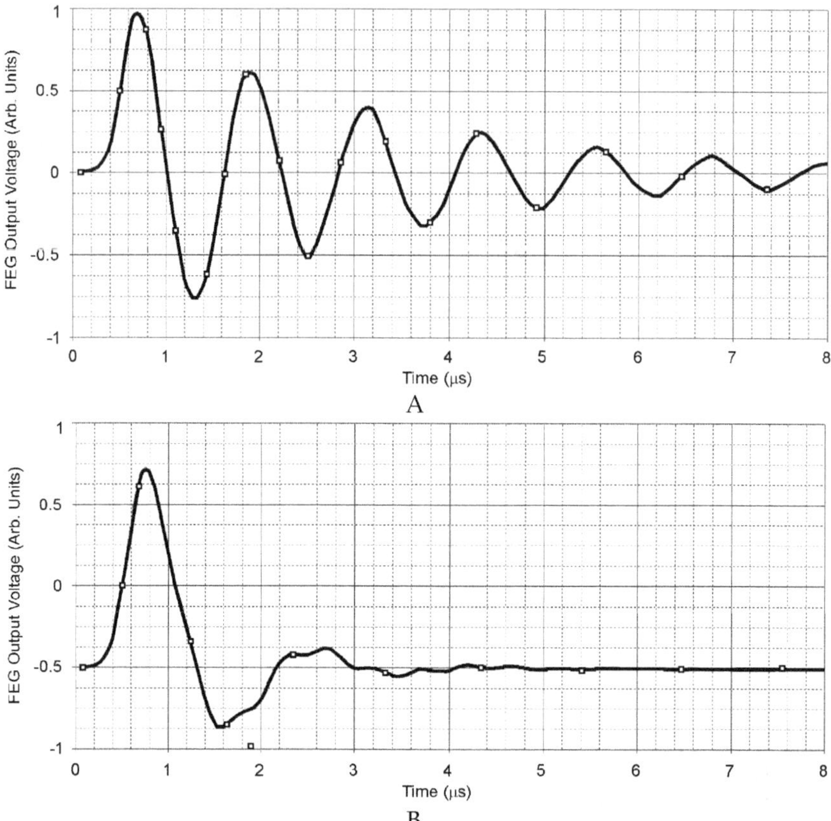

Fig. 18. Results of computer simulation of operation of FEG-Capacitor bank system.
A - capacitance of the capacitor bank of 18 nF; B - capacitance of the capacitor bank of 36 nF.

In the experiments described above, increasing the capacitance of the capacitor bank leads to decreasing the voltage produced by the FEG across the bank and, correspondingly, the voltage applied to shock-compressed ceramic disk. It effects significantly on the intensity of electrical breakdown in the shock-compressed PZT ceramics. Decreasing the voltage across the capacitor bank below threshold level results in lower impedance of conductive channels formed in the ceramics due to the electrical breakdown, and correspondingly aperiodic behavior of signals in FEG-Capacitor bank system.

6. FEG-Based Nanosecond Pulsed Power System

One of possible engineering applications of FEGs is to use it in combination with conventional pulsed power devices. A novel type of explosively driven combined pulsed power system was recently developed (Shkuratov et al., 2006a, Shkuratov et al., 2007a). The system is based on the FEG as a primary power source and the spiral vector inversion generator (VIG) as a power-conditioning stage. In this system, the amplitude of microsecond high-voltage pulses produced by FEGs can be amplified up to 20 times and the pulse width can be compressed to the nanosecond time range. A schematic diagram of the completely explosive FEG-VIG system is shown in Fig. 19.

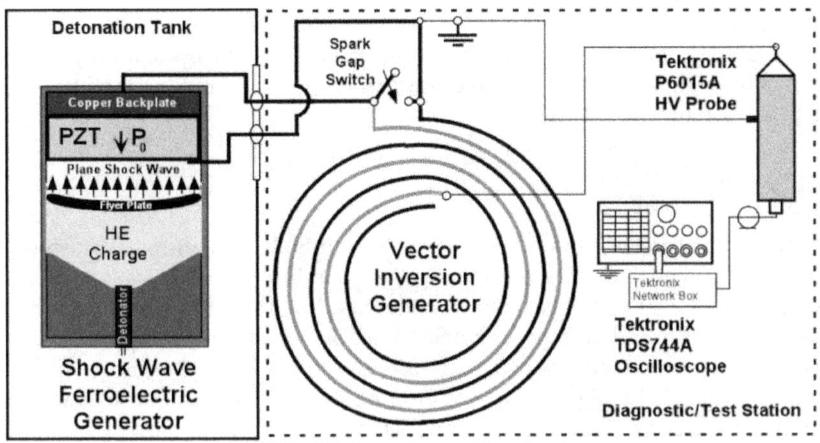

Fig. 19. Schematic diagram of a FEG-VIG nanosecond pulsed power system.

The FEG was placed inside the detonation tank. The output terminals of the FEG were connected to the input of the VIG. The negative terminal of the FEG was grounded. When fired, the FEG produced a positive high-voltage pulse that was applied to the input of the VIG spark gap switch. Detailed description of principles of operation of the FEG-VIG system is given in (Shkuratov et al., 2006a).

The waveform of a typical high-voltage pulse produced by an FEG-VIG system is shown in Fig. 20. The FEG contained PZT 52/48 disk with $D = 25.4$ mm/$h = 5.1$ mm.

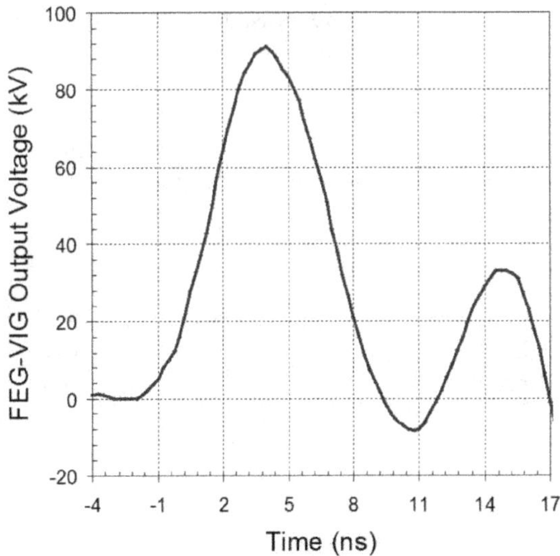

Fig. 20. Waveform of a typical high-voltage pulse produced by an explosive-driven FEG-VIG system.

This type of FEG is capable of producing output voltage up to 17 kV across the high-resistance load (see Section 3 of this Chapter). The FEG-VIG system (Figs. 19 and 20) produced high-voltage pulse with a peak voltage amplitude $U(t)_{max}$ = 91.4 kV, FWHM = 6.5 ns and τ = 5.25 ns.

Adding a VIG stage increases the voltage output of the FEG by a multiplication factor depending on the VIG's parameters, while simultaneously compressing the pulse width into the range of a few nanoseconds.

7. Summary

We designed, constructed, and systematically studied miniature explosively driven generators based on shock depolarization of polycrystalline ferroelectric elements. Results of our studies provided strong basis for understanding the operation of shock wave ferroelecric generators in different modes and its engineering applications. It was experimentally demonstrated that the amplitude of the output voltage produced by miniature FEGs across the high resistance load is directly proportional to the thickness of the ferroelectric element. The effect of almost complete shock wave depolarization of $Pb(Zr_{0.52}Ti_{0.48})O_3$ ferroelectric elements within miniature FEGs under longitudinal explosive shock pressure 1.5 GPa was detected experimentally. It was experimentally demonstrated that miniature FEGs can be used as an effective charging sources for capacitor banks. It follows from detailed parametric studies of FEG-Capacitor bank systems that there is an optimum capacitance of the bank for each type of ferroelectric element at which the maximum energy transfer is providing from the FEG to the bank. The charging power in the FEG-Capacitor bank system ranges from 0.25 to 0.30 MW. The specific energy density transferred from the PZT 52/48 element to the bank reaches 170 mJ/cm^3. It was experimentally demonstrated that the FEG-Capacitor bank system is capable of operating as a powerful oscillator with frequency up to few megahertz. Feasibility of construction of miniature autonomous 100-kV nanosecond pulsed power system based on miniature FEG and the vector inversion generator was experimentally demonstrated. Adding a VIG stage increases the voltage output of the FEG by a multiplication factor depending on the VIG's parameters, while simultaneously compressing the pulse width into the range of a few nanoseconds. This two-stage FEG-VIG pulsed power system produces an extremely high power, single-shot pulser that is unrivaled for specific power.

8. References

Altgilbers, L.L., Stults, A.H., Kristiansen, M., Neuber, A., Dickens, J., Young, A., Hold, T., Elsayed, M., Curry, R., O'Connor, K., Baird, J., Shkuratov, S., Freeman, B., Hemmert, D., Rose, F., Shotts, Z., Roberts, Z., Hackenberget, W., Alberta, E., Rader, M. & Dougherty, A. (2009). Recent advances in explosive pulsed power. *Journal of Directed Energy*, Vol. 3, No. 2, (Spring 2009) pp. 149-191.

Altgilbers, L.L., Baird, J., Freeman, B., Lynch, C.S., Shkuratov, S.I. (2010). *Explosive Pulsed Power*, Imperial College Press, ISBN 1-84816-322-3, London, U.K.

Bauer, F. & Vollrath, K. (1976a). Behaviour of non-linear ferroelectric ceramics under shock waves. *Ferroelectrics*, Vol. 12, No. 1-4, (Spring 1976) pp. 153-156, ISSN 0015-0193.

Bauer, F. & Vollrath, K. (1976b). New aspects in ferroelectric energy sources for impact fuses. *Propellants, Explosives, Pyrotechnics*, Vol. 1, No. 3, (August 1976) pp. 55-59, ISSN 0721-3115.

Cutchen, J.T. (1966). Polarity effects and charge liberation in lead zirconate titanate ceramics under high dynamic stress. *Journal of Applied Physics*, Vol. 37, No. 13, (December 1966) pp. 4745-4750, ISSN 0021-8979.

Dick, J.J. & Vorthman, J.E. (1978). Effect of electrical state on mechanical and the electrical response of a ferroelectric ceramic PZT 95/5 to impact loading. *Journal of Applied Physics*, Vol. 49, No. 4, (April 1978) pp. 2494-2498, ISSN 0021-8979.

Dungan, R.H. & Storz, L.J. (1985). Relation between chemical, mechanical, and the electrical properties of Nb_2O_5-modified 95 mol% $PbZrO_3$-5 mol% $PbTiO_3$. *Journal of the American Ceramic Society*, Vol. 68, No. 10, (October 1985) pp. 530-533, ISSN 0002-7820.

Duvall, G.E. & Graham, R.A. (1977). Phase transitions under shock-wave loading. *Reviews of Modern Physics*, Vol. 49, No. 3, (September 1977) pp. 523-579, ISSN 0034-6861.

Halpin, W.J. (1966). Currents from a shock-loaded short-circuited ferroelectric ceramic disk. *Journal of Applied Physics*, Vol. 37, No. 1, (January 1966) pp. 153-163, ISSN 0021-8979.

Halpin, W.J. (1968). Resistivity estimates for some shocked ferroelectrics. *Journal of Applied Physics*, Vol. 39, No. 8, (July 1968) pp. 3821-3826, ISSN 0021-8979.

Jaffe, B., Cook, W.R. & Jaffe, H. (1971). *Piezoelectric Ceramics*, Academic Press, ISBN 1-87890-710-7, London, U.K.

Kuznetsov, D.K., Shur, V.Ya., Baturin, I.S., Menou, N., Muller, C.H., Schneller, T., Sternberg, A. (2006). Effect of penetrating irradiation on polarization reversal in PZT thin films. *Ferroelectrics*, Vol. 340, No. 1, Part 1, (January 2006) pp. 161-167, ISSN 0015-0193.

Lysne, P.C. (1973). Dielectric breakdown of shock-loaded PZT 65/35. *Journal of Applied Physics*, Vol. 44, No. 2, (February 1973) pp. 577-582, ISSN 0021-8979.

Lysne, P.C. & Percival, C. M. (1975). The electric energy generation by shock compression of ferroelectric ceramics: Normal-mode response of PZT 95/5. *Journal of Applied Physics*, Vol. 46, No. 4, (April 1975) pp. 1519-1525, ISSN 0021-8979.

Lysne P.C. (1975). Kinetic effects in the electrical response of a shock-compressed ferroelectric ceramic. *Journal of Applied Physics*, Vol. 46, No. 9, (September 1975) pp. 4078-4079, ISSN 0021-8979.

Lysne, P.C. & Percival, C. M. (1976). Analysis of shock-wave-actuated ferroelectric power supplies. *Ferroelectrics*, Vol. 10, No. 1, (January 1976) pp. 129-133, ISSN 0015-0193.

Lysne, P.C. (1977). Resistivity of shock-wave-compressed PZT 95/5. *Journal of Applied Physics*, Vol. 48, No. 11, (November 1977) pp. 4565-4568, ISSN 0021-8979.

Mesayts, G.A. (2005). *Pulsed Power*, Kluwer Academic/Plenum Publishers, ISBN 0-306-48653-9, New York, U.S.A.

Mineev, V.N. & Ivanov, A.G. (1976). Electromotive force produced by shock compression of a substance. *Soviet Physics – USPEKHI*, Vol. 19, No. 5, (May 1976) pp. 400-419, ISSN 1063-7869.

Moulson, A.J. & Herbert, J. M. (2003). *Electroceramics: Materials, Properties, Applications*, John Wiley & Sons, ISBN 0-47149-748-6, West Sussex, England.

Neilson, S.W. (1957). Effect of strong shocks in ferroelectric materials. *Bulletin of the American Physical Society*, Vol. 2, No. 2, (March 1957) pp. 302-302, ISSN 0361-2228.

Novitskii, E.Z., Sadunov, V.D. & Karpenko, G.A. (1978). Behavior of ferroelectrics in shock waves. *Combustion, Explosion, and Shock Waves*, Vol. 14, No. 4, (July 1977), pp. 505-516, ISSN 0010-5082.

Novitskii, E.Z. & Sadunov, V.D. (1985). Behavior of ferroelectrics in shock waves. *Combustion, Explosion, and Shock Waves*, Vol. 21, No. 5, (May 1985), pp. 611-615, ISSN 0010-5082.

Online A. Available: www.teledynerisi.com

Online B. Available: www.defense.itt.com

Reynolds, C. E. & Seay, G. E. (1961). Multiple shock wave structures in polycrystalline ferroelectrics. *Journal of Applied Physics*, Vol. 32, No. 7, (July 1961) pp. 1401-1402, ISSN 0021-8979.

Reynolds, C. E. & Seay, G. E. (1962). Two-wave shock structures in the ferroelectric ceramics barium titanate and lead zirconate titanate. *Journal of Applied Physics*, Vol. 33, No. 7, (July 1962) pp. 2234-2241, ISSN 0021-8979.

Setchell, R.E. (2003). Shock wave compression of the ferroelectric ceramic $Pb_{0.99}(Zr_{0.95}Ti_{0.05})_{0.98}Nb_{0.02}O_3$: Hugoniot states and constitutive mechanical properties. *Journal of Applied Physics*, Vol. 94, No. 1, (July 2003) pp. 573-588, ISSN 0021-8979.

Setchell, R.E. (2005). Shock wave compression of the ferroelectric ceramic $Pb_{0.99}(Zr_{0.95}Ti_{0.05})_{0.98}Nb_{0.02}O_3$: Depoling currents. *Journal of Applied Physics*, Vol. 97, No. 1, (January 2005), Article number 013507, ISSN 0021-8979.

Setchell, R.E., Montgomery, S.T., Cox, D.E. & Anderson, M.U. (2006). Dielectric properties of PZT 95/5 during shock compression under high electric fields. *AIP Conference Proceedings*, Vol. 845, Part 1, (August 2006) pp. 278-281, ISBN 0-7354-0341-4.

Setchell R.E. (2007). Shock wave compression of the ferroelectric ceramic $Pb_{0.99}(Zr_{0.95}Ti_{0.05})_{0.98}Nb_{0.02}O_3$: Microstructural effect. *Journal of Applied Physics*, Vol. 101, No. 5, (March, 2007) Article number 053525, ISSN 0021-8979.

Shkuratov, S.I., Talantsev, E.F., Dickens, J.C. & Kristiansen, M. (2002a). Transverse shock wave demagnetization of $Nd_2Fe_{14}B$ high-energy hard ferromagnetics. *Journal of Applied Physics*, Vol. 92, No. 1, (July 2002) pp. 159-162, ISSN 0021-8979.

Shkuratov, S.I., Talantsev, E.F., Dickens, J.C. & Kristiansen, M. (2002b). Ultracompact explosive-driven high-current source of primary power based on shock wave demagnetization of $Nd_2Fe_{14}B$ hard ferromagnetics. *Review of Scientific Instruments*, Vol. 73, No. 7, (July 2002) pp. 2738-2741, ISSN 0034-6748.

Shkuratov, S.I., Talantsev, E.F., Dickens, J.C. & Kristiansen, M. (2002c). Compact explosive-driven generator of primary power based on a longitudinal shock wave demagnetization of hard ferri- and ferromagnets. *IEEE Transactions on Plasma Science*, Vol. 30, No. 5, (October 2002) pp. 1681-1691, ISSN 0093-3813.

Shkuratov, S.I., Talantsev, E.F., Dickens, J.C., Kristiansen, M. & Baird, J. (2003a). Longitudinal-shock-wave compression of $Nd_2Fe_{14}B$ high-energy hard ferromagnet: The pressure-induced magnetic phase transition. *Applied Physics Letters*, Vol. 82, No. 8, (February 2003) pp. 1248-1250, ISSN 0003-6951.

Shkuratov, S.I., Talantsev, E.F., Dickens, J.C. & Kristiansen, M., (2003b). Currents produced by explosive driven transverse shock wave ferromagnetic source of primary power in a coaxial single-turn seeding coil of a magnetocumulative generator. *Journal of Applied Physics*, Vol. 93, No. 8, (April 2003) pp. 4529-4535, ISSN 0021-8979.

Shkuratov, S.I., Talantsev, E.F., Menon, L., Temkin, H., Baird, J. & Altgilbers, L.L. (2004). Compact high-voltage generator of primary power based on shock wave depolarization of lead zirconate titanate piezoelectric ceramics. *Review of Scientific Instruments*, Vol. 75, No. 8, (August 2004) pp. 2766-2769, ISSN 0034-6748.

Shkuratov, S.I., Talantsev, E.F., Baird, J., Rose, M.F., Shotts, Z., Altgilbers, L.L. & Stults, A.H. (2006a). Completely explosive ultracompact high-voltage nanosecond pulse-generating system. *Review of Scientific Instruments*, Vol. 77, No. 4, (April, 2006) Article number 043904, ISSN 0034-6748.

Shkuratov, S.I., Talantsev, E.F., Baird, J., Altgilbers, L.L. & Stults, A.H. (2006b). Compact autonomous explosive-driven pulsed power system based on a capacitive energy storage charged by a high-voltage shock-wave ferromagnetic generator. *Review of Scientific Instruments*, Vol. 77, No. 6, (June, 2006) Article number 066107, ISSN 0034-6748.

Shkuratov, S.I., Talantsev, E.F., Baird, J., Temkin, H., Altgilbers, L.L. & Stults, A.H. (2006c). Longitudinal shock wave depolarization of $Pb(Zr_{0.52}Ti_{0.48})O_3$ polycrystalline ferroelectrics and their utilization in explosive pulsed power," *AIP Conference Proceedings*, Vol. 845, Part 2, (August 2006) pp. 1169-1172, ISBN 0-7354-0341-4.

Shkuratov, S.I., Baird, J., Talantsev, E.F., Rose, M.F., Shotts, Z., Altgilbers, L.L., Stults, A.H. & Kolossenok, S.V. (2007a). Completely explosive ultracompact high-voltage pulse generating system. *Digest of Technical Papers-IEEE International Pulsed Power Conference*, pp. 445-448, ISBN 978-078039190-1, Monterey, CA, June 2005, IEEE, Piscataway.

Shkuratov, S.I., Talantsev, E.F., Baird, J., Temkin, H., Tkach, Y., Altgilbers, L.L. & Stults, A.H. (2007b). The depolarization of a $Pb(Zr_{0.52}Ti_{0.48})O_3$ polycrystalline piezoelectric energy-carrying element of compact pulsed power generator by a longitudinal shock wave. *Digest of Technical Papers-IEEE International Pulsed Power Conference*, pp. 529-532, ISBN 978-078039190-1, Monterey, CA, June 2005, IEEE, Piscataway.

Shkuratov, S.I., Baird, J., Talantsev, E.F., Tkach, Y., Altgilbers, L.L., Stults, A.H. & Kolossenok, S.V. (2007c). Pulsed charging of capacitor bank by compact explosive-driven high-voltage primary power source based on longitudinal shock wave depolarization of ferroelectric ceramics. *Digest of Technical Papers-IEEE International Pulsed Power Conference*, pp. 537-540, ISBN 978-078039190-1, Monterey, CA, June 2005, IEEE, Piscataway.

Shkuratov, S.I., Talantsev, E.F., Baird, J., Ponomarev, A.V., Altgilbers, L.L. & Stults, A.H. (2007d). Operation of the longitudinal shock wave ferroelectric generator charging a capacitor bank: Experiments and digital model. *PPPS-2007 - Pulsed Power Plasma Science 2007*, pp. 1146-1150, ISBN 978-142440914-3, Albuquerque, NM, June 2007, IEEE, Piscataway.

Shkuratov, S.I., Baird, J., Talantsev, E.F., Ponomarev, A.V., Altgilbers, L.L., & Stults, A.H. (2008a). High-voltage charging of a capacitor bank. *IEEE Transactions on Plasma Science*, Vol. 36, No. 1, (February 2008) pp. 44-51, ISSN 0093-3813.

Shkuratov, S.I., Talantsev, E.F., Baird, J., Altgilbers, L.L. & Stults, A.H. (2008b). Pulse charging of capacitor bank by explosive-driven shock wave ferroelectric generator. *2006 International Conference on Megagauss Magnetic Field Generation and Related Topics, including the International Workshop on High Energy Liners and High Energy Density Applications, MEGAGAUSS*, pp. 325-330, ISBN 978-142442061-2, Santa Fe, NM, November 2006, IEEE, Piscataway.

Shkuratov, S.I., Baird, J., Antipov, V.G., Talantsev, E.F., Lynch, C.S. & Altgilbers, L.L. (2010). Depolarization of PZT 52/48: Longitudinal explosive shock versus quasi-static thermal heating. *IEEE Transactions on Plasma Science*, accepted for publication. ISSN 0093-3813.

Shur, V.Ya., Negashev, S.A., Subbotin, A.L. & Borisova, E.A. (1997). Crystallization kinetics of amorphous ferroelectric films. *Ferroelectrics*, Vol. 196, No. 1-4, (January 1997) pp. 183-186, ISSN 0015-0193.

Tkach, Y., Shkuratov, S.I., Talantsev, E.F., Dickens, J.C., Kristiansen, M., Altgilbers, L.L. & Tracy, P.T. (2002). Theoretical treatment of explosive-driven ferroelectric generators. *IEEE Transactions on Plasma Science*, Vol. 30, No. 5, (October 2002) pp. 1665-1673, ISSN 0093-3813.

CPSIA information can be obtained
at www.ICGtesting.com
Printed in the USA
LVHW061406010423
743231LV00003B/295